METHODS IN PHARMACOLOGY AND TOXICOLOGY

For further volumes:
http://www.springer.com/series/7653

Process Simulation and Data Modeling in Solid Oral Drug Development and Manufacture

Edited by

Marianthi G. Ierapetritou
and Rohit Ramachandran

Rutgers University, Piscataway, NJ, USA

Humana Press

Editors
Marianthi G. Ierapetritou
Rutgers University
Piscataway, NJ, USA

Rohit Ramachandran
Rutgers University
Piscataway, NJ, USA

ISSN 1557-2153 ISSN 1940-6053 (electronic)
Methods in Pharmacology and Toxicology
ISBN 978-1-4939-2995-5 ISBN 978-1-4939-2996-2 (eBook)
DOI 10.1007/978-1-4939-2996-2

Library of Congress Control Number: 2015949385

Springer New York Heidelberg Dordrecht London

Printed on acid-free paper

Humana Press is a brand of Springer
Springer Science+Business Media LLC New York is part of Springer Science+Business Media (www.springer.com)

Preface

The need for a more structured approach to process and product development has been recently identified in the pharmaceutical industry in order to consistently guarantee quality and value to processes and products. This need has been formally identified by the pharmaceutical industry as the Quality by Design (QbD) initiative and has been in the center of attention over the last few years. To enable the implementation of this concept, there is a need for quantitative characterization of process and product performance.

The goal of this handbook is to provide pharmaceutical engineers with an introduction to the current state of the art in modeling and simulation of pharmaceutical processes and to summarize a number of practical applications of such methodologies in drug product development. Chapters include reviews of the simulation and modeling methodologies, data collection and analysis, development of novel sensing techniques, development and integration of individual unit models, optimization approaches for data-based models, design space evaluation techniques, informatics-based methodologies, and emerging topics in pharmaceutical process development.

The first chapter focuses on examples of more mechanistic models developed in pharmaceutical manufacturing and specifically addresses the modeling of fluid bed granulation, tablet coating, and spray drying. The importance of the integration of such models in existing manufacturing workstreams is highlighted. Chapter 2 provides a detailed review of Population Balance Models (PBM) as one of the most promising alternatives to represent particulate systems. The distributed nature of the PBM makes it an appropriate modeling framework for pharmaceutical powder-based drug manufacturing.

Material properties and how they affect product and process performance are discussed in Chap. 3. In particular, specific properties of raw materials and the techniques used to measure them are described, and the state of the art of applying this information to define product formulation is also presented.

Chapter 4 reviews one of the most commonly used mechanistic approaches to model particular systems, which is Discrete Element Method (DEM). The main advantage of this approach is that it can capture mechanistic, particle-scale information such as velocity and collision profiles.

In pharmaceutical manufacturing, it is important to establish the right procedures to guarantee the consistency in process performance. Chapter 5 reviews the procedures used for the development of near infrared (NIR) spectroscopic methods and describes examples of the most recently developed approaches based on partial least squares (PLS) calibration methods that minimize the number of samples used. Following the ideas on latent-variable modeling techniques, Chap. 6 presents how those approaches can be used to support pharmaceutical development and manufacturing activities.

Chapter 7 is dedicated to control strategies required to support the switch from batch to continuous models of operations for the production of pharmaceutical products. A general methodology required to design and implement a control system is presented together with the required software and hardware of the control platform. Chapters 8 and 9 focus on the importance of mathematical modeling for process simulation, optimization, and the development of the design space for pharmaceutical process operations. In particular, Chap. 8 reviews the approaches that are used to define process feasibility and how those are applied for pharmaceutical manufacturing. Chapter 9 provides a review of the optimization methodologies that have been applied to achieve better performance in pharmaceutical manufacturing. Finally, Chapter 10 provides a very detailed step-by-step description of the manufacturing stages involved in the production of solid dosage forms. Process analytical technology (PAT) devoted to the continuous manufacturing of solid-based drugs is also described.

It is hoped that the collection of these papers will promote research into the process systems methodologies and their application in pharmaceutical product and process development, which will undoubtedly become an increasingly important area in the future.

Piscataway, NJ, USA

Marianthi G. Ierapetritou
Rohit Ramachandran

Contents

Contributors

MANEL ALCALÀ • *Grup de Quimiometría Aplicada, Departament de Química, Unitat de Química Analítica, Universitat Autònoma de Barcelona, Bellaterra, Barcelona, Spain*

MASSIMILIANO BAROLO • *Computer-Aided Process Engineering Laboratory (CAPE-Lab), Department of Industrial Engineering, University of Padova, Padova, PD, Italy*

DANA BARRASSO • *Department of Chemical and Biochemical Engineering, Rutgers, The State University of New Jersey, Piscataway, NJ, USA*

LIEVE BIJNENS • *Pharmaceutical Development and Manufacturing Sciences, Janssen Pharmaceuticals, Beerse, Belgium*

ANWESHA CHAUDHURY • *Department of Chemical and Biochemical Engineering, Rutgers, The State University of New Jersey, Piscataway, NJ, USA*

M. SEBASTIAN ESCOTET-ESPINOZA • *Department of Chemical and Biochemical Engineering, Rutgers, The State University of New Jersey, Piscataway, NJ, USA*

JEROEN GEENS • *Pharmaceutical Development and Manufacturing Sciences, Janssen Pharmaceuticals, Beerse, Belgium*

MARIANTHI G. IERAPETRITOU • *Department of Chemical and Biochemical Engineering, Engineering Research Center for Structured Organic Particulate Systems (ERC-SOPS), Rutgers, The State University of New Jersey, Piscataway, NJ, USA*

KRIZIA M. KARRY • *Department of Chemical and Biochemical Engineering, Engineering Research Center for Structured Organic Particulate Systems (ERC-SOPS), Rutgers, The State University of New Jersey, Piscataway, NJ, USA*

JOHANNES G. KHINAST • *Research Center for Pharmaceutical Engineering (RCPE), Graz, Austria; Institute for Process and Particle Engineering, Graz University of Technology, Graz, Austria*

SARA KOYNOV • *Department of Chemical and Biochemical Engineering, Engineering Research Center for Structured Organic Particulate Systems (ERC-SOPS), Rutgers, The State University of New Jersey, Piscataway, NJ, USA*

OLAV LYNGBERG • *Janssen Supply Chain, Technical Operations, Janssen Pharmaceuticals, Raritan, NJ, USA*

ALEX MARCHUT • *Janssen Supply Chain, Technical Operations, Janssen Pharmaceuticals, Raritan, NJ, USA*

STEVE MEHRMAN • *Pharmaceutical Development and Manufacturing Sciences, Janssen Pharmaceuticals, Beerse, Belgium*

FERNANDO J. MUZZIO • *Department of Chemical and Biochemical Engineering, Engineering Research Center for Structured Organic Particulate Systems (ERC-SOPS), Rutgers, The State University of New Jersey, Piscataway, NJ, USA*

MATTEO OTTAVIAN • *Manufacturing Science & Technology, Sandoz Industrial Products S.p.A., Rovereto, TN, Italy*

ROHIT RAMACHANDRAN • *Department of Chemical and Biochemical Engineering, Engineering Research Center for Structured Organic Particulate Systems (ERC-SOPS), Rutgers, The State University of New Jersey, Piscataway, NJ, USA*

AMANDA ROGERS • *Department of Chemical and Biochemical Engineering, Rutgers, The State University of New Jersey, Piscataway, NJ, USA*

ANDRÉS D. ROMÁN OSPINO • *Department of Chemistry, Recinto Universitario de Mayagüez, Mayagüez, PR, USA*
RODOLFO J. ROMAÑACH • *Department of Chemistry, Recinto Universitario de Mayagüez, Mayagüez, PR, USA*
STEPHAN SACHER • *Research Center for Pharmaceutical Engineering (RCPE), Graz, Austria*
ABHISHEK SAHAY • *Department of Chemical and Biochemical Engineering, Engineering Research Center for Structured Organic Particulate Systems (ERC-SOPS), Rutgers, The State University of New Jersey, Piscataway, NJ, USA*
ELISABETH SCHAFER • *Janssen Supply Chain, Technical Operations, Janssen Pharmaceuticals, Raritan, NJ, USA*
MAITRAYE SEN • *Department of Chemical and Biochemical Engineering, Rutgers, The State University of New Jersey, Piscataway, NJ, USA*
RAVENDRA SINGH • *Department of Chemical and Biochemical Engineering, Engineering Research Center for Structured Organic Particulate Systems (ERC-SOPS), Rutgers, The State University of New Jersey, Piscataway, NJ, USA*
EMANUELE TOMBA • *Technical Development Drug Product, GSK Vaccines, Siena, SI, Italy*
CARLOS VELAZQUEZ • *Department of Chemical Engineering, Engineering Research Center for Structured Organic Particulate Systems (ERC-SOPS), University of Puerto Rico Mayaguez, Mayaguez, PR, USA*

Chapter 1

Applications of Modeling in Oral Solid Dosage Form Development and Manufacturing

Olav Lyngberg, Lieve Bijnens, Jeroen Geens, Alex Marchut, Steve Mehrman, and Elisabeth Schafer

Abstract

Historically application of mechanistic modeling approaches to processing steps in pharmaceutical oral solid dosage form manufacturing have been limited compared to similar efforts for small molecule synthesis. One plausible explanation may be that there has been a lack of fundamental governing equations for drug product processes. The general similarity of oral solid dosage form processes from product to product make models in this area highly reusable and they can be used for purposes such as reducing scale-up and technology transfer times as well as reducing material usage for these activities. In this chapter we describe three mechanistic models used in oral solid dosage form process development and manufacturing from the perspective of how they are developed, used, and linked with supporting empirical data. The three models cover fluid bed granulation, tablet coating, and spray drying. The focus of the chapter is to illustrate the importance of embedding the models into existing development, scale-up, and manufacturing work-streams such as to create a model enhanced workstream that is more efficient, faster, and better routed in science than the fully empirical approach. Such workstreams require verification of their performance before being fully operational which is also discussed.

Key words Mechanistic models, Modeling enhanced workstreams, Fluid bed granulation, Spray drying, Tablet coating

1 Introduction

Historically application of mechanistic modeling approaches to processing steps in pharmaceutical oral solid dosage (OSD) forms manufacturing have been limited compared to similar efforts for small molecule synthesis. A search of articles for modeling of blending, milling, granulation, and tableting of pharmaceuticals turns up about 3000 references, whereas a similar search for modeling of reaction, extraction, and crystallization processes of pharmaceuticals turns up about 30,000 [1]. A plausible explanation may be that there has been a lack of fundamental governing equations for drug product processes [2]. Particularly the lack of constitutive

Marianthi G. Ierapetritou and Rohit Ramachandran (eds.), *Process Simulation and Data Modeling in Solid Oral Drug Development and Manufacture*, Methods in Pharmacology and Toxicology, DOI 10.1007/978-1-4939-2996-2_1, © Springer Science+Business Media New York 2016

equations for particulate systems and the lack of reproducible analytical powder characterization techniques have created significant difficulties. Similarly thermodynamic descriptions of active pharmaceutical ingredients APIs in solid or semisolid formulations has proven more challenging than the solution based systems commonly found in small molecule synthesis. Empirical approaches in OSD development therefore have been common. Current practice is often to test most lots of API batches produced during API process development and scale-up in the accompanying OSD process. There is only a partial reliance on material properties specification. Again if comparing with API processes the small molecule synthesis modeler can lean on several first principle fundamental equations in the areas of reaction kinetics, vapor liquid equilibrium, solid liquid equilibrium, chemical equilibrium, and fluid flow while similar well-established models have not been available for the drug product scientist. In this chapter we describe some examples of first principle modeling that are part of fully or partially developed modeling enhanced workstreams within the pharmaceutical development paradigm. Often the models do not describe directly the critical quality attributes of the final product, but instead key performance parameters or paths thereof. Operating spaces for these key parameters can be established independently of scale and equipment, as aids in selecting the most efficient scale-down studies to perform. The key performance parameters are linked through the development work to the final critical quality attributes of the product. Access to these workstreams provides strategic benefits to the organization by reducing work, timelines and materials needed to perform OSD process development, tech transfer, and validation.

1.1 Modeling of Critical Quality Attributes and Process Performance Parameters

A key goal of modeling is to establish a description of the critical quality attributes of the product as function of operating conditions. Ideally, the model should be able to capture effects that make it independent of equipment type and geometry, process scale, and material attributes. It should also be applicable to process control. With such a model it would, in theory, be possible to model the process ab initio before ever running it, purely based on material, equipment, and process parameters. In the area of small molecule processing this is reality in some cases [3] and recent papers have presented cases where the process conditions and equipment design was co-developed in silico in one simulation [4]. In the area of continuous OSD manufacturing great strides are being made in this direction and it is likely that first principle based control models will be deployed in industry within the next few years (or it may already have happened) [5]. Given the above mentioned lack of modeling capability one approach often taken is to model key process performance parameters or performance parameter paths. In short, performance parameters are process parameters where the value or paths are indicative of the end product quality such as content

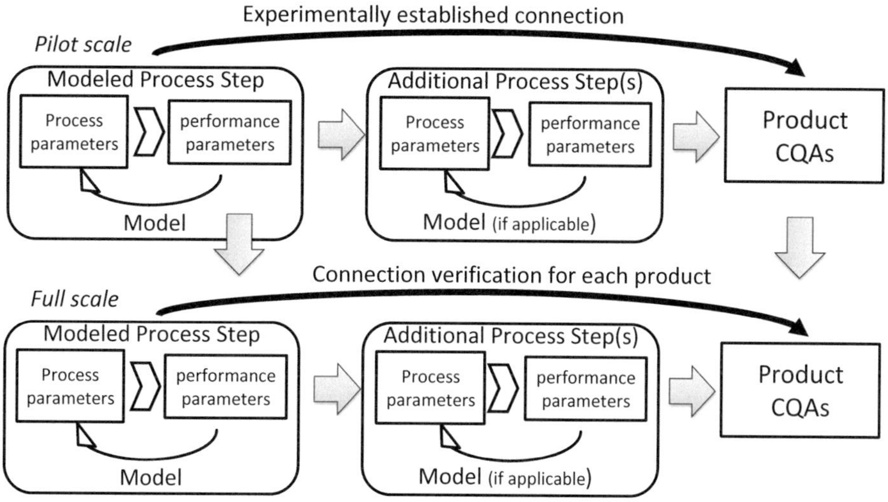

Fig. 1 Illustration of the link between unit-operations, process parameters, key performance parameters, and product CQAs

uniformity, assay, dissolution and appearance. If considering a fluid bed granulation, for example, air temperature, spray rate, air flow, fill level and atomization pressure, all impact the final granulated product in some way. The ultimate goal of the granulation is to create a powder that allows for a well-controlled compression step such that tablets consistently meet quality criteria. Although direct modeling of tablet weight variation or content uniformity as function of fluid bed granulation operation is difficult, modeling the LOD trajectory, PSD, or granule density may be tractable surrogates that provide as much value as long as it is possible to establish empirically the connection between the key performance parameter and the CQAs of the tablet. Figure 1 shows the connection between process parameters, key performance parameters, and tablet CQAs. Ranges may be established for these surrogate parameters that allow for successful operation of the tablet press. In this chapter we provide a few examples of this approach to modeling.

1.2 Modeling Enhanced Workstreams for Process Development and Scale-Up

Before presenting examples of individual model types used in OSD development, scale-up and manufacturing, it is important to touch on the aspects of the process development and tech transfer cycle that impact the effectiveness of modeling efforts. For the casual observer the pharmaceutical development cycle for a product may appear long—years—and hence the time allowed for model development and utilization would appear sufficient for most process types. However, to maintain a lean development environment, development activities are highly structured in time and effort and do not start until the last possible time. This approach is cost conscious and necessary in an environment where only a fraction of the initial development candidates become marketed drugs.

In practice this means that the process development time is as short as it can be made with existing technology, and it is within this paradigm that modeling has to function.

Model enhanced workstreams are necessary to allow for successful strategic deployment of models in this environment. The creation of a modeling enhanced workstream generally takes substantially longer than the time available for performing its intended purpose for any one product. As an example, one might have a modeling enhanced workstream around solubility of API in various solvents and buffers. Setting up such a workstream and validating it will take longer than the 1 week turnaround allowed for delivering solubility data whether they are empirical or modeled. In the case of scale-independent models the data needed for workstream validation requires both at scale and development scale data. Use of existing products to obtain this data is a possibility where such similar products exist. Once established, the modeling enhanced workstreams provide strategic value through enhanced efficiency, speed and reduced material usage. In the following we cover four separate examples of modeling enhanced workstreams that drive strategic advantages. They are spray drying, fluid bed granulation, tablet coating, and API properties in pharmaceutical grade polymers. In each case we demonstrate how a simple but validated model coupled with experimental data and linked to final CQAs can remove a substantial amount of work commonly done at scale and at times as part of the critical path timeline. Removing this type of work saves the organization API and allows for faster changes in plans in development. Later in the lifecycle, the strategic value is provided by allowing faster tech-transfer from one plant to the next. Being able to align capacity quickly and based on market conditions can eliminate stock-outs and make available capacity from plants with lower demand.

1.3 Modeling Focus from Early Development to Launch and Manufacturing

During early development of a new pharmaceutical entity the drug product development goal is often simply to produce a dosage form fit for purpose for conducting the initial toxicology work and phase I clinical trials. These may be drug-in-a-bottle, drug in a capsule, or standardized tablet formulations. To aid in enabling this effort, modeling approaches for early development focus on modeling product properties or performance more so then the relative limited effort around actual product manufacture. Typical properties to model may be oral bioavailability, properties indicative of process-ability for various unit-operations, stability in formulation solutions, and solubility in various aqueous or polymer systems.

The reduced investment paradigm executed in early development continues into mid and late development. Hence the marketed product oral solid dosage form development does not initiate until there are process development timeline based requirements to do so.

This creates opportunities and constraints for model based approaches. As described above, fully developed modeling enhanced workstreams can be executed in very short timeframes and with significantly reduced material requirements thereby providing strategic advantages. In late stage development these advantages may be reductions in the number batches planned as contingency, scale-up, or engineering batches for manufacture of phase II or phase III clinical trials. Reductions in batches for full scale characterization, registrational filings, or process validation may also be realized. Later in manufacturing, significant reduction in work for site to site tech transfer as well as efficiency optimizations may be possible.

During late development it is not uncommon for the majority of the total drug substance demand to be needed for drug product development efforts. Timelines associated with the delivery of 10–20 step synthetic schemes involving pilot or plant scale equipment are often as long as 6–12 months. These long delivery timelines reduce clinical flexibility or require material buffers to be built ahead of time as hedges against changing plans. Reductions in requirements for this API supply also reduce or eliminate a host of upstream activities, from API process development efforts to managing complex multi-vendor supply chains.

1.4 Regulatory Implications of Process Modeling

The FDA in its Q8 guidance [6] expressed with clarity that drug product development based on Quality by Design concepts are of public interest and that Design Spaces for DP processes should be established. From a modeling and engineering perspective, first principle modeling and Design Spaces go hand in hand. That is, first principle models inherently establish Design Spaces for the parameters in the model. In most DP processes there are no mechanistic models that cover all Critical Quality Attributes and a combination approach between modeling and an empirical approach has to be deployed.

It is difficult to authoritatively describe the extent to which mechanistic process modeling approaches have been used as part of the filing strategy for DP processes. Few if any publications describe in detail what was filed and the regulatory outcome of the modeling effort. This may be rooted in regulatory filing interactions generally being treated as confidential, but also because the modeling may provide strategic benefits to the sponsor. The regulatory expectations for model validation and documentation are generally covered by the FDA in Q8, Q9, and Q10 [6–8] with additional clarification provided later in 2011. The subject of first principle or mechanistic modeling and uses in filing strategies is an area that likely would benefit significantly from additional publications.

1.5 Future of Modeling of Oral Solid Dosage Form Processing Steps

The exponential growth in computing power in the last several decades has enabled many computational intensive numerical models to be deployed in engineering disciplines. Approaches such as Finite Element Method (FEM), Population Balance Models (PBM), Discreet Element Method (DEM), Computational Fluid Dynamics (CFD), and combinations of these are starting to make significant contributions to process understanding and equipment design in mining, agriculture, and construction when dealing with particulate systems [9–11]. It is likely that in the next decade these types of models potentially in combination with reduced order approaches will change the modeling landscape of drug product process design towards first principles. At the moment these techniques are still numerically too intensive to directly replace reduced complexity modeling approaches or even empirical approaches for validation of individual process steps. An example would be a standard blending step where a manufacturing scale bin may contain 800 kg powder of 3–10 components with particle sizes ranging from single microns to sub-millimeter granules. For process validation a scientist would seek to vary all relevant process parameters impacting blend quality to demonstrate process robustness and establish ranges. In practice this is not possible to model fully by DEM. While this is a near term limitation, likely to be overcome in the not too distant future, at the moment these models still provide insights for the purpose of general process understanding or product development and can be of significant aid if combined with process analytical technology (PAT) data. In addition to computation limitations the lack of direct analytical methodologies and understanding of powders to fully characterize their properties is another important aspect making powder systems highly challenging. Recently developed instrumentation such as powder rheometers and continued reduction in cost and complexity of NIR PAT methodology to determine inline powder density and composition have improved upon this problem. However the application of measured powder property values directly to scale-up models has still not been established.

2 Example: Modeling of Fluid Bed Granulation

2.1 Introduction

Fluidized bed granulation (FBG) is a common pharmaceutical drug product processing technology. The main advantage is that several ingredients can be granulated and dried in the same vessel thereby reducing material handling and shortening process times. Additionally fluidization is efficient at drying the product when compared with other wet granulation processes [12–14]. Granulated materials are desirable and enable common oral solid dosage form downstream processing in several ways. They have improved flow characteristics and compressibility which can be important for efficient and reproducible tablet compression. Granulation also reduces or eliminates the propensity for segregation hence improves content uniformity and has a low propensity for dust, improving operator and environmental safety [15].

2.2 Process Variables

The granulator has relatively few process variables available to the operator for control. In a top-spray granulator the controllable parameters are spray rate, inlet air temperature and flow rate, process air humidity (if controllable), and the atomization air pressure or flow. Atomization airflow is not directly controlled, but a function of atomization pressure and liquid viscosity [16]. The effects of process parameters on the physical properties of granules are studied during drug product development. As would be expected a faster rate of liquid addition and higher LOD trajectory results in larger granule sizes and less-friable granules. At high LOD trajectories formation of large agglomerated masses and bed collapse is possible, limiting the LOD at the upper end. At low LOD trajectories the material may be under granulated resulting in poor flow and compressibility which in turn leads to inability to form tablets meeting quality criteria. During granulation the granule size increases initially as the binder is applied and then decreases during the final drying phase due to erosion and breakage. Too much binder may result in hard agglomerates that do not compress well, too little can lead to too friable granules. Finally the operation type (either discontinuous or continuous airflow mode) also affects the granulation process and the upper and lower LOD limits [13, 17–26].

2.3 Process Components of Fluid Bed Granulation

For the purpose of scale-up and process transfer the FBG process can be divided into three fundamental process components: (a) the heat and moisture trajectory, (b) the granule fluidization and collision trajectory, and (c) the spray distribution profile. If these process components or trajectories are kept constant during scale-up and tech-transfer the process will remain stable. The process component trajectory concept is outlined in Fig. 2 below.

2.4 Fluid Bed Granulation Modeling

In the literature there are several reports of modeling and control of fluidized bed granulation processes. Modeling has been applied to determine LOD trajectories, particle size distributions (PSD) as function of time and process conditions and particle densities as function of processing conditions [23–25, 27–30].

In pharmaceutical development and manufacturing where the final product is a tablet, the critical quality attributes (CQA) are those associated with the tablet. Commonly they are Content Uniformity (CU), Assay, Hardness, Dissolution, and Appearance and are determined as part of the criticality assessment for each product. The goal of modeling the FBG process is therefore to establish a connection between the CQAs and the modeled FBG

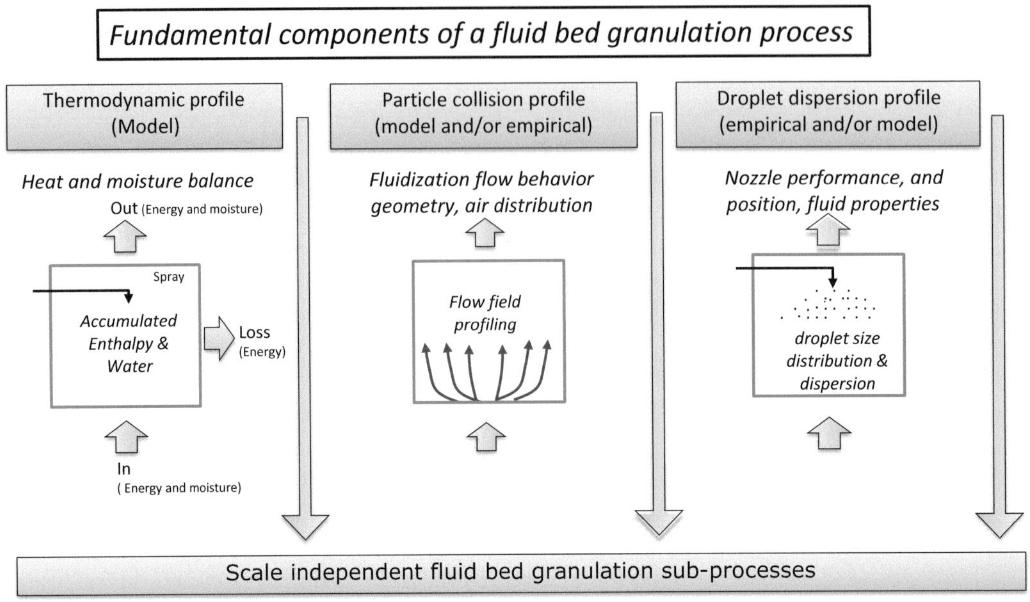

Fig. 2 The three pillar trajectory approach to FBG scale-up

performance parameters and or intermediate material attributes, e.g., final LOD, LOD trajectory, temperature trajectory, particle size distribution, and granulate bulk and tapped density. In a model based development and tech-transfer scenario the goal is to establish an operating space for the performance parameters by linking these to the tablet CQAs. Having established such as space it is possible through the model to link this performance parameter space back to a scale and equipment independent process parameter space and ensure that quality is maintained.

It has been proposed in the literature that development of a robust fluid bed process depends on control of the moisture profile [23, 24, 31–40]. Moisture profile control of a fluid bed granulation process requires an operator or controller to manipulate three interdependent control "levers" available during processing. The control levers are Spray Rate, Air Flow, and Inlet Air Temperature but in reality there is only one which is the overall LOD trajectory. Figure 3 shows the connection between Process Parameters, Performance Parameters and Final Product CQAs.

Modeling PSD in FBG processes is done either through population balance models or directly by modeling individual particles in DEM or DEM-CFD coupled simulations. In population balance models, the key terms are the growth and breakage kernels and their dependence on operating and powder conditions. Making independent measurements of growth and breakage kinetics as function of conditions is a key challenge when performing PSD modeling. Given the multitude of effects that impact growth and breakage of granules it can be difficult to establish scale-independent parameters. This can make the model difficult to use for a priori design and

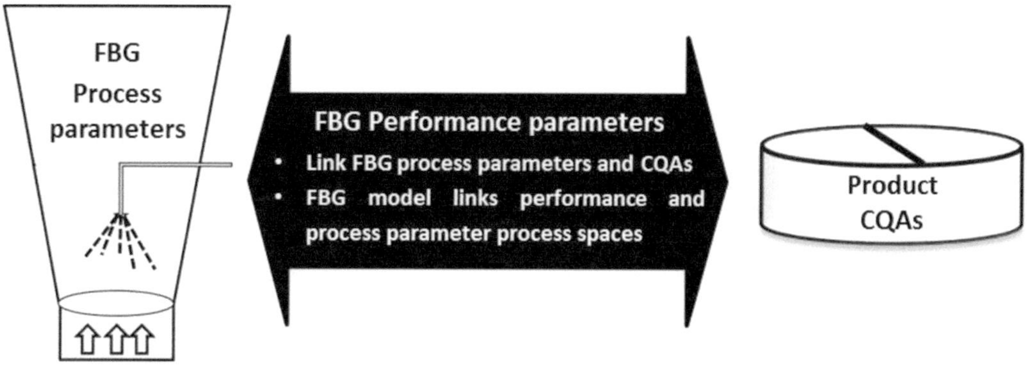

Fig. 3 Linking FBG product trajectories to final product properties

scale-up [41]. DEM-CFD or similar based PSD approaches are first principle based models and can be scale independent if the particles are either not scaled or alternatively their properties are scaled to ensure that bulk behavior remains constant. With these approaches it is possible to capture effects such as particle flow and material stresses which are not possible with less rigorous approaches. Insights provided by CFD/DEM simulations can be very helpful as trouble shooting tools or for general understanding. The main limitation at this point is that they are computationally very demanding and hence difficult to execute with time scales equal to those of actual FBG processes [42].

2.5 A Practical Fluid Bed Granulation Moisture and Heat Balance Model

A number of approaches have been developed to model bed humidity during fluid bed granulation and spray coating operations based on heat and mass balances [30, 43, 44]. Figure 4 displays one possible approach that has been used in Janssen development and manufacturing to convert the functionalities and characteristic features of the real FBG into a model of the equipment. The model is an adapted version of another model from the literature [43]. The reduced complexity approach allows a simplified mathematical description of the process. The underlying assumptions in the approach are as follows. The FBG can be approximated as a stirred tank reactor and there are no gradients of temperature or moisture. The spray first hits the bed and only a negligible amount evaporates prior to deposition on the particles. The model has two compartments: (1) the wet particles with humidity being only in the liquid state and (2) the air and water vapor phase consisting of air blown through the granulator and the evaporated liquid. Consequently, there exist separate heat and mass balances for the bed and the gas/vapor compartment which are linked via heat and mass transfer equations. Additionally a third set of heat and mass balances as well as additional expressions for heat transfer are employed to include the metal wall surfaces of the granulator. This is used to enable modeling of condensation conditions during start-up. The particles are treated

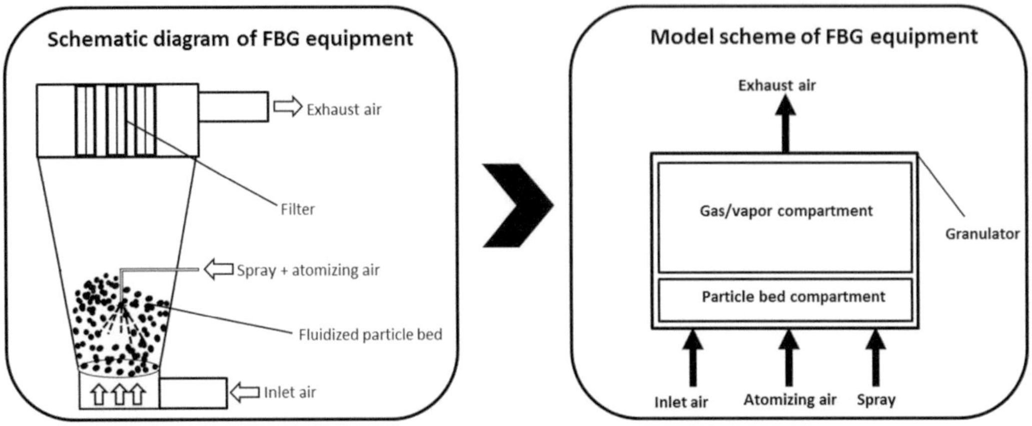

Fig. 4 Illustration of the modeling concept for the fluidized bed granulation moisture and heat model

as uniform spheres whereby micro level phenomena with respect to mass transfer such as pore diffusion are not captured explicitly.

The central part of the model is the expression describing the evaporation of liquid from the granules. In the literature, one work used a first principle derivation approach [30], whereas another [43] used a fitted term that was linear with partial pressure differential between surface and bulk gas phase. For the model presented here the driving force (DF) for the evaporation rate is assumed to be linear with the partial pressure differential between the surface and moisture saturated air and can be described as shown in Eq. (1).

$$\mathrm{DF} = M_{\mathrm{L}} \cdot (c_{\mathrm{sat}} - c) \tag{1}$$

M_{L} is the molar mass of the evaporating liquid, L, c_{sat} is the concentration of the evaporated liquid in the gas/vapor compartment at saturation conditions, and c is the concentration at the particle's surface.

For FBGs the evaporation rate is also a function of the wetted surface area of the particles which can change as the particles grow. A principal difficulty that arises here is to estimate the continuously changing area without having a separate model for the particle size distribution. In addition, it is at least equally challenging to determine the ratio of wetted to un-wetted surface even if the total area is known. Consequently, assessing the correctness of the calculated values applying total area and relative wetted area expressions are problematic. One approach around this problem is therefore to include a simple empirical term accounting for the LOD dependent size of the wetted particle surface area WSA in the model (Eq. 2)

$$\mathrm{WSA} = A_{\mathrm{p}} \cdot \mathrm{DW} \tag{2}$$

with A_{p} being the total particle surface area, and DW the degree of wetness. Depending on the total number of particles n_{p} and the particle diameter D_{p} the total area can be written as (Eq. 2).

$$A_{\mathrm{p}} = n_{\mathrm{p}} \cdot \pi \cdot D_{\mathrm{p}}^2 \tag{3}$$

The degree of wetness of the total particle surface area, DW, depends on LOD and a wetness parameter WP

$$\mathrm{DW} = \exp(\mathrm{LOD})/\mathrm{WP} \quad \text{with} \quad 0 \leq \mathrm{DW} \leq 1 \tag{4}$$

This leaves three fitting parameters which are n_{p}, D_{p} and WP. Applying the model to different types of products, process conditions and scales as will be demonstrated in the following section which shows that D_{p} and WP can be fitted with one data set consisting of process data and LOD. This means that D_{p} and WP can be treated as being product specific and constant throughout the granulation process and independent of process conditions and scale. N_{p} is dependent on the amount of material added to the granulator and therefore not a fitting parameter but linked to batch size.

A third term in an expression for the evaporation rate is the mass transfer coefficient k. Following an approach suggested in the literature [45], the following expression for calculating k is used (Eq. 5).

$$k = \frac{\alpha}{\rho_{\mathrm{gas}} \cdot \mathrm{cp}_{\mathrm{gas}}} \tag{5}$$

Equation (5) is based on the similarities between heat and mass transfer and the inherent assumption is that these are linearly related. Here, α is the heat transfer coefficient between the particles and the surrounding gas, $\mathrm{cp}_{\mathrm{gas}}$ the heat capacity and ρ_{gas} the density of the gas. In the model presented the heat transfer coefficient is in turn determined via the Nusselt number.

The average evaporation rate of the entire bed can be calculated by using the equation below.

$$\dot{m}_{\mathrm{evap}} = k \cdot A_{\mathrm{p}} \cdot M_{\mathrm{L}} \cdot (c_{\mathrm{sat}} - c) \cdot \exp(\mathrm{LOD})/\mathrm{WP} \tag{6}$$

The remaining heat and mass balance equations in the model are identical to those described in the literature [43]. In the following section we describe the testing and validation of the model presented above.

2.6 Model Testing and Validation

To ensure that the model was relevant and fit for the purpose of development, scale-up, tech transfer and troubleshooting, a rigorous model testing and validation program was initiated. The model was specifically tested to ensure that it was applicable to predict in the following areas:

- Granulation @ different scales
- Granulation of different products
- Granulation @ different process conditions

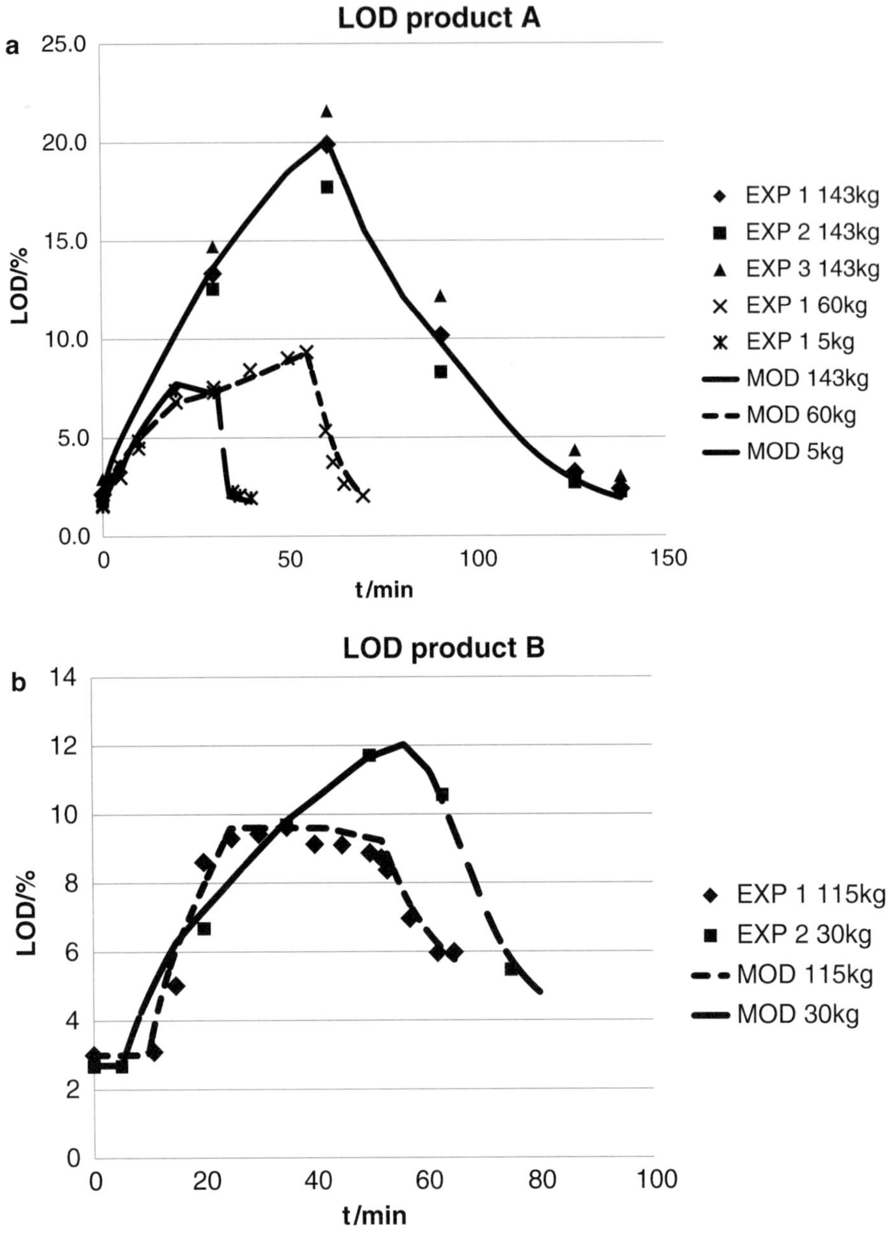

Fig. 5 Modeling LOD trajectories for product A (**a**) and product B (**b**) for various scales with single set product parameters

2.6.1 Model Validation for Different Scales and Products

First, the applicability of the model for predicting LOD as function of scale was tested. Figure 5a depicts modeled and experimental LOD trajectories of a granulation processes for product A at different batch sizes (5, 60, and 143 kg). The granulators used were Glatt 5, 60, and 120 respectively. In applying the model, only equipment specific and process parameter values were changed between the scales. The product related parameters n_p, D_p, and

WP (Eq. 6) were fitted using one experimental LOD and process data set at 60 kg scale. The same set of product related parameter values were then also employed at 5 and 143 kg scale. Experimentally measured and model predicted LOD trajectories for product A agree well for the scales investigated. In Fig. 5b, a second example for product B is given. Applying the same modeling procedure for 30 and 115 kg batch sizes gave very good agreement between experimental and modeled data. Based on this comparison and additional analysis for other products at different scales it can be concluded that the modeling approach is sufficiently robust to allow prediction as function of scale with constant product parameters determined only once at a single scale.

Having a validated and scale-independent model to calculate LOD data across scales is a first step towards implementing a model enhanced workstream. In scaling FBG processes it is common practice to follow general guidelines. These rules of thumb are outlined below:

- Apply similar air velocity values, calculated based on base plate area
- Apply similar or only slightly scaled bed heights
- Use similar relative humidity and inlet air temperatures
- Use similar material or bed temperature trajectories (T_{bed})

As function of scale-up the parameters that are left to be determined are:

- Spray rate
- Type, number, position, etc., of nozzles, e.g., Top/Bottom/etc. spray
- Atomizing air flow rate or pressure

As described in the tablet coating and spray drying sections, various scale-up approaches exist for nozzle selection to ensure that the spray quality remains constant. However, it is also common to select the nozzle settings and ratio of spray rate to atomization air flow rate experimentally or based on experience depending on the process sensitivity. The spray rate can be determined by the FBG model if using the LOD trajectory as the scaling approach.

2.6.2 Model Validation for Complex Process Conditions

The LOD trajectory model was further validated by testing its predictive capability against varying process conditions during the granulation process. For a product C the process conditions and the spray binder solution were varied. A single set of product related parameter values were employed and used to predict the LOD as function of time. Figure 6 shows the predicted and experimentally determined LOD trajectories. Using the model to predict LOD data for more complex process conditions confirmed the capability of the model to correctly represent the effects of varying process

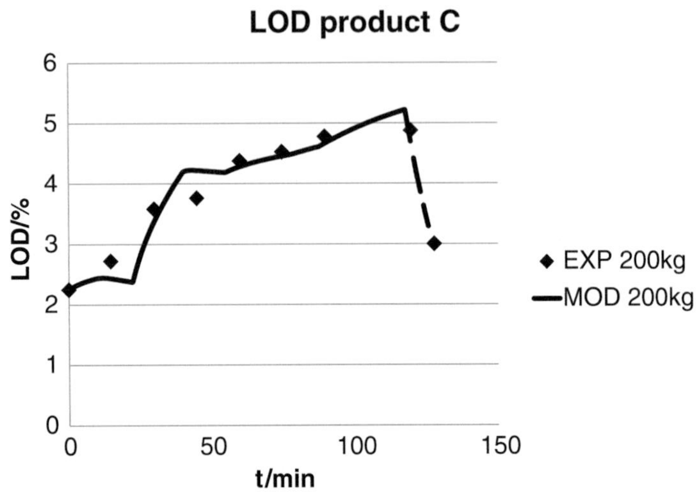

Fig. 6 Modeling LOD trajectory of Product C for a complex set of process conditions

parameters over time on the performance parameters. This again was essential in order to establish that the model was sufficiently fundamental to allow use for building process control strategies and for troubleshooting scenarios.

Additionally, the model was tested against the effect of changing RH of the incoming air. In this troubleshooting case where RH was not controlled, the question was if different RH could explain the significant variation in the LOD trajectory observed. Figure 7 depicts experimental and modeled LOD trajectories for a product D. With the help of the model it was established that the model was capable of predicting the effect of variations in RH from batch to batch (data not shown). Additionally the model could then be used to investigate significant outlier batches where possible changes in material properties or processing execution methodology were also playing a role. The observed effects in LOD trajectory could not be explained by the concurrent changes in RH of the incoming air. Figure 7 shows two batch granulation trajectories (Exp 1 and Exp 2) where the model establishes that RH cannot be the sole reason for the variations observed in the LOD profiles. This shows that the model can predict variations in processing conditions using only one set of material parameters.

2.7 Experimental Design for Pilot Granulation Experiments

As mentioned in the introduction, the development team's goal is to establish the relationship between the process parameters and the critical quality attributes of the product (the tablet). Establishing such relationships requires significant experimental efforts as there are no general models that link these directly. To have sufficient confidence in the generated data the experimental approach has to be executed at a reasonable scale which is commonly set in the range

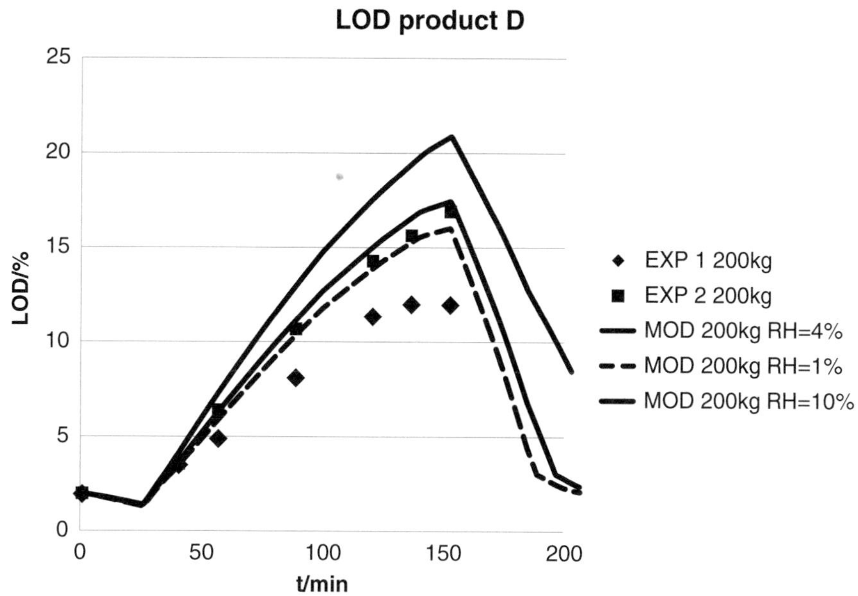

Fig. 7 Modeling of LOD trajectory to deconvolute of the effects of multiple process, material and operational changes between batches

of 15–60 kg. A partial or full factorial DOEs design may be used for these studies depending on the number of parameters selected. In using a model enhanced approach the process parameters are all linked through the LOD trajectory, and hence the DOE is selected such that is covers a sufficiently large LOD space. For most products the approximate range for the LOD profile is initially tested at small scale to establish a starting point. For the pilot scale DOEs, if taking the approach outlined above where air flow is determined based on the geometry of the granulator, it is possible to execute a full factorial design involving just spray rate and temperature, with RH being controlled by the equipment air handler during the test and later established from the model. A typical full factorial design of Spray Rate and Inlet Temperature is shown in Fig. 8. In some cases using constant spray rates or airflow rates is not possible during the entire granulation process as these would lead to excessively high LODs and over wetting. In these cases the model can be helpful to narrow options while still maintaining distinct LOD trajectories.

2.7.1 Linking LOD Trajectory to Granulate Properties and Tablet CQAs

Typically, the granulation material produced in designed experiments is well characterized to maximize the information and knowledge gained from designing and executing pilot scale DOE's. Commonly this involves flow and sieve analyses as well as executing several tableting runs for each granulate produced. It is not uncommon to produce five (5) different compression points for each material. The resulting tablets are then further characterized with standard physical measures such as weight, thickness, and hardness followed by content uniformity, dissolution and NIR spectral analysis. PLS or PCA type multivariate analysis is commonly employed to study the impact of the process parameters on the performance parameters and finally product quality. Figure 9 below shows the impact of the fluid bed granulation process conditions on intermediate product quality (granulate) and final product quality (tablets). Each point in the plot below represents a completed batch from a DOE executed as in Fig. 8 where each granulation batch was then tableted at multiple conditions. The *x*-axis represents the dissolution test output measuring percent API dissolved (in solution) at 20 min. The *y*-axis represents the in-line particle size of each granulation as measured by the Lasentec at end of drying (i.e., end of batch). Tablet hardness is indicated by the color scale on the Y2 axis. Red is a harder tablet and blue a softer one.

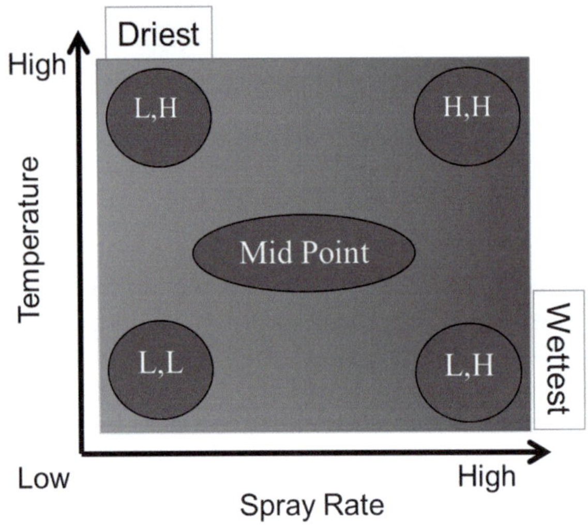

Fig. 8 Full factorial DOE design based on LOD trajectory development approach

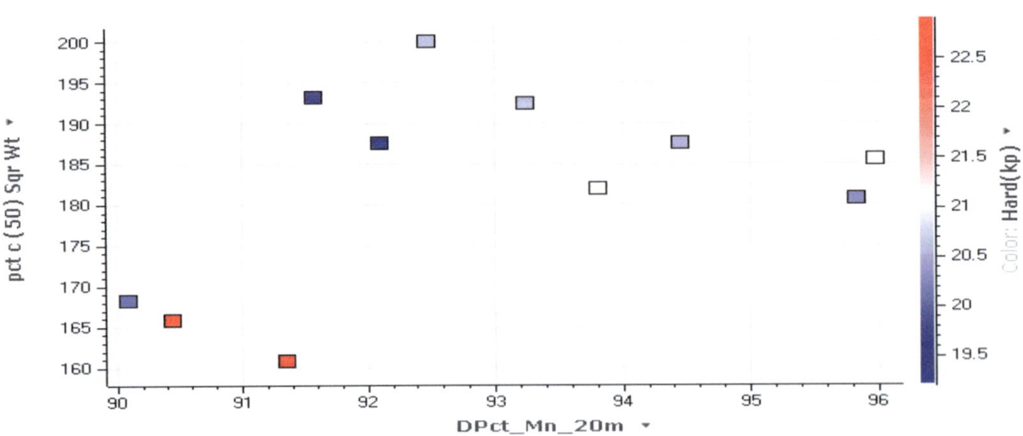

Fig. 9 Linking granulation profile to tablet dissolution results

From the plot is it clear that there is a significant impact of particle size and hence LOD trajectory on the dissolution properties for each tablet and granulation. Higher c50 values lead to softer tablets that in-turn have a higher dissolution percentage. The optimal dissolution properties are around 96 % API dissolved with hardness around 20.5 kPa and particle size between 180 and 185 μm. Conversely, one can see that smaller granules lead to harder tablets and lower dissolution percentages while still meeting specifications. One exception to this trend is the blue box (softer tablet) in the lower left which has poor dissolution. This exception was related to a significant interruption event that occurred during processing leading to poor granulation, affecting content uniformity and in-turn leading to low dissolution percentages for the fraction of tablets chosen for this dissolution test.

Building a predictive model for this relationship is also quite helpful to quantitate relationships between variables. An example of this is shown below where data from the previous in-line particle size measurement is correlated to the table dissolution percentage using a PLS model. The scores plot below shows the relationships between the *X*—particle size data block and *Y*—dissolution and hardness data block as indicated by the loadings plot below (Fig. 10). In this model c10, c50, and c90 were included to represent the particle size. One can see that higher particle size leads to higher dissolution (i.e., correlation seen in left of scores plot) and that lower particle size leads to harder tablets and lower dissolution.

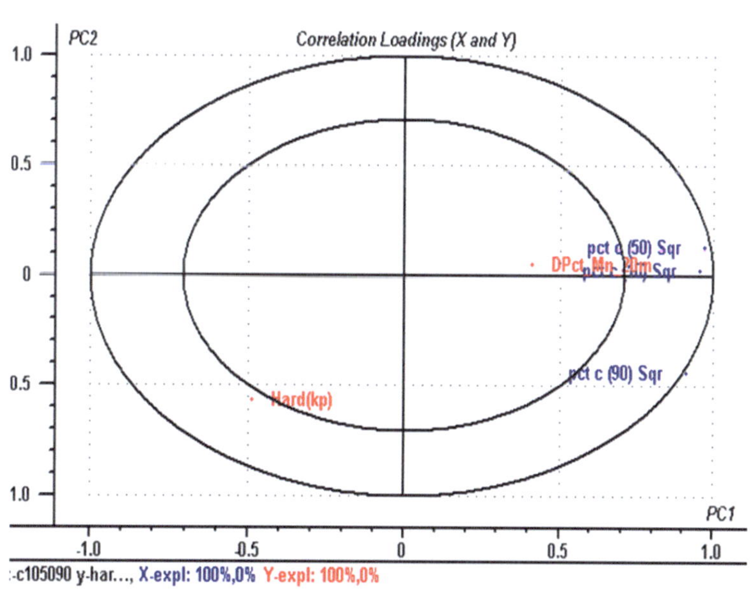

Fig. 10 Correlation by PCA of granule particle size (c10, c50, c90) to tablet dissolution

Another opportunity for control in the design space for solid dosage manufacturing is to understand the relationship of in-line particle size measurement to compression force and then relating both of these to their impact on the final CQA for the product—dissolution. Figure 11 below shows the relationship between in-line particle size (*x*-axis) and dissolution (*y*-axis). The Y2 axis indicates the categorical compression force used to tablet the granulation batches. Blue indicates a lower main compression force and Red a higher compression force. Two very distinct groupings can be seen where by increasing the main compression force leads to higher dissolution results and a much tighter grouping (less impact of granulation on dissolution). This behavior is often seen where high compression forces can fracture the granulation [46–48]. Compression force adjustments can be valuable as part of an overall control strategy to maintain product quality and consistency.

In conclusion, we have demonstrated an approach to utilize modeling to help in development, scale-up and transfer of FBG processes. The FBG process is reduced to its core components (LOD trajectory, particle collision trajectory, and spray distribution). In the case of FBG the LOD trajectory can be modeled while the collision trajectory and spray distribution can be maintained relatively stable through selection of similar equipment types as the process is scaled up. Through modeling of the LOD it is now possible to link the pilot scale processing parameter values to full scale manufacturing parameter values, getting the most out of the pilot scale runs and consuming less material.

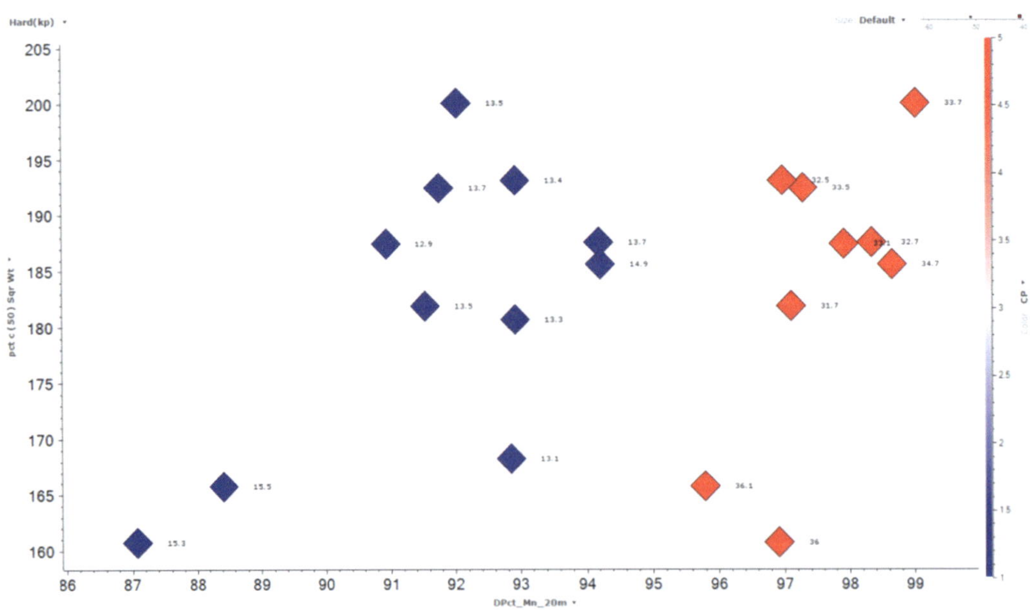

Fig. 11 Impact of tablet press compression force setpoint on dissolution of DOE batches from Fig. 8

3 Example: Modeling of Spray Drying for Small Molecule Active Pharmaceutical Ingredients

3.1 Introduction

Over the last 10 years, spray drying has become an effective technique for pharmaceutical applications where the properties of the molecule either preclude crystallization or the amorphous material provides specific advantages [49]. Within the pharmaceutical industry, amongst other technologies such as bead coating, nano-milling or hot melt extrusion, spray dry technology has been applied as a so-called enabling technology for the manufacturing of amorphous solid dispersions. Amorphous material can provide higher bioavailability, and hence improve the pK-profile, efficacy or reduce the pill burden [49]. Besides manufacturing of amorphous solid dispersions, spray dry technology can also be used as a pure API-isolation technology, e.g., in case the target product is difficult to crystallize in a controlled way [49]. Other important advantages of the technology are the ability to design the particle size distribution and isolate thermally instable products [49].

Spray drying is a method of producing a "dry material" from a "liquid" by rapidly drying using a hot gas stream (air, nitrogen). In the spray drying process, there are four basic steps. (a) Preparation of spray solution/suspension in water or an organic solvent. (b) Spray atomization of the prepared liquid into fine droplets. (c) Drying of the droplet in a drying chamber by a heated gas stream. (d) Separation of the dried droplets/particles from the heated gas/vapor into a collection vessel.

Spray dryers can be operated in an open loop system (in/out), or in a closed-loop configuration using a condenser and a recycle loop for the drying gas. Figure 12 shows a typical spray dryer configuration when operating in closed-loop.

Spray dried material collected in the collection vessel often have residual solvent levels that exceeds ICH-limits. As a consequence, secondary drying (post-drying) is needed. This is typically done in a conventional dynamic dryer. The resulting product can then be processed through conventional oral solid dosage form processing into a final tablet or capsule formulation.

3.2 Critical Quality Attributes and Critical Process Parameters

For pharmaceutical applications and as outlined in ICH-Q10 [7] a Criticality Analysis is performed on the drug product to identify the critical quality attributes (CQA). The CQAs are then through additional development and analysis related back to the critical process parameters (CPP) that impact the CQAs. In most cases for spray dried drug products the amorphous state, the particle size distribution and residual solvent level(s) are identified as CQAs. Other attributes like bulk/tapped density or specific surface area are often found to be not-critical-to-quality, but can be critical-to-manufacturability. As describe above particle formation in spray drying consists of two

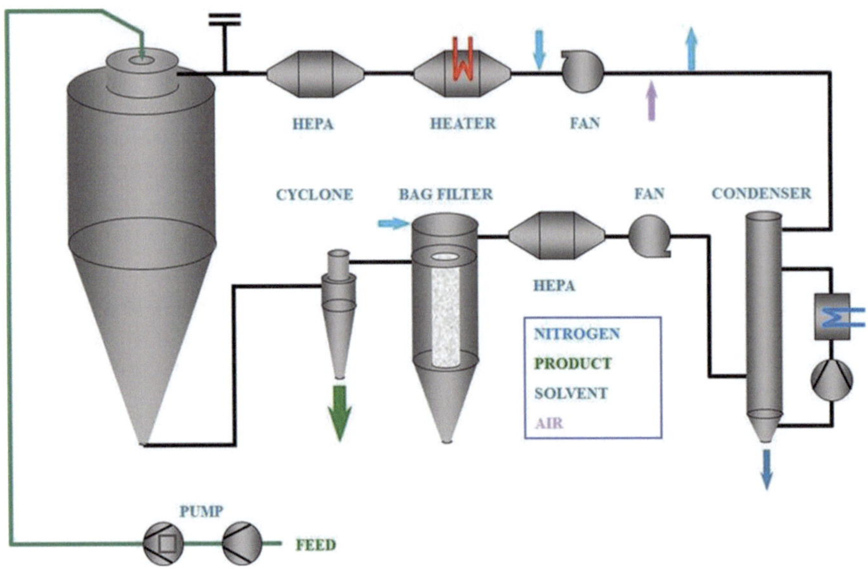

Fig. 12 Closed-loop spray dryer configuration

critical sub-processes (a): the atomization (droplet formation), and (b) the drying of the droplets. Both of these sub-processes typically impact the CQAs and hence both subprocesses contain CPPs. The droplet size is determined by feed properties such as viscosity, density, and surface tension as well as the nozzle configuration and the atomization pressure (or the related flow rate). The evaporation of the droplets is controlled by the thermodynamic conditions and is therefore affected by air/nitrogen flow rate, drying temperature and condenser temperature to mention the primary parameters.

3.3 Spray Dry Modeling

Different types of modeling can be applied to spray dry technology, and these can be found in the literature [49]:

- Process Modeling
- Chemometric Modeling
- Predictive Modeling

In the following we demonstrate how these types of models can be applied to spray drying design, scale-up, and control.

3.4 Process Modeling

As spray drying is fundamentally a drying technology the same principles applied to conventional drying can be applied to the spray drying process. Drying kinetics and vapor liquid equilibrium are determined by the droplet size (at time zero), the temperature, the drying time, and the relative saturation in the carrier gas. Most often temperature is the only process parameter measured on pilot-scale or commercial scale spray dryers. The other critical fundamentals of the process are indirectly controlled by measuring the atomization pressure (or liquid flow rate), the main gas flow rate and the condenser temperature. For the purpose of development, scale-up, and control it is beneficial to separately model atomization and evaporation.

3.4.1 Atomization: Nozzle Characterization and Modeling

The goal of atomization is to produce finely dispersed droplets of the liquid feed stream. On a commercial scale, this is typically done using a high pressure (HP)-configuration. The HP-nozzle consists of different components such as an orifice/swirl or an orifice/insert. Each part contributes to the unique correlation between pressure, flow rate, and droplet size which is again dependent on the specific liquid feed properties, e.g., density, viscosity, surface tension. Furthermore, the feed system is a once-through design, on which normal fluid mechanics apply. This means that for given nozzle and fluid property, pressure and flow are linked directly such that the flow cannot be changed without also affecting the droplet size or likewise the other way around.

$$F = f(P, h, r, d_i) \tag{7}$$

$$d_{32} = f(F, P, \phi) \tag{8}$$

Rather than the exact description of these theoretical correlations (formulation specific), it is a well-accepted practice to characterize nozzle configurations by offline measurement of the droplet size as function of the flow rate (or pressure) for a specific reference liquid. Droplet size measurement can in this way be done by advanced analytical technologies such as laser diffraction or phase Doppler. Figure 13 shows a representation of a typical spray nozzle characterization experiment.

In this way, the scale-up/process modeling approach does not rely on theoretical values, but on experimentally determined droplet size values. In scale-up, the goal is to select a suitable nozzle

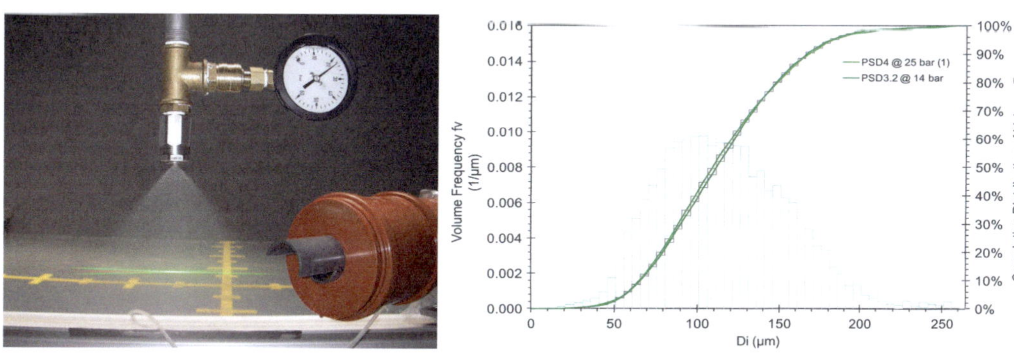

Fig. 13 Droplet size measurement (offline)

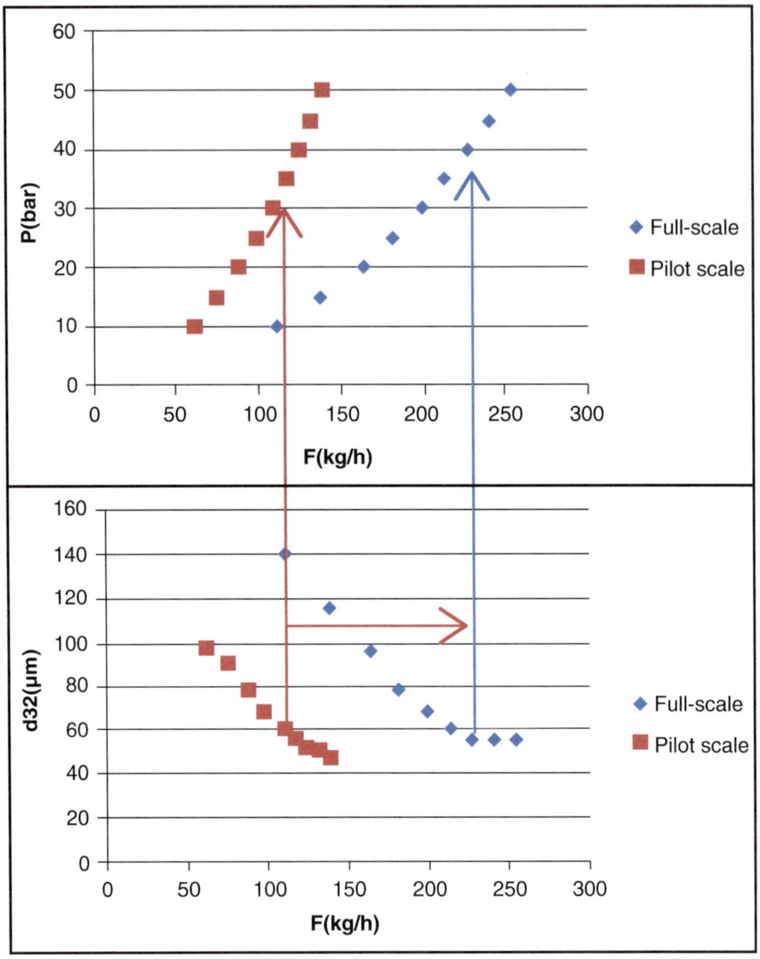

Fig. 14 Nozzle-selection for scale-up

configuration, which maintains the same droplet size, while increasing the throughput. By scaling up the process but keeping the key process fundamental constant the output can be maintained. In this case as long as the atomization provides a similar droplet size distribution at the different scales, the drying kinetics and thermodynamics are kept constant. This approach is conceptually shown in Fig. 14.

3.4.2 Drying Thermodynamics

After successful atomization, drying thermodynamics take over as the fundamentals of the process. The key concept to successful scale-up is to establish the same drying conditions at different scales, while throughput is increased through correct selection of the nozzle type and operating parameters.

Drying kinetics and thermodynamic equilibrium are determined by:

- Droplet size (time zero)
- Drying time

- Drying temperature
- Relative saturation (%RS)

The outlet temperature is recommended to be kept constant in scale-up, while drying time is linked to the residence time inside the spray dry chamber. As for any process, average residence time is defined as the ratio between the process volume (i.e. of the PSD-chamber) and the throughput (i.e. main gas flow + vaporized feed flow rate). The residence time inside the spray dry chamber can in many cases be considered as an equipment specific value, since the main gas flow rate is recommended to be used at a default design value (per scale), which is related to the process volume to maintain constant residence time during scale-up. It is important during scale-up to keep the liquid/gas ratio constant. This will result in a linear scale-up, ensuring that all process conditions can be kept constant at the same time. Nonlinear scale-up is possible but will result in deviating percent relative saturation (%RS) values when dealing with multi-solvent systems in the liquid feed. The amount and composition of solvent left in the solids affect the glass transition temperature and hence the stability and processability of the powder. The relative saturation can be calculated using chemical thermodynamics fundamentals. As the flash evaporation occurs very fast—in milliseconds—after droplet formation is it an acceptable approximation to consider the entire spray dry chamber as an isothermal process-unit operating at the (outlet) temperature and relative saturation. With this assumption, the relative saturation can be expressed as a function of the inlet liquid feed, the inlet main gas flow rate, the vapor concentration in the main gas flow (determined by the condenser temperature), and the outlet temperature of the chamber.

$$\%\mathrm{RS} = f(F_\mathrm{L}, F_\mathrm{G}, T_\mathrm{o}, T_\mathrm{c}) \tag{9}$$

Where, F_L is liquid flow, F_G is gas flow, T_o is outlet temperature and T_c is the condenser temperature.

For the modeling of the thermodynamic equilibria for multi-solvent systems the Antoine equation can conveniently be used. For the condensation step it is assumed that the outlet vapor stream remains fully saturated at the temperature and pressure of the condenser. With this as a starting point, a straightforward mass/energy balance can be compiled across the spray dry chamber enabling calculation of the outlet vapor pressures of the individual components.

3.4.3 Scale-Up Process Models

Having the tools in place for nozzle characterization, and for modeling thermodynamic conditions, the small-scale, experimentally identified design space (using P, T_o, T_c) can be translated into a scale-independent design space using d32, %RS, and P spray dryer. This space can then be used to define/predict the design space on a larger scale, which can be verified by a reduced number of experiments. This approach is illustrated in Fig. 15.

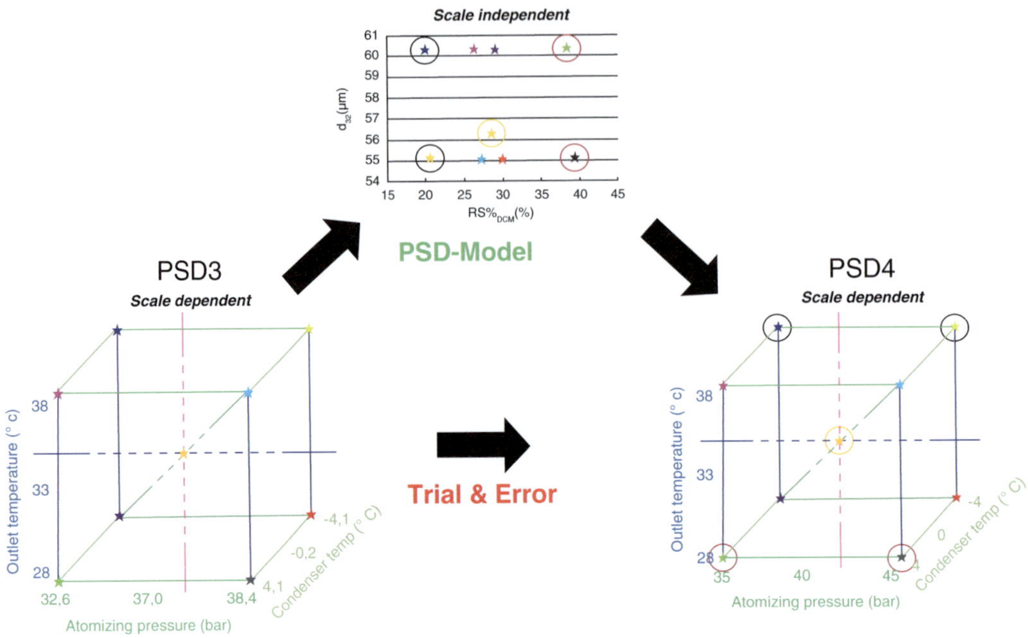

Fig. 15 Scale-up approach for spray drying

3.5 Chemometric Modeling

As mentioned previously during development, scale-up and tech-transfer of pharmaceutical processes the sponsor should investigate process robustness to ensure that product quality is maintained. For this it is desirable to adopt a QbD-approach, which focuses on the multi-variate design space of the critical process parameters. The investigation of the design space is typically done by executing a Design-of-Experiment (DoE) at pilot or even lab scale. Having the previously described models for scale-up ensures that the studied parameter ranges can be translated into relevant parameter ranges at full scale. Different types of DoE are possible, depending of the number of critical process parameters, as well as the desired level of robustness [49]. For spray drying processes, a central composite design is commonly used (Fig. 16). Critical process parameters can for instance be atomization pressure, outlet temperature, and condenser temperature.

DoE-analysis can be performed by different statistical tools, e.g., Design Expert. Figure 17 shows an example of a summary-of-analysis of design.

Based upon experimental data in a multivariate design space, a statistical correlation can be compiled between critical quality attributes (e.g., particle size) and critical process parameters (e.g., P, T_o, T_c). These correlations and its regression parameters show the sensitivity to each individual process parameter. They are purely statistically built and only valid within the experimentally assessed

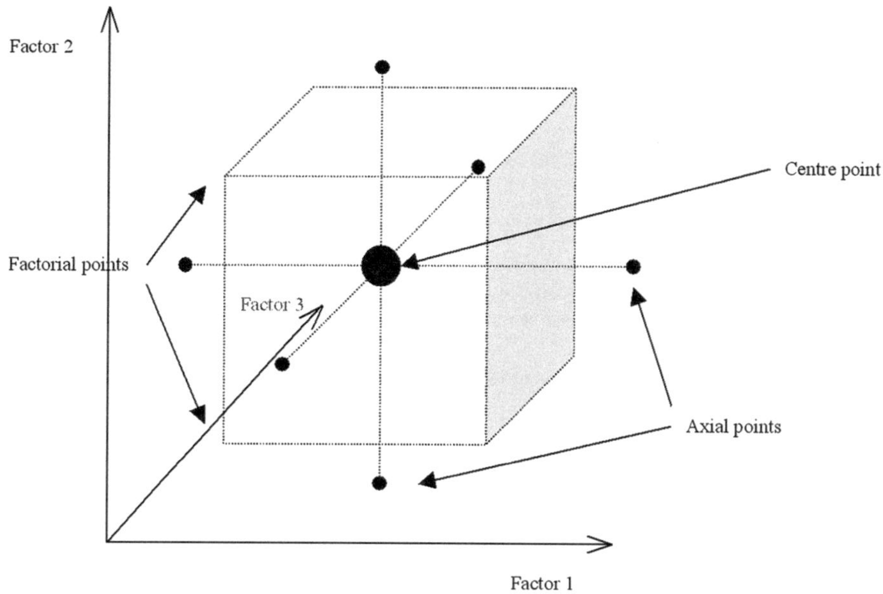

Fig. 16 Central Composite DoE

design space, without any fundamental physics behind. Equation (10) is an example of a typical relation.

$$D50 = a + b \times P + c \times T_{\mathrm{o}} + d \times T_{\mathrm{o}} \times T_{\mathrm{c}} \qquad (10)$$

Once the robustness of the design space has been established, despite the lack of physical fundamental basis, these correlations can be used for statistical based process control as long as the correlations are used within the initial parameter space studied.

3.6 Future Avenues for Spray Drying Predictive Modeling

Ultimately the aim of spray drying modeling is to establish a fully predictive relationship between process and products input (formulation, scale, process conditions) and final product critical quality attributes (e.g., particle size distribution, density, purity). In the academic literature [49], multiple attempts are ongoing to describe the correlation between droplet size and particle size. These approaches combine all elements of atomization and drying thermodynamics into one model. This research area is referred to as "single droplet drying" [49]. Although this is an area of potentially high interest for academia with respect to process understanding and optimization it has so far had limited impact on industrial applications. The main reason is most likely the availability of scale-up models as presented above which when combined with small-scale experiments are relatively accurate while at the same time removing the need to perform sizable at-scale process development, characterization and validation batches. In the pharmaceutical business environment where small molecule properties changes dramatically from compound to compound, small-scale

Range	Min = 39µm Max = 63µm Range = 24µm
Model	P<0.0001 Lack of fit P= 0.1339
Significant terms	Atomisation pressure P=0.0035 Outlet temperature P <0.0001 Condenser temperature P=0.0002 Outlet temperature² P=0.0429
Goodness of fit	Std. Dev.= 2.53 R²= 0.8341 R²adj= 0.8009 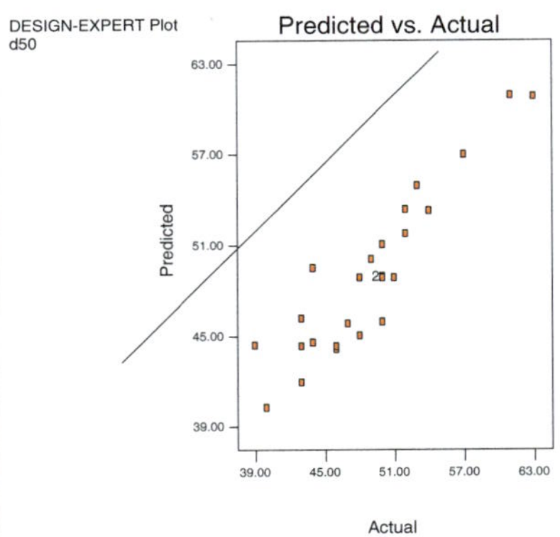
Model	D50 = +172.91372 -0.38919* atomisation pressure -5.58106* outlet temperature +0.74117* condenser temperature +0.068197* outlet temperature²

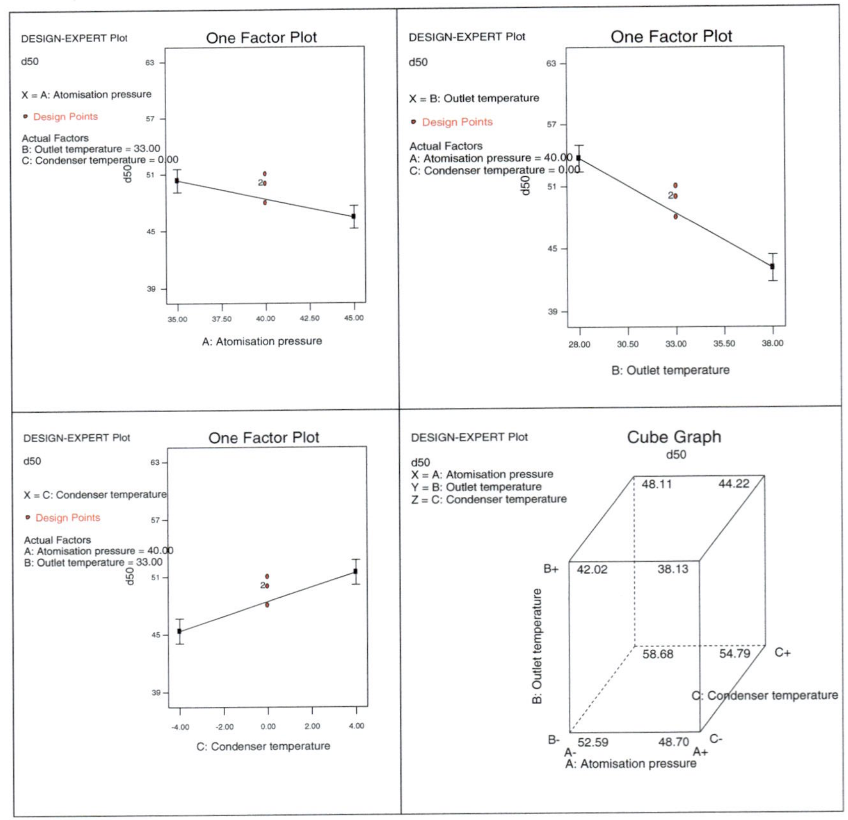

Fig. 17 Summary of analysis design (example)

experiments provide an amount of experimental reality check of sorts at a relatively limited expense. It should be noted that downstream process development, toxicological studies, and pharmacokinetic studies often require small (g to kg) amounts of material. This makes it necessary to maintain physical space to perform small scale batch runs; hence the savings in a fully theoretical approach would likely be limited. Small scale experiments are also often invaluable when dealing with degradation kinetics and powder properties.

4 Example: Modeling of Active Pharmaceutical Ingredient Tablet Coating

4.1 Introduction to Tablet Coating

Coating of tablets post compression is a standard OSD operation. The coating affects organoleptic performance, it can be functional in terms of drug release profile as well as help in ease of identification and be a tool in marketing [50]. Cases where the active drug substance (API) is coated onto the tablet surface of a blank tablet core or as part of a combination product remain a subset of OSDs. While there can be no exact limit for how much coating can be applied, and sugar coatings may comprise as much as 30–50 % of tablet weight, reported total mass increase as function of surface area for various API coated tablets have been around 10–12 mg/cm^2 or in the range of 0.5–5 % of tablet weight [51]. To achieve the required percent relative standard deviation in coat content the overall process time can be substantial at full scale, on the order of several hours up to more than 12 h per batch depending on tablet size and actual scale and the RSD required. As with all OSDs the required %RSD for API in the coating will have to meet standard pharmacopeia criteria [52].

In cases where the unit dose is low in the microgram to low milligram range and where the API may be difficult to combine with other APIs due to stability or be adversely affected by standard tableting operations, coating of the API on to the tablet remains a very attractive option. Once developed it is relatively easy to apply the coating process unto a tablet containing another API, making the API tablet coating a good platform process for products where many combination products of existing products can be envisioned [51, 53].

As discussed previously in this chapter the timeline for completing the drug product development is compressed. The approach to modeling to help reduce the time required for scale-up and tech-transfer will depend significantly on what model-enhanced workstreams are available at the start of the development effort and to what extend these have been verified for the intended purpose such as scale-up or design space determination. For pan coating, one can divide the process and the auxiliary operations into four main pillars for which control of the operational space is

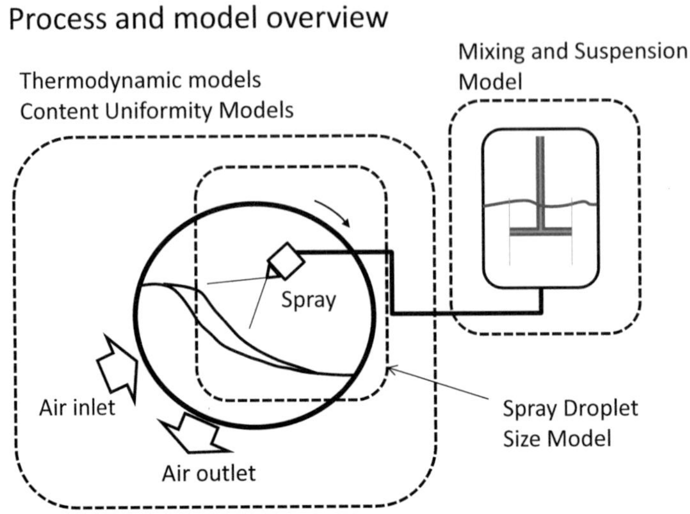

Fig. 18 Model boundaries for coater and auxiliary equipment

required. Coating operations for applying the API solution to the tablets [54]. Heating operation for removal of moisture during coating [43, 44]. Spray gun operations for controlling the delivery of a quality spray [55] and finally control of the spray suspension [56]. Each of these four areas can be modeled separately, either for general understanding or selecting the final process space later on. Figure 18 shows a schematic of where the four models apply.

Table 1 shows the main output of each of these models, the model type and the impact to quality attributes as well as process efficiency. Table 1 shows that only the Content Uniformity model directly predicts a quality attribute. The other three models are all indirect models of quality or process attributes. As described in the introduction this is common since it is often very difficult to obtain models that directly predict final product attributes. Commonly it may be more tractable to determine a ranges or trajectories of certain key process attributes that are quality indicating and scale-independent and allows for scale-up and tech-transfer. An example of this would be spray droplet size. Directly linking droplet size to one or more quality attributes is difficult, but maintaining consistent droplet size distributions throughout development and scale-up to different spray guns and spray rates allows the impact of this attribute on %RSD or yield to be minimized and controlled. Similarly it should be noted that the output of most models of indirect process attribute impacts several final product attributes. To illustrate this consider the thermodynamic model. The main purpose of the thermodynamic model is to allow modeling of scale-independent LOD and temperature trajectories. A LOD or temperature trajectory by themselves does not predict any final product attributes. However by modeling LOD it is possible to select scale

Table 1
Model types and capabilities

Model type	Thermodynamic model	Content uniformity model	Spray droplet size model	Mixing suspension model
Output	LOD profile Product Temp profile	%RSD	Droplet size distribution	Agitation speed for suspension
Model types	ODE based heat and moisture balance	Analytical expression, DEM or Monte Carlo	CFD or correlational	CFD or correlational
Model capability	Scale-up or product/ equipment specific	Product/equipment specific	Scale-up	Scale-up
Quality and processing impact				
• %RSD	Indirect minor	Direct major	Indirect minor	Indirect major
• Potency	Indirect major	Indirect minor	Indirect major	Indirect minor
• Appearance	Indirect major	None	Indirect major	None
• Yield	Indirect major	None	Indirect major	None
• Process efficiency	Indirect minor	Indirect major	Indirect minor	None

Direct = Direct output of model, Indirect = Indirect output of model, Major = Major impact to quality or process attribute, Minor = Minor impact to quality or process attribute. Scale-up model = direct comparison of different equipment sizes, Equipment model = modeling of each equipment based on direct equipment measurements

independent boundaries where product quality is maintained. At the higher LOD trajectory limit the bed may become too wet impacting appearance and %RSD. At the lower LOD trajectory limit, process yields and potency could be impacted. Similarly higher temperatures may lead to degradation or appearance changes and lower temperatures may impact processing efficiency.

4.2 Applications Scenarios

There are a number of potential methods to apply modeling as part of the development, scale-up and tech-transfer process. Which approach is taken will depend significantly on the confidence behind each model and its ability to be used in scale-up, availability of applicable data, prior technical and regulatory experience as well as timeline or financial necessities that may be taken into consideration. Figure 19 below outlines three common approaches.

The minimal approach is to use the model as a means to inform—in the background—a fully empirical approach. The development and scale-up data is still obtained as if there was no modeling. In such cases the model may serve as a means for reducing the risk of failure either for at scale DOE studies or characterization and validation batches. This approach may also be utilized during initial model development before sufficient model base prediction experience has been obtained.

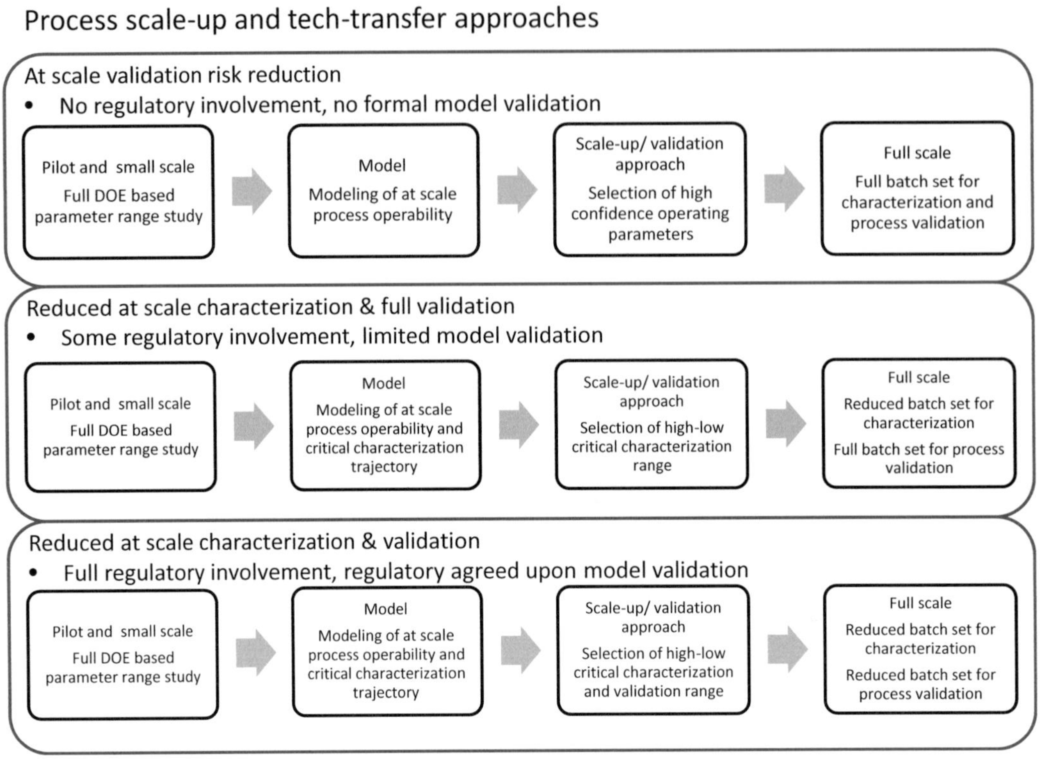

Fig. 19 Application approaches to modeling of coating processes

In the intermediate approach the model is used to reduce at-scale work either just before validation or as part of intermediate scale-up campaigns, but does not impact process validation. The model will serve to inform on at-scale process performance based on pilot or small scale studies. To enable such prediction it is important that the model is based on first principals to the extent possible and it will likely be required that the model scale-up performance was tested in one or more products before being used in this manner. The predictive power of the model should be documented and discussed in collaboration with the regulatory function. In the fully modeled scale-up approach the model is applied not only to process characterization work, but also to validation batches. In such a scenario the use of the model will likely need to be agreed upon with health authorities prior to use to avoid having insufficient data from the validation batches. In this scenario the model will have to be well documented and likely validated on its own, either from prior work or as part of the development and scale-up path. If the model is to be validated as part of the development process, this may require additional experimental work involving PAT technology or other in-line measurements that can help validate that the models predictive power. In the remaining of this chapter we focus mostly on the Content Uniformity Model to demonstrate how this model can be applied to aid in process development, scale-up, and validation regardless of the path taken above.

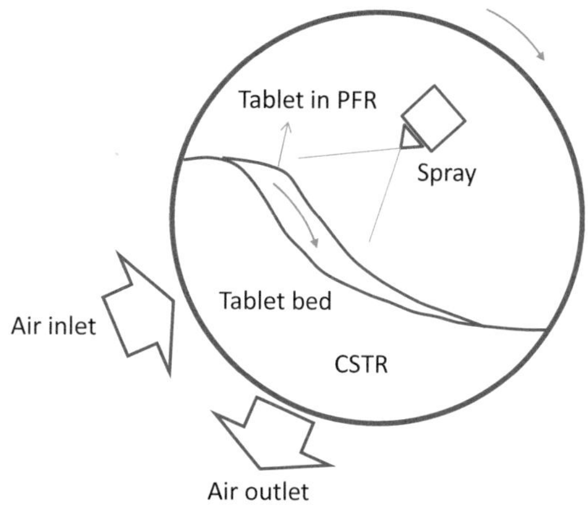

Fig. 20 Schematic of a coating pan with modeled CSTR and PFR shown

4.3 Pan Coating Operations

During a pan coating process tablets are placed in a horizontally mounted cylindrical pan such that they will form a tumbling or cascading flow when the pan is rotated. The pan is equipped with spray gun(s) which are stationary inside the pan and directed downward towards the bed. See Fig. 20 for illustration.

During rotation the spray guns direct a spray of droplets towards the bed to distribute the coating liquid onto the tablets at the surface. Perforations in the pan surface or sides or through baffles allow drying air to enter and exit the coating bed while it is coating. Baffles placed on the pan surface or held into the bed help move tablets axially during rotation to enhance mixing. The purpose of the spray is to distribute the coating liquid onto the tablets evenly and with minimal losses. During the coating process sufficient air and heat has to be applied to remove the applied liquid (which is usually water) and maintain acceptable moisture levels.

4.4 Background on Models for RSD Determination in OSD Coating Processes

In case of coating active pharmaceutical ingredient one of the primary goals during development is to understand the critical process parameters that affect %RSD. These parameters can be divided into operating, product, and equipment parameters. One approach to describe %RSD would be to seek to accurately describe the path of every tablet in the pan coater as it operates. Recent applications of Discreet Element Modeling have allowed such models to be built [57, 58]. At this time the computational effort for such undertakings are still too great to allow modeling of the full process. However it is possible to extract significant information around tablet flows within the coater that this can be used in conjunction with simplified modeling approaches or for parameterization of such models. Another way to model %RSD is to follow one tablet in the coating pan as it moves through the coater. This approach was developed using Monte Carlo methods where by one tablet is tracked repeatedly through a series of random walks [59, 60]. This approach is less computationally intensive than DEM, but here it can be difficult to link the random walk parameters directly to physical parameters in the actual system. A third approach which we present in more detail here is to model the coater as two distinct zones, one being the coating zone and another being a mixing zone [54].

4.5 A Simple Relative Standard Deviation Coating Model

Percent RSD is defined by the following Eq. (11):

$$\%\text{RSD} = \frac{\sqrt{(-m_i - \overline{m})/n}}{\overline{m}} \tag{11}$$

Where m_i is the mass of API on the ith tablet in the sample, $\overline{m}$ is the average mass of API and n is the number of tablets in the sample.

If we represent the pan as two zones one being the coating zone and one being the mixing zone, it is possible to see the process as a series of transitions back and forth between these two zones. If we assume that the travel through the coating zone is plug flow like and the mixing zone behaves as a CSTR it is possible to apply the same principles of residence time distribution that has been applied previously throughout the chemical process industry [57, 61, 62]. Figure 20 shows a radial view of the pan coater with the CSTR-PFR concept schematically outlined.

The model's key attribute is that all of its input parameters can be measured directly with commercially available methodologies. The model is not a scale-up model in that the inputs have to be measured directly for each coater, tablet, and process parameter combination. In practice the amount of experimental input needed can be significantly reduced by use of correlations developed for tablet velocity, number density, and coating zone width.

In the following section we briefly describe the main parts of the derivation of the %RSD as function of input parameters as presented earlier [54]. The model is an extension of earlier work by others [58, 63–66]. The model was developed based on the following assumptions:

1. the spray is uniform within the spray zone axially and horizontally.
2. the tablets are free flowing and well mixed in the CSTR zone and unmixed in the PFR zone. It should be noted that axial mixing in the pan is assumed to be sufficient to characterize the pan as a 2D model as shown in Fig. 20.
3. the coating zone (PFR) is much smaller than the mixing zone (CSTR).
4. tablets only acquire coated material from the spray and not through contact with each other.

The %RSD as defined in Eq. (11) can be represented as shown in Eq. (12) either expressed in terms of a mass standard deviation or based on time through the coating zone.

$$\%\mathrm{RSD} = \frac{\sigma_m}{\overline{m}} = \frac{\sigma_t}{\overline{t}} \tag{12}$$

Where σ_{m} and σ_{t} are the mass and time based standard deviations and where m_{bar} and t_{bar} are the mean weight gain or time spent in the coating zone. Using the methodology of standard residence time distribution based on a series of CSTR-PFR it is possible to arrive at the following equation for %RSD.

$$\%\mathrm{RSD} = \sqrt{\frac{\Gamma(2+N)}{N^2(N-1)!} - 1} = \sqrt{\frac{1}{N}} \tag{13}$$

Where N is the real number of times the tablets passed through the coating zone. For practical applications where N can be approximated by an integer we arrive at a final simple expression for $\%\mathrm{RSD} = N^{-1/2}$.

Given that the coating zone is small and that tablets spend the majority of their time in the mixing zone, N can be calculated as the ratio of the total coating time t and the mean residence time in the mixing zone t_{m}. From a practical perspective this result, while mathematically impressive still is not yet fully useful in that t_{m} is still difficult to estimate directly. If we invoke the mass balance that all tablets are either in the mixing zone or in the coating zone, and the coating zone is small compared to the mixing zone we see that t_{m} may be estimated based on the flow rate f through the coating zone and the total number of tablets in the coater V_{total}. See Eq. (14).

$$N = \frac{t}{t_{\mathrm{m}}} = \frac{tf}{V_{\mathrm{m}}} \approx \frac{tf}{V_{\mathrm{total}}} \tag{14}$$

Finally if we extend the expression of the flow rate in the coating zone based on our initial assumptions of plug flow, we can write the flow as the tablet velocity v [cm/s], coating zone width w [cm], and tablet density ρ [tablets/cm^2] (Eq. 15). The last three quantities are all properties that can be measured directly by suitable camera systems mounted in the coater or maybe modeled using Discrete Element Modeling approaches. Finally it is also possible to estimate these based on correlational approaches and data provided in the literature.

$$f = v\rho w \tag{15}$$

We arrive at the final result being that we can express our critical quality attribute %RSD as follows (16):

$$\%\mathrm{RSD} = \sqrt{\frac{1}{N}} = \sqrt{\frac{V_{\mathrm{m}}}{t v \rho w}} \tag{16}$$

Where V_{m} is the total number of tablets in the coater [#], t is the coating time [s], v is the tablet velocity in the coating zone [cm/s], ρ is the tablet density in the coating zone [#/cm^2], and w is the coating zone width [cm].

4.6 Model Usage and Incorporation with DOE Experimentation

DOE either full or partial factorial designs are generally the main method employed to demonstrate process performance and determine process parameter ranges. In the case of tablet coating, one issue is that common parameters such as spray rate, drum rpm and air flow are scale dependent and do not scale directly with pan or batch size. The %RSD model is therefore used in combination with the DOE to enable scale-independent determination of parameter ranges. Prior to executing the DOE the model may also be used to execute a sensitivity analysis against known equipment capability ranges. This allows selection of the right parameters and the correct range to utilize in the DOE. In many cases the selection of DOE parameters and ranges for process definition are not the best settings of parameters for testing the predictive power and capability of the model itself. It is therefore important to validate the model for it intended usage prior to the DOE.

When considering the coating pan as shown in Fig. 20, it is important to distinguish between parameters that affect the thermal efficiency of the process, such as spray rate, air temperature, and air flow versus parameters that affect coating uniformity such as pan loading, rotational speed, coating time, and spray zone width, and those that affect spray gun performance such as atomization and pattern air pressure, nozzle type and spray rate. If combined into a single DOE these parameters will form the basis for a very large experimental plan that may not be practical to perform on a reasonable scale. A more targeted approach would be to build plan for understanding the performance of each of these sub-processes separately and then select conditions that provides for control of each. The plan can be either a purely empirical DOE approach or a model aided DOE approach.

If we compare the effort required for process scale-up between a model aided DOE approach with that of a purely empirical approach the work involved with model building and validation is significant. The return of this investment is based on the value of the product required to perform large scale studies and the value of the potential time savings. In the case of coating processes where it is hard to select scale-independent process parameters to study in the DOE, the issue becomes that the DOE's have to be repeated upon scale-up to some extent to demonstrate control. For cosmetic coatings where placebo tablets may be use and the coating quality range is less stringent than when coating API's, this effort may not be warranted. For new molecular entities where the availability of the API may be limited during the development phase the situation can be very different. As an example of this consider the study from 2002 [67]. In this study an efficient Plackett–Burman design was used to study 6 factors with a total of 18 runs at a relatively modest scale of 8 kg. The factors they chose were atomization pressure, fan pressure, pan speed, inlet air flow, spray time and inlet air temperature. They concluded that it was possible to obtain %RSDs below 6 % in perforated pans and identified atomization pressure, pan speed and coating time as critical process parameters. Given that pan speed, pan loading and coating zone width will change with scale, the continuation of the development would likely require additional DOE's at larger scale. Also considering the modeling work performed after this study was conducted, the finding that atomization pressure was critical seems to suggest that there were secondary effects going on in the coater either affecting the thermal state or the spray zone width.

Another study [68] used a prioritization matrix approach to reduce seven initial factors down to five final factors. These factors were spray rate, inlet temp, air flow rate, solids content, and pan speed. Based on the remaining 5 factors they conducted a 32 run study at 15 kg scale each. They concluded that based on their study they were able to determine a process operating space. Furthermore they remarked it is prudent to reevaluate the DOE when studying scale-dependent process parameters and when there are significant increases in scale or changes in equipment. Again we see the same pattern where the presence of a model would likely have resulted in a different and more fundamental choice of DOE factors.

A third study [69] used a thermodynamic based model to assist the scale-up of from a Glatt GC1250 and IMA 200 (115–160 kg scale) to a Glatt GC1500 (350 kg) scale. They used the model to provide a scientific basis for establishing the process parameters in the new equipment while using a QbD approach. They used an approach of environmental similarity inside the pan to set inlet air temperature set points. An atomization droplet size model was used to determine atomization air pressures. They concluded that fundamental process models can be applied in a QbD approach while determining the design space during scale-up. In terms of experimental effort this study was highly efficient due to the utilization of a model based approach. In the study presented above [54] it was remarked that in a traditional approach, DOE experiments are used to study the impact of process variables on product quality. If parameters that affect only one of four sub-processes (thermal efficiency, coating uniformity, spray performance mixing) are lumped into a larger DOE, then these will contribute to a significant enlargement of the required experimental design. Furthermore, if some parameters such as pan speed and loading contribute to a single model parameter that may be independently measured, it can be possible to have even further reductions in the experimental design.

While the model presented here [54] is elegant it is not a direct scale-up equation, in that the parameters have to be determined for each scale, product and operating parameter set. This is a key limitation, but it can partially be overcome due to all the parameters being directly measureable a priori. They further estimated that compared to partial or full factorial designs it is possible to combine the experimental runs with model predictions such that one may realize further reduction in experimental effort. The reductions observed were on the order of 30–70 %. At small to modest scale such an approach is likely to be only marginally beneficial given the model effort required, but once batch sizes reach hundreds of kg the monetary value of the batch alone may justify the model based approach.

5 Examples of Modeling of Physical Properties of Oral Solid Dosage Forms

Formulation of new Active Pharmaceutical Ingredients (API's) with poor aqueous solubility is a challenge since delivering them to the gut in the thermodynamically stable crystal form will lead to little dissolution and poor bioavailability. One way to overcome this challenge is to deliver these medicines as a solid solution of amorphous API molecularly dissolved in a glassy polymer. Although some amorphous formulations can have API concentrations over the thermodynamic solubility limit and not crystallize over pharmaceutically relevant timescales due to kinetic stabilization (prevention of crystallization due to restricted mobility of the API in the glassy polymer matrix) it is desirable to have formulations below the thermodynamic solubility limit in order to avoid the risk that the crystallization kinetics may not always be sufficiently inhibited [70, 71].

5.1 Thermodynamics Applied to OSD Development

Thermodynamic models for the prediction of solubility are widely used in the pharmaceutical industry for designing unit operations such as crystallization [72–77]. This is because these models are accurate enough to save development time by guiding researchers in the right direction towards desired solubility profiles. Typically, the models are used to explore a wide design space and then the most interesting model-predicted results are selected for follow-up verification in the laboratory, eliminating the need to do round after round of laboratory experiments in order to explore the design space. Given the success of thermodynamic models in this area and the need for thermodynamic data to design amorphous API/polymer solid solution formulations, there is currently much interest in applying thermodynamic models to designing these formulations [78–90].

Validation of models used to predict if the API will be under the thermodynamic solubility limit is extremely important in order to enable decisions to be made based on these models in the pharmaceutical industry. A first step in model validation is collection of the data that the model predictions will be compared to in order to test their accuracy. In the case of solid solutions of API's dissolved in glassy polymers, this can be much more challenging than, for example, measuring the solubility of an API in a solvent or mixture of solvents to be used in a crystallization [71, 91, 92]. However, researchers in the field of amorphous API/polymer solid solutions have risen to the challenge and developed methods capable of determining the thermodynamic solubility of API's in these polymer blends. Typically, Differential Scanning Calorimetry (DSC) is used, sometimes in combination with other supporting techniques such as Powder X-Ray Diffraction (PXRD), Hot Stage Microscopy (HSM), and Infrared spectroscopy (IR) [71, 79, 88, 91–95]. Use of DSC in combination with improved sample preparation via cryomilling [71, 91] and in some cases annealing the sample [92] has been demonstrated to produce results that are sufficiently accurate to be used in design of the amorphous API/polymer solid solution formulations [71]. Therefore, these techniques should also be sufficiently accurate to produce the data needed for model validation, enabling development of accurate thermodynamic models to predict the necessary solubility data. It should be noted that these experimental techniques can be very time consuming, making it desirable to validate the models with a minimum amount of data but also increasing the power of the models to save time in pharmaceutical formulation development. Additionally, these techniques work by measuring the phase behavior above room temperature with extrapolation back to room temperature, typically with an empirical correlation having a very simple form [71]. Using a rigorous thermodynamic model for the extrapolation to room temperature would no doubt improve the accuracy of that extrapolation.

5.2 Historical Approaches to API-Polymer Solubility Modeling

The desire to use thermodynamic models to predict the equilibrium solubility of API's in the polymers for use in these formulations is not a new one and various approaches have been tried by researchers in the field. There has been extensive use of Hansen and Hildebrand solubility parameters to attempt to predict the phase behavior of these systems [78–81, 87, 88]. Unfortunately use of these solubility parameters has many limitations including the lack of an ability to account for specific and/or directional association interactions between molecules such as hydrogen bonding, and the fact that it is fundamentally based in a "like dissolves like" assumption which in addition to lacking thermodynamic rigor, completely ignores entropic contributions to the interactions between the molecules in the system. Indeed, no doubt due to these reasons, researchers have concluded that these models have either very poor predictive power [78] or very limited predictive power [79, 80, 88].

Flory-Huggins theory has also been used in various attempts at predicting the solubility of these systems [85, 86, 89, 90]. These approaches do add some necessary thermodynamic rigor (and corresponding improvements in predictive power) with improved estimates of the mixing entropy in these systems, but ultimately the question remains of how to get at the all-important chi parameter describing the mixing enthalpy of the system. Ways to estimate the chi parameter used in the literature include group contribution methods [85, 89, 90] melting point depression [85, 86, 89] measurements of the solubility of the API in a solvent similar to the monomer that the polymer is made of [85, 86], and atomistic simulations [84]. Atomistic simulations require significant computational resources and therefore have limited applicability in pharmaceutical modeling enhanced workstreams. The group contribution or experimentally based approaches involve more rapid calculations and data that may already have been collected for other pharmaceutical development research reasons so they fit into pharmaceutical development more readily. Unfortunately, specific and/or directional association interactions are not accounted for in the entropic term and are typically not accounted for when calculating the chi parameter. These models, while a vast improvement on approaches that consider only solubility parameters, can therefore still be limited in their predictive power [85, 86, 90].

5.3 New Developments Using SAFT

One approach that has appeared in the literature recently involves applying Perturbed Chain Statistically Associating Fluid Theory (PC-SAFT) to the problem yielding a method capable of modeling the specific association interactions (such as hydrogen bonding) that some of the other models cannot incorporate and a model that is firmly rooted in rigorous thermodynamics. This model involves six parameters which are based on the fundamental physics of the molecules in question: the number of segments that make up the molecule, the size of those segments, their interaction energy, the number of sites used to describe the association (hydrogen bonding) interactions, the size of the association sites, and the interaction energy of the association sites. Since these parameters are based on the molecular physics in question, some (such as the number of association sites) may be fixed based on the molecular structure, reducing the number of parameters to be regressed. The necessary parameters for homopolymers can typically be found in the literature [83] and the parameters for API's can be regressed based on solubility data for the API's in solvents. Since the solubility data for API's in solvents is typically gathered during crystallization process development, this model fits into current pharmaceutical development work streams quite readily. Good qualitative and quantitative accuracy has been observed for the prediction of solubility of API's in homopolymers [83] and the model has been extended to describe copolymers with good qualitative and quantitative accuracy for the prediction of solubility of API's in these copolymers [82].

We can conclude that there is a bright future for the use of thermodynamic models to predict the solubility of API's in polymers for use in amorphous solid solutions. The data necessary to validate the model predictions can be measured, enabling development of models that will be sufficiently accurate to speed up pharmaceutical development time lines. Simplistic models that lack accuracy and thermodynamic rigor such as the Hansen and Hildebrand solubility parameter based approaches have been examined enough to know that they are not fit for this purpose and the field is moving on to more fundamentally sound and accurate approaches such as PC-SAFT.

References

1. Scholar G. Search terms: pharmaceutical granulation, milling, tableting process modeling or pharmaceutical extraction reaction crystallization process modeling. In *Google Scholar* 2014
2. Mikami T, Kamiya H, Horio M (1998) Numerical simulation of cohesive powder behavior in a fluidized bed. Chem Eng Sci 53:1927–1940
3. Grossmann C, Amanullah M, Erdem G, Mazzotti M, Morbidelli M, Morari M (2008) 'Cycle to cycle' optimizing control of simulated moving beds. AIChE J 54:194–208
4. Flavien N. Application of QbD in continuous processing. In *Dynochem Seminar* 2014
5. Singh R, Sahay A, Muzzio F, Ierapetritou M, Ramachandran R (2014) A systematic framework for onsite design and implementation of a control system in a continuous tablet manufacturing process. Comput Chem Eng 66:186–200
6. Guideline IHT. Pharmaceutical development. Q8(2R). As revised in August 2009
7. Guideline IHT. Quality risk management. Q9. Current Step 2005, 4: 408
8. Guideline IHT. Pharmaceutical quality system. Q10. Current Step 2008, 4
9. Liu H, Kang Y, Lin P (2013) Hybrid finite-discrete element modeling of geomaterials fracture and fragment muck-piling. Int J Geotech Eng 9:115
10. Hastie DB, Wypych P (2010) Experimental validation of particle flow through conveyor transfer hoods via continuum and discrete element methods. Mech Mater 42:383–394
11. Norton T, Sun D-W (2006) Computational fluid dynamics (CFD) – an effective and efficient design and analysis tool for the food industry: a review. Trends Food Sci Technol 17:600–620
12. Räsänen E, Rantanen J, Mannermaa JP, Yliruusi J (2004) The characterization of fluidization behavior using a novel multichamber microscale fluid bed. J Pharm Sci 93:780–791
13. Faure A, York P, Rowe RC (2001) Process control and scale-up of pharmaceutical wet granulation processes: a review. Eur J Pharm Biopharm 52:269–277
14. Sauer D, Cerea M, DiNunzio J, McGinity J (2013) Dry powder coating of pharmaceuticals: a review. Int J Pharm 457:488–502
15. Srivastava S, Mishra G (2010) Fluid bed technology: overview and parameters for process selection. Int J Pharm Sci Drug Res 2:236–246
16. Gu L, Liew C, Heng P (2004) Wet spheronization by rotary processing-a multistage single-pot process for producing spheroids. Drug Dev Ind Pharm 30:111–123
17. Vuppala MK, Parikh DM, Bhagat HR (1997) Application of powder-layering technology and film coating for manufacture of sustained-release pellets using a rotary fluid bed processor. Drug Dev Ind Pharm 23:687–694
18. Leuenberger H, Luy B, Studer J (1990) New development in the control of a moist

agglomeration and pelletization process. STP Pharma Sci 6:303–309
19. Burggraeve A, Van den Kerkhof T, Hellings M, Remon JP, Vervaet C, De Beer T (2011) Batch statistical process control of a fluid bed granulation process using in-line spatial filter velocimetry and product temperature measurements. Eur J Pharm Sci 42:584–592
20. Burggraeve A, Silva AF, Van den Kerkhof T, Hellings M, Vervaet C, Remon JP, Vander Heyden Y, De Beer T (2012) Development of a fluid bed granulation process control strategy based on real-time process and product measurements. Talanta 100:293–302
21. Burggraeve A, Monteyne T, Vervaet C, Remon JP, De Beer T (2013) Process analytical tools for monitoring, understanding, and control of pharmaceutical fluidized bed granulation: a review. Eur J Pharm Biopharm 83:2–15
22. Ehlers H, Liu A, Raikkonen H, Hatara J, Antikainen O, Airaksinen S, Heinamaki J, Lou H, Yliruusi J (2009) Granule size control and targeting in pulsed spray fluid bed granulation. Int J Pharm 377:9–15
23. Watano S, Takashima H, Miyanami K (1997) Control of moisture content in fluidized bed granulation by neural network. J Chem Eng Jpn 30:223–229
24. Watano S, Sato Y, Miyanami K (1995) Application of fuzzy logic to moisture control in fluidized bed granulation. J Chem Eng Jpn 28: 282–287
25. Watano S, Fukushima T, Miyanami K (1994) Application of fuzzy logic to bed height control in agitation-fluidized bed granulation. Powder Technol 81:161–168
26. Narvanen T, Lipsanen T, Antikainen O, Raikkonen H, Yliruusi J (2008) Controlling granule size by granulation liquid feed pulsing. Int J Pharm 357:132–138
27. Lister J, Smit D, Hounslow M (1995) Adjustable discretized population balance for growth and aggregation. AIChE J 41:591–603
28. Hounslow M, Ryall R, Marshall V (1988) A discretized population balance for nucleation, growth, and aggregation. AIChE J 34: 1821–1832
29. Kumar J, Peglow M, Warnecke G, Heinrich S, Mörl L (2006) Improved accuracy and convergence of discretized population balance for aggregation: the cell average technique. Chem Eng Sci 61:3327–3342
30. Hu X, Cunningham J, Winstead D (2008) Understanding and predicting bed humidity in fluidized bed granulation. J Pharm Sci 97: 1564–1577
31. Troup GM, Georgakis C (2013) Process systems engineering tools in the pharmaceutical industry. Comput Chem Eng 51:157–171
32. Laínez JM, Schaefer E, Reklaitis GV (2012) Challenges and opportunities in enterprise-wide optimization in the pharmaceutical industry. Comput Chem Eng 47:19–28
33. Stephanopoulos G, Reklaitis GV (2011) Process systems engineering: from Solvay to modern bio- and nanotechnology: a history of development, successes and prospects for the future. Chem Eng Sci 66:4272–4306
34. Bogomolov A (2011) Multivariate process trajectories: capture, resolution and analysis. Chemom Intell Lab Syst 108:49–63
35. Hartung A, Knoell M, Schmidt U, Langguth P (2011) Role of continuous moisture profile monitoring by inline NIR spectroscopy during fluid bed granulation of an Enalapril formulation. Drug Dev Ind Pharm 37:274–280
36. Chablani L, Taylor MK, Mehrotra A, Rameas P, Stagner WC (2011) Inline real-time near-infrared granule moisture measurements of a continuous granulation-drying-milling process. AAPS PharmSciTech 12:1050–1055
37. Li W, Bruce R, Fermier A, Mehrman S (2009) PAT-monitoring fluid-bed wet granulation and tablet production by NIR and NIR chemical imaging. Am Pharma Rev 12:30
38. Lipsanen T, Narvanen T, Raikkonen H, Antikainen O, Yliruusi J (2008) Particle size, moisture, and fluidization variations described by indirect in-line physical measurements of fluid bed granulation. AAPS PharmSciTech 9: 1070–1077
39. Li W, Cunningham J, Rasmussen H, Winstead D (2007) A qualitative method for monitoring of nucleation and granule growth in fluid bed wet granulation by reflectance near-infrared spectroscopy. J Pharm Sci 96:3470–3477
40. Abberger T (2001) The effect of powder type, free moisture and deformation behaviour of granules on the kinetics of fluid-bed granulation. Eur J Pharm Biopharm 52:327–336
41. Goldschmidt M, Weijers G, Boerefijn R, Kuipers J (2003) Discrete element modelling of fluidised bed spray granulation. Powder Technol 138:39–45
42. Grima AP, Wypych PW (2011) Investigation into calibration of discrete element model parameters for scale-up and validation of particle–structure interactions under impact conditions. Powder Technol 212:198–209
43. Page S, Baumann K-H, Kleinebudde P (2006) Mathematical modeling of an aqueous film coating process in a Bohle Lab-Coater, Part 1:

Development of the model. AAPS PharmSciTech 7:E79–E86

44. Page S, Baumann K-H, Kleinebudde P (2006) Mathematical modeling of an aqueous film coating process in a Bohle Lab-Coater: Part 2: Application of the model. AAPS PharmSciTech 7:E87–E94
45. Hoffmann L, Greiter I, Wagner A, Weiss V, Alefeld G (1996) Experimental investigation of heat transfer in a horizontal tube falling film absorber with aqueous solutions of LiBr with and without surfactants. Int J Refrig 19:331–341
46. Bodmeier R (1997) Tableting of coated pellets. Eur J Pharm Biopharm 43:1–8
47. Jarosz PJ, Parrott EL (1983) Comparison of granule strength and tablet tensile strength. J Pharm Sci 72:530–535
48. Johansson B, Wikberg M, Ek R, Alderborn G (1995) Compression behaviour and compactability of microcrystalline cellulose pellets in relationship to their pore structure and mechanical properties. Int J Pharm 117:57–73
49. Broadhead J, Edmond Rouan S, Rhodes C (1992) The spray drying of pharmaceuticals. Drug Dev Ind Pharm 18:1169–1206
50. Fourman GL, Hines CW, Hritsko RS (1995) Assessing the uniformity of aqueous film coatings applied to compressed tablets. Pharm Technol 19:70–76
51. Tobiska S, Kleinebudde P (2003) Coating uniformity and coating efficiency in a Bohle Lab-Coaterusing oval tablets. Eur J Pharm Biopharm 56:3–9
52. Pharmacopoeia J (2002) United States' Pharmacopeia
53. Tobiska S, Kleinebudde P (2003) Coating uniformity: influence of atomizing air pressure. Pharm Dev Technol 8:39–46
54. Chen W, Chang SY, Kiang S, Marchut A, Lyngberg O, Wang J, Rao V, Desai D, Stamato H, Early W (2010) Modeling of pan coating processes: prediction of tablet content uniformity and determination of critical process parameters. J Pharm Sci 99:3213–3225
55. Aliseda A, Hopfinger E, Lasheras JC, Kremer D, Berchielli A, Connolly E (2008) Atomization of viscous and non-Newtonian liquids by a coaxial, high-speed gas jet. Experiments and droplet size modeling. Int J Multiphase Flow 34:161–175
56. Buurman C, Resoort G, Plaschkes A (1986) Scaling-up rules for solids suspension in stirred vessels. Chem Eng Sci 41:2865–2871
57. Yamane K, Sato T, Tanaka T, Tsuji Y (1995) Computer simulation of tablet motion in coating drum. Pharm Res 12:1264–1268
58. Fichana D, Marchut AJ, Ohlsson PH, Chang S-Y, Lyngberg O, Dougherty J, Kiang S, Stamato H, Chaudhuri B, Muzzio F (2009) Experimental and model-based approaches to studying mixing in coating pans. Pharm Dev Technol 14:173–184
59. Pandey P, Katakdaunde M, Turton R (2006) Modeling weight variability in a pan coating process using Monte Carlo simulations. AAPS PharmSciTech 7:E2–E11
60. Nakamura H, Abe E, Yamada N (1998) Coating mass distributions of seed particles in a tumbling fluidized bed coater Part II. A Monte Carlo simulation of particle coating. Powder Technol 99:140–146
61. Levenspiel O (1972) Chemical reaction engineering, vol 2. Wiley, New York, NY
62. Danckwerts P (1953) Continuous flow systems: distribution of residence times. Chem Eng Sci 2:1–13
63. Wnukowski P, Setterwall F (1989) The coating of particles in a fluidized bed (residence time distribution in a system of two coupled perfect mixers). Chem Eng Sci 44:493–505
64. Sherony DF (1981) A model of surface renewal with application to fluid bed coating of particles. Chem Eng Sci 36:845–848
65. Mann U (1983) Analysis of spouted-bed coating and granulation. 1. Batch operation. Ind Eng Chem Process Des Dev 22:288–292
66. Joglekar A, Joshi N, Song Y, Ergun J (2007) Mathematical model to predict coat weight variability in a pan coating process. Pharm Dev Technol 12:297–306
67. Rege BD, Gawel J, Kou JH (2002) Identification of critical process variables for coating actives onto tablets via statistically designed experiments. Int J Pharm 237:87–94
68. Teckoe J, Mascaro T, Farrell TP, Rajabi-Siahboomi AR (2013) Process optimization of a novel immediate release film coating system using QbD principles. AAPS PharmSciTech 14:531–540
69. Prpich A, am Ende MT, Katzschner T, Lubczyk V, Weyhers H, Bernhard G (2010) Drug product modeling predictions for scale-up of tablet film coating—a quality by design approach. Comput Chem Eng 34:1092–1097
70. Vasconcelos T, Sarmento B, Costa P (2007) Solid dispersions as strategy to improve oral bioavailability of poor water soluble drugs. Drug Discov Today 12:1068–1075
71. Kyeremateng SO, Pudlas M, Woehrle GH (2014) A fast and reliable empirical approach for estimating solubility of crystalline drugs in polymers for hot-melt extrusion formulations. J Pharm Sci 103:2847

72. Tung HH, Tabora J, Variankaval N, Bakken D, Chen CC (2008) Prediction of pharmaceutical solubility Via NRTL-SAC and COSMO-SAC. J Pharm Sci 97:1813–1820
73. Spyriouni T, Krokidis X, Economou IG (2011) Thermodynamics of pharmaceuticals: prediction of solubility in pure and mixed solvents with PC-SAFT. Fluid Phase Equilib 302: 331–337
74. Ruether F, Sadowski G (2009) Modeling the solubility of pharmaceuticals in pure solvents and solvent mixtures for drug process design. J Pharm Sci 98:4205–4215
75. Kokitkar PB, Plocharczyk E, Chen C-C (2008) Modeling drug molecule solubility to identify optimal solvent systems for crystallization. Org Process Res Dev 12:249–256
76. Chen C-C, Crafts PA (2006) Correlation and prediction of drug molecule solubility in mixed solvent systems with the nonrandom two-liquid segment activity coefficient (NRTL-SAC) model. Ind Eng Chem Res 45: 4816–4824
77. Chen C-C, Song Y (2004) Solubility modeling with a nonrandom two-liquid segment activity coefficient model. Ind Eng Chem Res 43:8354–8362
78. Nair R, Nyamweya N, Gönen S, Martınez-Miranda L, Hoag S (2001) Influence of various drugs on the glass transition temperature of poly (vinylpyrrolidone): a thermodynamic and spectroscopic investigation. Int J Pharm 225: 83–96
79. Forster A, Hempenstall J, Tucker I, Rades T (2001) Selection of excipients for melt extrusion with two poorly water-soluble drugs by solubility parameter calculation and thermal analysis. Int J Pharm 226:147–161
80. Just S, Sievert F, Thommes M, Breitkreutz J (2013) Improved group contribution parameter set for the application of solubility parameters to melt extrusion. Eur J Pharm Biopharm 85:1191–1199
81. Hancock BC, York P, Rowe RC (1997) The use of solubility parameters in pharmaceutical dosage form design. Int J Pharm 148:1–21
82. Prudic A, Kleetz T, Korf M, Ji Y, Sadowski G (2014) Influence of copolymer composition on the phase behavior of solid dispersions. Mol Pharm 11:4189–4198
83. Prudic A, Ji Y, Sadowski G (2014) Thermodynamic phase behavior of API/polymer solid dispersions. Mol Pharm 11:2294
84. Maus M, Wagner KG, Kornherr A, Zifferer G (2008) Molecular dynamics simulations for drug dosage form development: thermal and solubility characteristics for hot-melt extrusion. Mol Simul 34:1197–1207
85. Marsac PJ, Shamblin SL, Taylor LS (2006) Theoretical and practical approaches for prediction of drug–polymer miscibility and solubility. Pharm Res 23:2417–2426
86. Marsac PJ, Li T, Taylor LS (2009) Estimation of drug–polymer miscibility and solubility in amorphous solid dispersions using experimentally determined interaction parameters. Pharm Res 26:139–151
87. Gupta J, Nunes C, Vyas S, Jonnalagadda S (2011) Prediction of solubility parameters and miscibility of pharmaceutical compounds by molecular dynamics simulations. J Phys Chem B 115:2014–2023
88. Greenhalgh DJ, Williams AC, Timmins P, York P (1999) Solubility parameters as predictors of miscibility in solid dispersions. J Pharm Sci 88:1182–1190
89. Tian Y, Booth J, Meehan E, Jones DS, Li S, Andrews GP (2012) Construction of drug–polymer thermodynamic phase diagrams using Flory–Huggins interaction theory: identifying the relevance of temperature and drug weight fraction to phase separation within solid dispersions. Mol Pharm 10:236–248
90. Thakral S, Thakral NK (2013) Prediction of drug–polymer miscibility through the use of solubility parameter based Flory–Huggins interaction parameter and the experimental validation: PEG as model polymer. J Pharm Sci 102:2254–2263
91. Tao J, Sun Y, Zhang GG, Yu L (2009) Solubility of small-molecule crystals in polymers: d-mannitol in PVP, indomethacin in PVP/VA, and nifedipine in PVP/VA. Pharm Res 26: 855–864
92. Sun Y, Tao J, Zhang GG, Yu L (2010) Solubilities of crystalline drugs in polymers: an improved analytical method and comparison of solubilities of indomethacin and nifedipine in PVP, PVP/VA, and PVAc. J Pharm Sci 99: 4023–4031
93. Leuner C, Dressman J (2000) Improving drug solubility for oral delivery using solid dispersions. Eur J Pharm Biopharm 50:47–60
94. Forster A, Hempenstall J, Rades T (2001) Characterization of glass solutions of poorly water-soluble drugs produced by melt extrusion with hydrophilic amorphous polymers. J Pharm Pharmacol 53:303–315
95. Rumondor AC, Ivanisevic I, Bates S, Alonzo DE, Taylor LS (2009) Evaluation of drug-polymer miscibility in amorphous solid dispersion systems. Pharm Res 26:2523–2534

Chapter 2

Population Balance Models for Pharmaceutical Processes

Anwesha Chaudhury, Maitraye Sen, Dana Barrasso, and Rohit Ramachandran

Abstract

The pharmaceutical industry is predominantly dominated by the handling of particulate matter in the form of solids and emulsions. With the enforcement of the Quality by Design (QbD) initiative by the Food and Drug Association (FDA), a process systems engineering based case toward particulate process design is advantageous. This suggests the need for mechanistic modeling approaches that can be used for an accurate representation of the process dynamics. The inherent discrete nature of population balance models (PBM) makes it an appropriate framework for modeling particulate processes. With the representation of the particulate processes used for pharmaceutical product manufacturing using various modeling frameworks, advancements can be made to improved control and optimization of the process. This chapter provides a detailed review on the applicability and significance of PBMs in drug product manufacturing and is aimed to provide greater insight into the field of process systems engineering.

Key words Pharmaceutical manufacturing, Population balance model, Flowsheet modeling, Particulate processes, Quality by Design, Process systems engineering

1 Introduction

Particulate processes are ubiquitous in various fields of engineering. Such processes, which are essentially multiphase in nature, are crucial and extensively used in industries, such as pharmaceuticals, detergent manufacturing, polymer processing, and food processing. Within the pharmaceutical industry, these unit operations involve handling of powder or emulsions and transforming their inherent attributes, improving flowability, uniformity, modifying their bioavailability/dissolution characteristics or changing the polymorphic form. Historically, manufacturing costs contributed little to the overall cost of bringing a product to the market. However, due to rising research and development expenditures, rising manufacturing costs, expiration of drug patents, and competition from generic manufacturers, the industry has recognized the need to alter its manufacturing practices to improve operational

Marianthi G. Ierapetritou and Rohit Ramachandran (eds.), *Process Simulation and Data Modeling in Solid Oral Drug Development and Manufacture*, Methods in Pharmacology and Toxicology, DOI 10.1007/978-1-4939-2996-2_2, © Springer Science+Business Media New York 2016

efficiency. The proper regulation of such particulate processes in the pharmaceutical industry is of immense importance due to the imposition of a tight quality criteria [75]. The Quality by Design (QbD) initiative launched by the US FDA suggests that quality should be built into a product based on a thorough understanding of the product and process by which it is developed and manufactured, along with an in-depth knowledge of the risks involved in manufacturing the product and procedures to mitigate those risks. Thus, it is beneficial for a pharmaceutical company to implement a process systems engineering-based approach with heavy focus on modeling in order to attain a sustainable position in the market [31, 75, 89]. A model-based systems approach can play a crucial role in the design, operation, evaluation, and analysis of processes because of its potential to provide better process understanding, leading to a reduction in the time and cost for the operation of the process [31, 39].

Processes involving particulate matter cannot be modeled via the continuum approach (e.g., Navier Stokes/momentum balance equation for fluid) since these systems are discrete in nature and can be described using the population balance modeling (PBM) framework. Size, porosity, and composition are some of the critical properties that characterize the end product obtained from these processes. PBMs can be successfully considered for modeling particulate processes, owing to its discrete nature. PBMs are a class of hyperbolic partial differential equations that represent the mesoscopic framework. They involve classification of the particulate population within a process or a system based on discrete set of classes describing their inherent attributes by tracking the number of particles in each class as they are subjected to rate processes. It can be successfully used to relate macroscopic properties based on the information obtained at the microscale. Figure 1 shows the overall pharmaceutical drug manufacturing process starting from the synthesis of the drug. It can be seen that a number of unit operations within the flowsheet (circled) can be conveniently modeled using PBMs, such as crystallization, granulation, milling, mixing, and coating.

PBMs have been very popular for modeling the unit operations involved in the pharmaceutical industry. Significant progression in the application of PBMs for improved pharmaceutical operation has been observed over the past decade [11, 46, 63]. Several researchers have implemented the PBM framework to model the crystallization process [1, 11]. More detailed discussion is presented in the following sections. Granulation has also been modeled using PBMs, but there still exists a lack of available mechanistic knowledge [80]. Milling is yet another poorly understood process which has been described using PBMs [3, 6]. The mixing process has also recently been studied in detail using PBMs [86]. PBMs have also been used to represent coating in order to control and track the

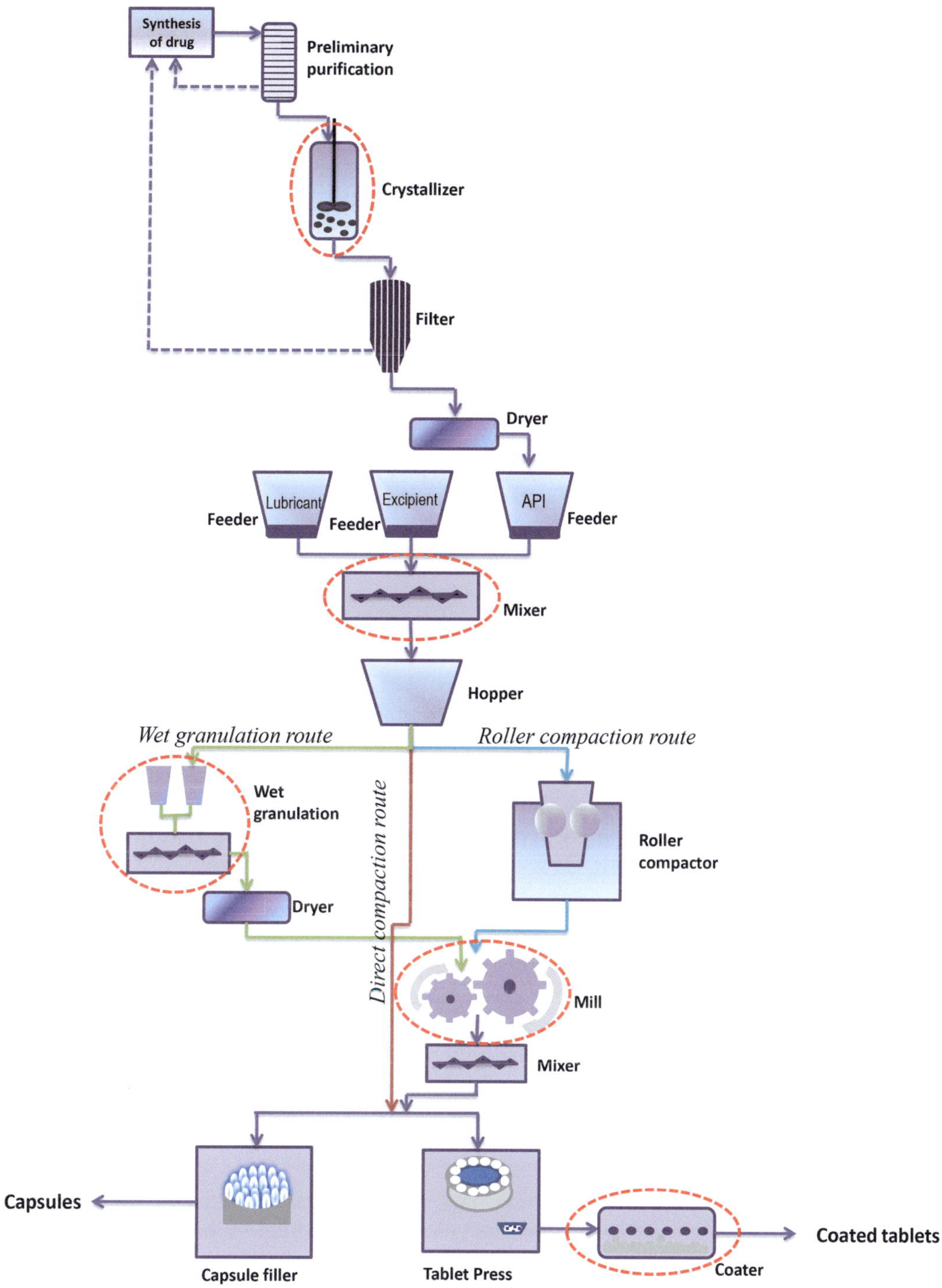

Fig. 1 Schematic of process for continuous drug manufacturing (PBM applicable on the circled unit operations)

coating variability. Coating investigations pertaining to polydisperse particles are limited in the literature. Experimental observations in a fluid bed coater suggested that coating distribution of the particles is primarily governed by the available surface area [61, 71]. Wnukowski and Setterwall [98] identified two distinct regions within the coater which suggests the existence of an *active* zone (where particles are coated and dried) and the *less active* zone (away from the coating fluid nozzle and near the walls). Later, Maronga and Wnukowski [58] identified three distinct domains within a fluid bed coater suggesting the existence of an *active spray zone*, an *active drying zone* and a *non-active domain*. With this observation, the coating process has been extensively modeled using a compartment based PBM which accounts for the inhomogeneities associated with the existence of distinct regions within the coater [30, 46, 57, 79]. A surface renewal based model has been implemented to model coating processes in order to track the motion of particles between the bulk and spray zone [24, 88]. This approach is useful to identify how particles get replenished in the spray zone with the progress of the operation. Some work has also been performed to model continuous coating processes [42]. However, utilization of PBMs for that purpose has not been explored significantly.

Using a mathematical model that captures the system behavior, various model based control [72] and optimization techniques [32] can be performed which improve the operation of the process. In this article we aim to provide the readers with a detailed discussion on the application of PBMs in modeling particulate processes, followed by solution techniques, implementation of reduced order models in order to combat the high computational overheads associated with solving PBMs and various advanced studies that can be performed using PBMs.

2 Mathematical Model Development

Particulate processes are very common in the pharmaceutical industry although they are very poorly understood due to the lack of mechanistic knowledge in terms of solid handling. The discrete nature of the particulate processes makes it difficult to understand the obscure behavior of such systems. Over the past few decades population balance has been successfully utilized for modeling processes such as crystallization, granulation, coating, milling, and mixing. The PBM in its most general form can be written as [74]:

$$\frac{\partial F}{\partial t}(\mathbf{x}, t) + \frac{\partial}{\partial \mathbf{x}}\left[F\frac{d\mathbf{x}}{dt}\right](\mathbf{x}, t) = \mathfrak{R}(x, t, F(x, t)) \qquad (1)$$

where F is the number density of the particulate matter, $\mathbf{x}$ is a vector of internal coordinates whose variation is tracked along the progression of the process, the partial derivative terms in the left-hand side of the equation represent the growth terms whereas the terms in the right-hand side ($h(x, t, F(x, t))$) depict the source terms involving an overall change in the number of particles. The source terms comprise of the rate of aggregation, breakage, and nucleation. In the following sections further discussion is provided with regard to using PBMs for describing various pharmaceutically relevant processes.

2.1 Crystallization

The crystallization process is used in the initial stages after the synthesis of the drug product. It can be used for various purposes, such as separation, purification, or stabilization of the molecule (through polymorph transformation). This step primarily decides the particle size distribution of the API crystals formed. It is desirable to obtain a uniform size distribution so that the properties of the final pharmaceutical blend remain uniform throughout. Crystallization occurs in two stages. The first stage is nucleation, where several nuclei of a distinct solid phase are formed. The second stage is crystal growth, during which the nuclei grow until a critical crystal size is reached.

When a supersaturated solution is cooled, nucleation and crystal growth takes place, causing the compound concentration in the solution to decrease. Hence, crystallization can be carried out in the manufacturing by manipulating the temperature profile to maintain a specific cooling schedule or by controlling the addition rate of an anti-solvent. When a crystal is formed, it retains some mother liquor that gets occluded within the crystal mass. PBMs are helpful to relate the shape and product size distribution to process parameters.

1-D PBMs have been most commonly used to model crystallization [1, 35, 55],until recently, when some 2-D models have also been developed to describe the crystallization processes [16, 27, 34, 54]. Certain 2-D models also employ the volume and surface area in order to account for the shape of particles [47]. The 1-D and 2-D PBMs are, respectively, shown as

$$\frac{\partial F(L,t)}{\partial t} + \frac{\partial\left(G(L,C,t)F\right)}{\partial L} = B_0(C,t)\delta(L) \tag{2}$$

$$\frac{\partial F(L_1,L_2,t)}{\partial t} + \frac{\partial (G_1(L_1,L_2,C,t)F)}{\partial L_1} + \frac{\partial (G_2(L_1,L_2,C,t)F)}{\partial L_2} = B_0(C,t)\delta(L_1)\delta(L_2) \tag{3}$$

Here, the number density, F, is a function of the two length directions and the concentration of solute in the solution. $G/\ G_1$,

G_2 represent the growth terms in the length direction (L/L_1,L_2), and B_0 represents the nucleation term. The growth and the nucleation terms in Eq. (3) can be written as

$$G_1 = k_{g1}\left(\frac{C - C_{sat}}{C_{sat}}\right)^{g_1} \tag{4}$$

$$G_2 = k_{g2}\left(\frac{C - C_{sat}}{C_{sat}}\right)^{g_2} \tag{5}$$

$$B_0 = k_b\left(\frac{C - C_{sat}}{C_{sat}}\right)^{b} \tag{6}$$

Here, C is the concentration of solute in the solution, C_{sat} is the solubility of the solute at the particular thermodynamic conditions, and k_{g1}, k_{g2}, k_b, g_1, g_2, b are the various empirical parameters used to represent growth and nucleation in the form of power law expressions.

Some works have also been observed that consider mother liquor occlusion into the crystal by representing it using a separate dimension [84, 87]. An empirical expression has been obtained for the growth with respect to the additional liquid dimension in order to address the occlusion of liquid into the crystals. From the works of Miki et al. [60], the calculations for mother liquor inclusion in a KDP crystal has been adopted, which was then used to extend the dimensionality of the PBM. Figure 6 in the work of Miki et al. [60] shows the relation between the amount of mother liquor included in the crystal and the size of the crystal. Although the size coordinate considered in the paper was unidimensional, Sen et al. [84] considered two length dimensions. An averaged equivalent length was utilized, which was obtained by calculating the diameter of a sphere with an equal amount of volume as the cuboidal crystal in case of the latter. Considering the depth of the crystal to be L_1, the equivalent length of the crystal can be expressed as

$$L_{eq}(L_1, L_2) = \left(\frac{6}{\pi} \times L_1^2 L_2\right)^{\frac{1}{3}} \tag{7}$$

From Eqs. (2) and (3), it can be seen that the number density of particles is a function of the solute concentration, thus indicating the need for a mass balance equation to update the solute concentration in the solution over time. The mass balance equation can be written as

$$\frac{dC}{dt} = -\rho_c \int_0^\infty \int_0^\infty F(L_1, L_2, t)\left(2G_1(L_1L_2 - L_1^2) + G_2L_1^2\right) dL_1 dL_2 \tag{8}$$

where ρ_c is the density of the crystal. The solubility of the solute in the solution is also a function of the temperature of the crystals. Hence, an energy balance equation is also necessary to quantify the change in the temperature, T over time. Aggregation of crystals has also been addressed in several works involving 1-D PBMs [7, 22, 43], however addressing aggregation in the case of multi-dimensional models (while preserving the shape) has not yet been explored. This is primarily due to the lack of mechanistic or statistical information suggesting the propensity of agglomeration based on the length/surface area of the colliding surfaces/length scales. Recently, PBMs have also been employed to study the continuous crystallization process by various researchers [70].

One of the most significant purposes of developing sophisticated models for crystallization processes is the ability to adopt the QbD approach. Using the detailed PBM, the outcome of a crystallization process can be predicted. The crystallization PBM can be calibrated using standard optimization techniques in order to be utilized for predictive purposes. In the work by Sen et al. [85], a 2-D PBM has been employed to validate the model against experimental results. A 2-D PBM enables the crystal shape to be tracked over time and provides more information regarding the progression of the crystallization process. The objective function for the parameter estimation is the minimization of the sum of squared error between the experimental and model predicted concentration. This objective function has been formulated as shown in Eq. (9):

$$\Omega^{\text{Concentration}}(\mathbf{p}) = \sum_{i=1}^{n} \left\| C_{predicted}(t_i, \mathbf{p}) - C_{experimental}(t_i) \right\|^2 \tag{9}$$

where Ω represents the objective function, $C_{predicted}$ and $C_{experimental}$ represent the predicted and the experimental concentrations, respectively, t_i represents the ith time instant, and $\mathbf{p}$ represents the set of empirical parameters within the PBM.

The shape of the crystal is tracked using the aspect ratio as expressed in Eq. (10):

$$\text{AR}(t) = \frac{\sum_{L_1=0}^{L_1\,\max} \sum_{L_2=0}^{L_2\,\max} F(L_1, L_2, t)\frac{L_1}{L_2}}{\sum_{L_1=0}^{L_1\,\max} \sum_{L_2=0}^{L_2\,\max} F(L_1, L_2, t)} \tag{10}$$

The optimization was solved using the *fminsearch* function in MATLAB which employs the Nelder–Mead algorithm. Figure 2 presents the validation results using a proprietary molecule. The model reveals good agreement with the experimental

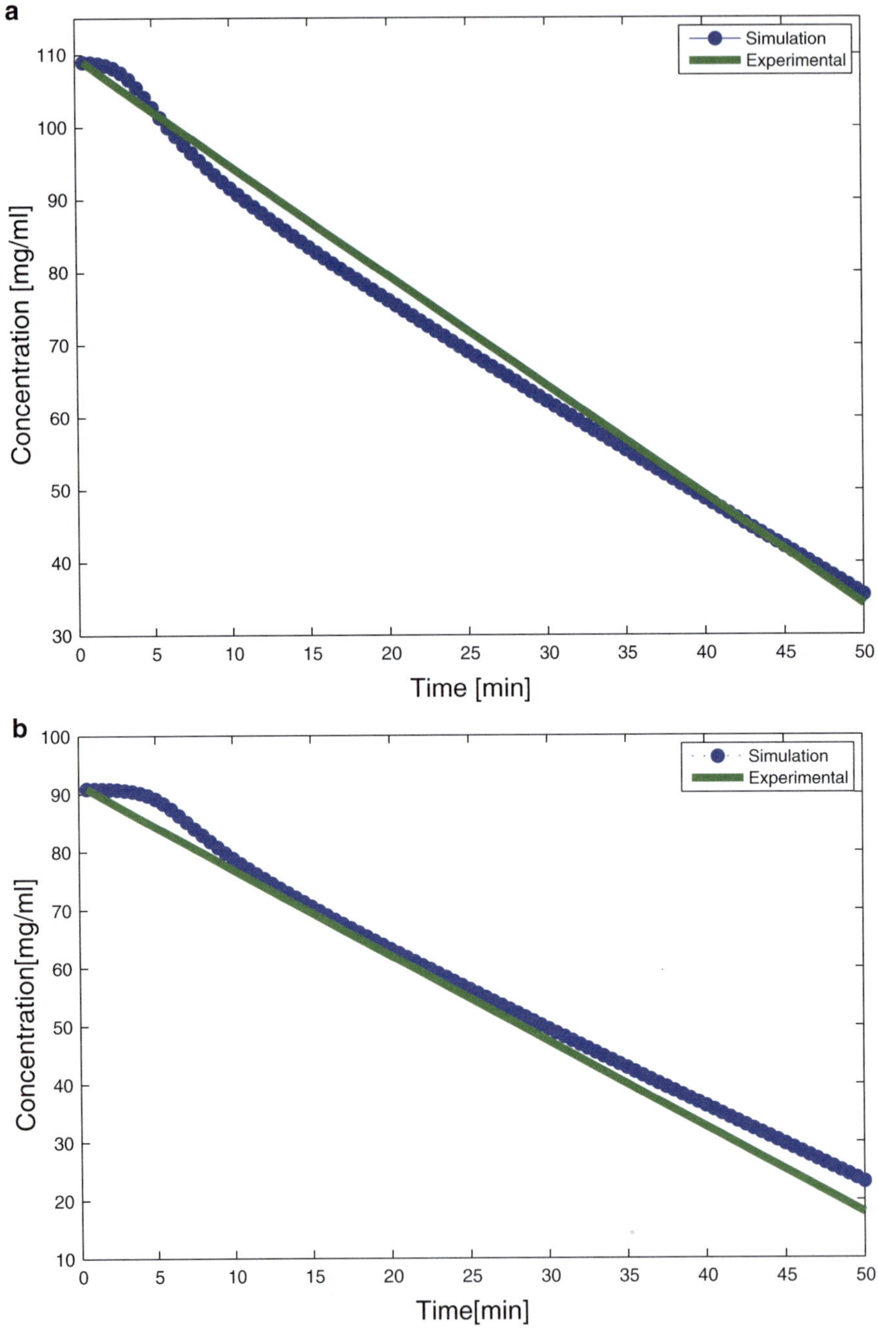

Fig. 2 Validation results from parameter estimation for the evolution of concentration with time for different cooling profiles

observations. The concentration of the solute was fitted with the simulated results using experimental data. The validation results show good agreement between the PBM and the experimental results. Typically, the crystallization process can be calibrated using experimental data consisting of crystal size and shape, the concentration of solute in the mother liquor, temperature or supersaturation of the system. For more advanced studies, other quantities such as polymorph composition or the mother liquor occlusion volume can also be considered.

2.2 Mixing

Mixing is one of the most important unit operations in the downstream tablet manufacturing process. In the pharmaceutical industry, an active pharmaceutical ingredient (API) is mixed with one or more excipients prior to tabletting, and the blend quality is primarily determined by the mixing operation. Poor mixing can lead to segregation and result in potency variations in the final dosage form. The outcome of this is a poor quality product which has a higher chance of being out of specification. Mixing is brought about by the particle velocity gradient within the mixer when two or more distinct bulk material particles come into intimate contact. Variability in the mixture is induced by segregation which occurs mainly when particles of different densities tend to settle in different layers [76]. The population balance equation for the mixing process can be written as shown in Eq. (11) [82]. Both internal and external coordinates have been considered, hence a multi-dimensional formulation of the PBM is required.

$$\frac{\partial}{\partial t}F(n,x,y,r,t)+\frac{\partial}{\partial x}\left[F(n,x,y,r,t)\frac{dx}{dt}\right]+\frac{\partial}{\partial y}\left[F(n,x,y,r,t)\frac{dy}{dt}\right]$$
$$+\frac{\partial}{\partial r}\left[F(n,x,y,r,t)\frac{dr}{dt}\right]=\Re_{\text{formation}}(n,x,y,r,t)$$
$$+\Re_{\text{depletion}}(n,x,y,r,t)+\text{Inflow}-\text{Outflow} \tag{11}$$

Here, x is the spatial co-ordinate in the axial direction, y is the spatial co-ordinate in the radial direction, and r is the internal co-ordinate of particle size. The counter n stands for the number of components. For example, if the model deals with the mixing of two components (component A and component B), $n = 1$ represents component A and $n = 2$ represents component B. The terms $\frac{dx}{dt}$ and $\frac{dy}{dt}$ represent the velocities in the axial and radial directions, respectively. $F(n, x, y, t)$ is the particle number density (number of particles per unit volume), which varies with the spatial location inside the mixer and the type of particle. Inflow is the rate at which the components are fed to the system, held at a constant value over time. Outflow is the rate at which the components exit the mixer.

Within the pharmaceutical industry, certain API particles are cohesive and tend to stick to each other, forming aggregates. As a result, cohesion leading to aggregation has been studied both experimentally [92] and via DEM simulations [77]. The aggregation rate process is defined in Eqs. (12)–(14):

$$\mathfrak{R}_{\text{aggregation}} = \mathfrak{R}_{\text{formation}} - \mathfrak{R}_{\text{depletion}} \tag{12}$$

where

$$R_{\text{formation}} = 0.5\int_{r_{\min}}^{r-r_{\min}} \beta(n,x,y,r',r-r')F(n,x,y,r',t) \times F(n,x,y,r-r',t)dr \tag{13}$$

$$R_{\text{depletion}} = \beta(n,x,y,r,r')F(n,x,y,r,t)F(n,x,y,r',t)dr \tag{14}$$

$\beta(n, x, y, r, r')$ is the aggregation kernel defined in Eq. (15). The aggregation kernel is based on the kinetic theory of granular flow (KTGF) [90].

$$\beta(n,x,y,r,r') = K\sqrt{\frac{3\theta}{\rho}(r+r')^2}\sqrt{\frac{1}{r^3}+\frac{1}{r'^3}} \tag{15}$$

Here K is a constant, ρ is the particle density, and θ is a pseudo-temperature termed granular temperature (Eq. (16)). By definition, v is the random fluctuation of the velocity within a continuous granular medium; however, in this study v is assumed to be the actual particle velocity depicted by its magnitude [33]. This assumption has been deemed valid since the relative differences in the aggregation model are important, as opposed to the absolute differences.

$$\theta = \frac{1}{3} < v.v > \tag{16}$$

In certain cases, formation of aggregates during mixing is not desired. For example, in the pharmaceutical industry, the API is rendered free flowing by addition of suitable lubricants [91]. In the case of free flowing powders, the internal coordinates can be dropped and the population balance equation can be written as shown in Eq. (17):

$$\frac{\partial}{\partial t}F(n,x,y,t)+\frac{\partial}{\partial x}\left[F(n,x,y,t)\frac{dx}{dt}\right]+\frac{\partial}{\partial y}\left[F(n,x,y,t)\frac{dy}{dt}\right]$$
$$=\text{Inflow}-\text{Outflow} \tag{17}$$

One of the approaches for modeling a mixer is by dividing it both axially and radially into several compartments. If there are $x_{\max} \times y_{\max}$ compartments, then outflow (as seen in Eq. (11)) can be represented as $\sum_{n=1}^{2}\sum_{y=1}^{y_{\max}} F(n, x_{ma}x, y, t)\,V_f$, which is the total number of particles exiting from the end compartments. V_f is the forward axial velocity. Mixing can occur in both the axial and radial directions by convection and dispersion. In a continuous mixer, mixing takes place when the particles are moved about by the motion of the blades with the dispersive component being negligibly small relative to the convective one. Such assumptions have been justified in literature [66]. Particles can be treated as discrete entities and their exchange between any two compartments is simulated. It is assumed that homogeneous mixing occurs in each of the compartments. The axial and radial velocities move the particles from one compartment to another in both the axial and radial directions. The exchange of mass between the compartments has been represented as a number of particles. Particles can either move forward to the compartment ahead of it or backward to the compartment behind it. On the other hand, radial mixing conserves the total number of particles at a fixed axial location at any given point in time. Hence the mass exchange (in terms of particle density) of a single component among the compartments can be simplified according to Eq. (18), given below:

$$\frac{\partial F(n,x,y,t)}{\partial t}=\frac{V_f[F_{n,x-1,y,t}-F_{n,x,y,t}]}{\Delta x}+\frac{V_b[F_{n,x+1,y,t}-F_{n,x,y,t}]}{\Delta x}$$
$$+V_r\frac{[F_{n,x,y+1,t}+F_{n,x,y-1,t}-2F_{n,x,y,t}]}{\Delta y} \tag{18}$$

The above equation can be written for each component present in the mixer. Here, V_f refers to the forward velocity in the axial direction, V_b refers to backward velocity in the axial direction, and V_r refers to the radial velocity.

The inputs to this model are V_f, V_b, and V_r. The velocity values can be obtained either experimentally or from a previously run detailed numerical simulation (e.g., DEM simulation). DEM calculates the velocity values from the particle properties (e.g., particle diameter, density). Once velocity values are extracted, the simulation can be used to provide information about the dynamics and the outputs of the process.

Consider a case where two components (component A and component B) are being mixed, where component A can be taken as the API and component B as the excipient. At inlet, the flowrate of the component A (API) is 16.3 % of the total flowrate (i.e., the sum of component A and component B inlet flowrates). The input parameters V_f, V_b, and V_r have been obtained from a previously run DEM simulation, where a continuous mixer has been simulated using EDEM™ (DEM Solutions). A commercial mixer (Gericke GCM250™) with impeller blades in alternating forward and backward orientation has been simulated. Equal number of particles each of component A and B have been introduced into the mixer using two feeders discharging particles on either side of the inlet. A feed rate of 1990 particles per second and an impeller speed of 250 RPM have been maintained. The details of the DEM simulation and velocity extraction have been previously reported by the authors [86]. The mixing performance is defined in terms of certain critical quality attributes (CQAs) such as relative standard deviation (RSD), composition of A, which can be active ingredient of interest for a given process (C_A) and residence time distribution (RTD). These CQAs should be regulated and controlled in order to achieve the desired mixing efficiency. These parameters can be found as described below:

$$C_A = \frac{\sum_{y=1}^{y_{\max}} F(1, x_{\max}, y, t)}{\sum_{n=1}^{n_{\max}} \sum_{y=1}^{y_{\max}} F(n, x_{\max}, y, t)} \tag{19}$$

In the above equation, the numerator stands for the total number of particles of component A which exit from the last compartments at any point in time. The denominator represents total number of particles of both components A and B exiting from the last compartments at any point of time. Since this model involves two components, the value of $n_{\max}$ is 2. $x_{\max}$ and $y_{\max}$ stand for the maximum number of grids in the axial and radial directions, respectively. Figure 3 shows how the API composition at the mixer outlet varies over time. It can be seen that the API composition at the outlet increases with time and finally reaches a steady value of 0.163 (which is same as the inlet composition).

The homogeneity of samples retrieved from the outflow is measured by calculating the variability in the concentration. The RSD of the tracer concentration measures the degree of homogeneity of the mixture and is given by

$$RSD = \frac{\sqrt{\sum_{i=1}^{N} \frac{(c_i - c_{\text{avg}})^2}{N-1}}}{c_{\text{avg}}} \tag{20}$$

where N represents the total number of compartments ($N = x_{\max} \times y_{\max}$). i is the index to represent the compartment. c_i is the

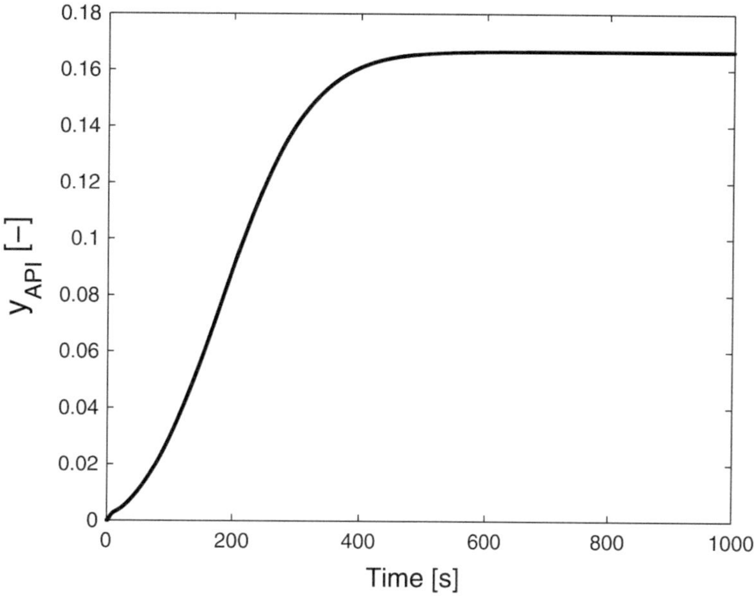

Fig. 3 Evolution of API composition

concentration of component A at any compartment i. c_{avg} is a spatial average of the concentration of component A. Figure 4 represents how the overall RSD decreases over time as well as over the mixer length. This is because the variability in the mixture is reduced over time. In Fig. 4, the mixer length is represented by the compartment number. The spikes present in the graph depict back mixing in the respective compartments.

The RTD, $E(t)$, is a measure of the time spent by the particles within the mixer, capturing the non-ideality associated with the flow. The RTD can be found by

$$E(t) = \frac{c(t)}{\int_0^\infty c(t)dt} \tag{21}$$

In the above equation, $c(t)$ stands for the concentration of component A in the outlet stream at any time t. It is important to make the following assumptions in order to determine the RTD: (1) the flow in the mixer is well mixed; (2) the powder elements entering the mixer simultaneously flow with constant velocity and leave the mixer at same time. Figure 5 depicts the RTD of the API in the mixer upon injecting a pulse of API at $t = 100$ s. The RTD is a measure of the time spent by the API particles in the mixer and correlates to mixing performance. The width of the RTD curve can

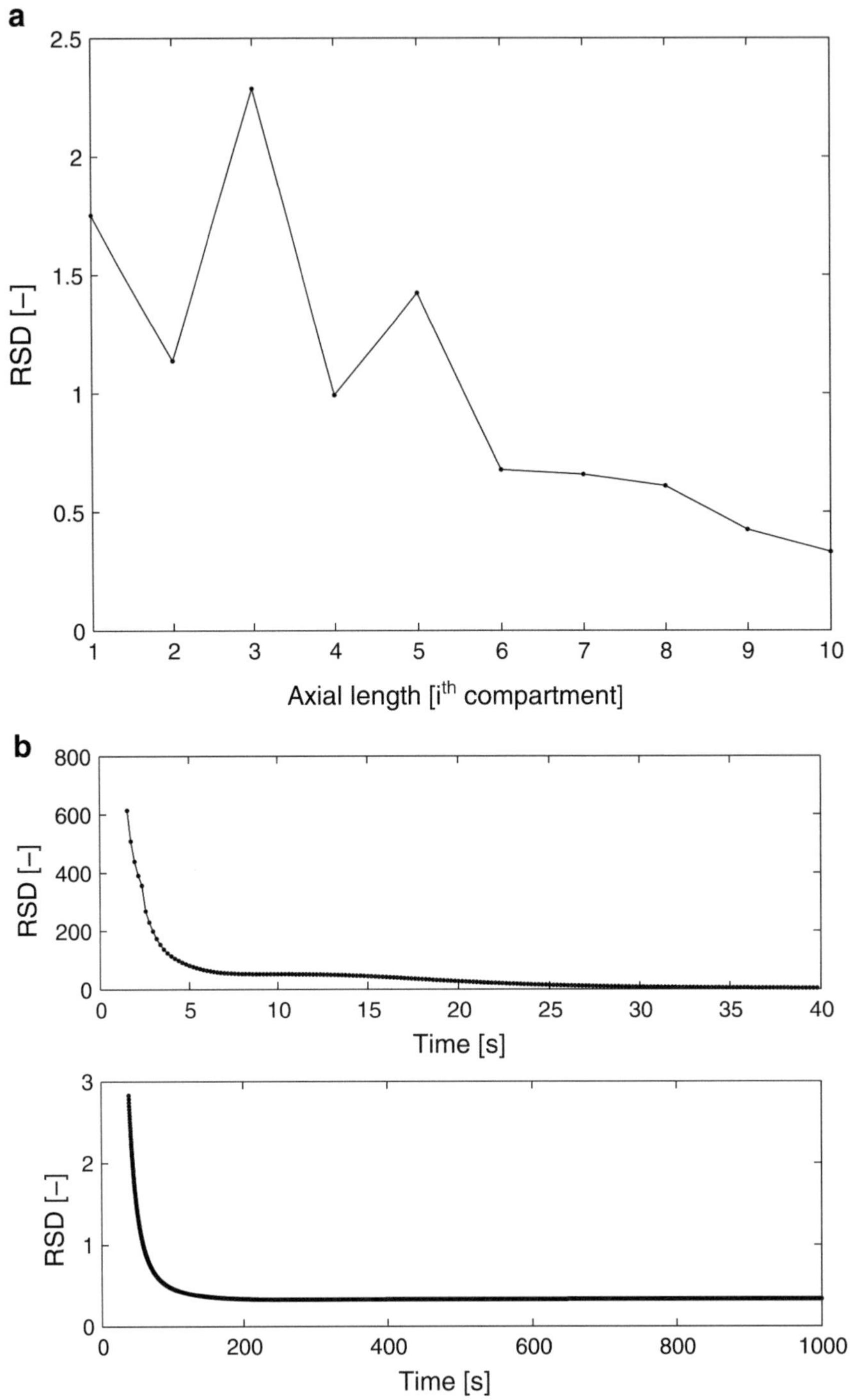

Fig. 4 Plot of (**a**) RSD versus axial length at end-point and (**b**) RSD versus time at mixer outlet for 0:40 s and 40:1000 s

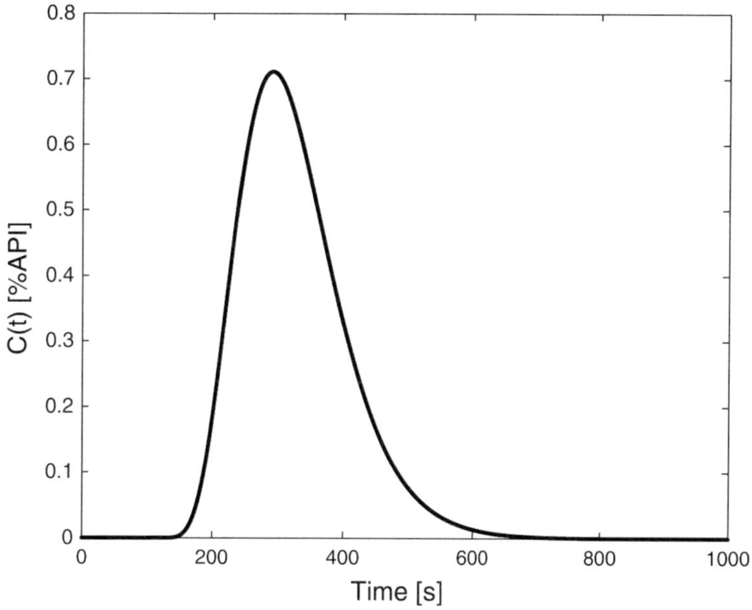

Fig. 5 RTD of API upon injection of pulse at $t = 100$ s

be set as a metric to optimize process performance as a function of formulation properties and processing conditions.

The mathematical model has been validated with experimental results. Experiments have been designed to obtain the RSD and API composition at the mixer outlet as a function of time. Materials, which have been used in the experiment are Avicel PH-200 (FMC biopolymer), which is the excipient and Acetaminophen (Mallinckrodt), which is the API. The mixing has been carried out in a commercial mixer (Gericke GCM250^{TM}). One experimental data set which has been obtained for a feed rate of 20 kg/h and mixer speed of 40 RPM has been considered. API concentration in the mixture fed to the mixer inlet is 10 %. More details on the experimental setup can be obtained from [83]. The API composition and RSD values have been obtained as a function of time at the mixer outlet. A linear trendline has been fitted to the experimental data points and the model has been validated with respect to the trendline. The objective function can be formulated as shown below (Eqs. (22)–(24)) [83]:

$$\Omega^{\mathrm{APIComposition}}(p) = \sum_{i=1}^{n} \left\| y_{\mathrm{API,predicted}}(t_i, \mathbf{p}) - y_{\mathrm{API,experimental}}(t_i) \right\|^2 \tag{22}$$

$$\Omega^{\mathrm{RSD}}(p) = \sum_{i=1}^{n} k \left\| \mathrm{RSD}_{\mathrm{predicted}}(t_i, \mathbf{p}) - \mathrm{RSD}_{\mathrm{experimental}}(t_i) \right\|^2 \tag{23}$$

$\mathbf{p}$ is the set of estimated parameters, which, in this case, are the forward, backward, and radial velocities (V_f, V_b, and V_r). The objective function for the RSD has been multiplied by a constant (k) to facilitate convergence since the API composition and RSD have values of different orders. The overall objective function was then formulated as follows:

$$\Omega^{\text{Total}}(p) = \Omega^{\text{RSD}}(p) + \Omega^{\text{APIComposition}}(p) \tag{24}$$

An RTD study has been performed, as well. Since the RTD measurement requires a different set of experiment, the objective function has been formulated and optimized separately, as shown in Eq. (25):

$$\Omega^{\text{RTD}}(p) = \sum_{i=1}^{n} \left\| E_{\text{predicted}}(t_i, \mathbf{p}) - E_{\text{experimental}}(t_i) \right\|^2 \tag{25}$$

Figure 6 shows that there is a good agreement between the model predicted and experimental (linear trendline) values for both RSD and API composition.

2.3 Granulation

Granulation is yet another crucial process that is very relevant to the pharmaceutical industry. It involves the size enlargement of particles in order to improve flowability and dissolution and enhance homogeneity of finer particulate solids. It is a highly complex process that is poorly understood and is currently operated in the industry under high recycles [80]. Due to the discrete nature of the process, PBMs are highly useful in modeling granulation [48]. In early days, 1-D models were more popular for modeling granulation due to their reduced complexity [53, 81]. However, Iveson [38] presented a detailed case study emphasizing the need for multi-dimensional models for accurately modeling granulation. Eventually, Wauters [97] suggested an easier yet effective approach of framing a multi-dimensional population balance model for granulation by re-casting it in terms of their individual volumes of solid (s), liquid (l), and gas (g), which enables the decoupling of the integrated process with respect to the individual meso-scopic sub-processes. The overall 3-D PBM for granulation can be written as

$$\begin{aligned}\frac{\partial}{\partial t}F(s,l,g,t) &+ \frac{\partial}{\partial s}\left[F(s,l,g,t)\frac{ds}{dt}\right] + \frac{\partial}{\partial l}\left[F(s,l,g,t)\frac{dl}{dt}\right] \\ &+ \frac{\partial}{\partial g}\left[F(s,l,g,t)\frac{dg}{dt}\right] = \Re_{\text{nuc}} + \Re_{\text{agg}} + \Re_{\text{break}}\end{aligned} \tag{26}$$

where $F(s, l, g, t)$ represents the population density function such that $F(s, l, g, t)dsdldg$ is the number of moles of granules with a

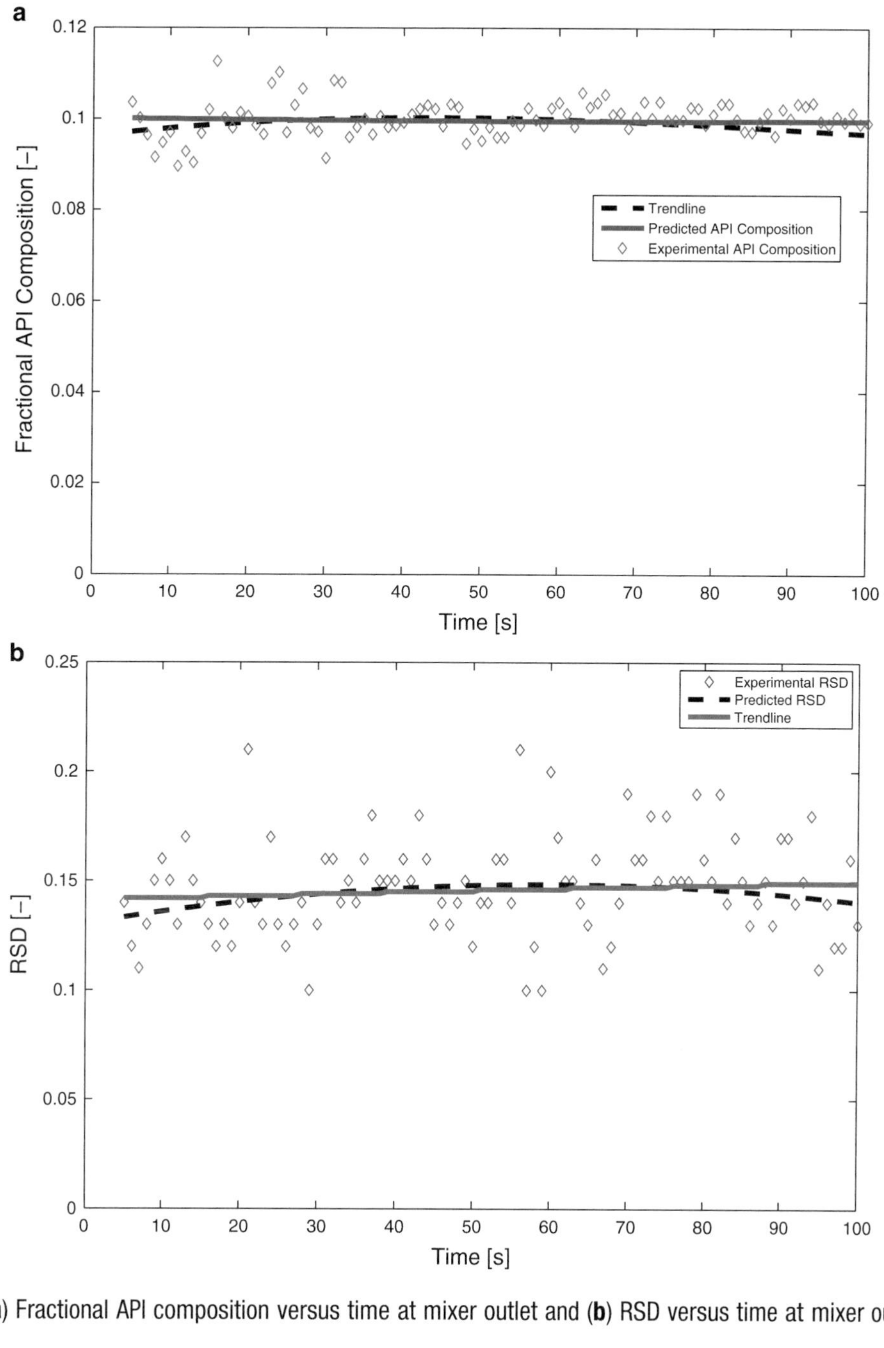

Fig. 6 (**a**) Fractional API composition versus time at mixer outlet and (**b**) RSD versus time at mixer outlet

solid volume between s and $s + ds$, a liquid volume between l and $l + dl$, and a gas volume between g and $g + dg$. The partial derivative term with respect to s accounts for the layering of fines onto the granule surfaces, the partial derivative term with respect to l accounts for the drying of the binder or the re-wetting of granules and the partial derivative with respect to g accounts for consolidation, which, due to compaction of the granules, results in an increase in the pore saturation and decrease in the porosity.

Various submodels can be used to describe the different growth mechanisms involved in granulation. Wang et al. [96] modeled the layering of fines using a Monod based growth model as shown in Eq. (27):

$$G = G_{\max} \frac{M_{\text{powder}}}{k_i \sum M_i + M_{\text{powder}}} e^{[-\alpha(x_w - x_{wc})^2]} \tag{27}$$

where $G_{\max}$ is the maximum growth rate, M_{powder} is the mass of fine powder, M_i is the mass of particles in the ith size class, x_{wc} is the critical moisture, and k and α are fitting parameters (based on experimental data-particle size distribution, porosity, moisture content). Layering is highly affected by the amount of fines present in the system.

Drying and rewetting are associated with the gain or loss of liquid into or from the granulation system due to evaporation or liquid addition. The liquid growth rate can be obtained from the mass balance as

$$\frac{dL}{dt} = \frac{\dot{m}_{\text{spray}}(1 - c_{\text{binder}}) - \dot{m}_{\text{evap}}}{m_{\text{solid}}}, \tag{28}$$

where

$$m_{\text{solid}} = m_{\text{solid fraction}} + \dot{m}_{\text{spray}} c_{\text{binder}} \Delta t, \tag{29}$$

In the above equations, $\dot{m}_{\text{spray}}$ is the rate at which liquid is being sprayed in the system, c_{binder} is the concentration of solid binder in the liquid added, m_{evap} is the rate of liquid being evaporated, $m_{\text{solid fraction}}$ is the volume of solid of the particles in each bin and L is the liquid in the powder bed. In the case of a high shear mixer, the assumption of $\dot{m}_{\text{evap}} = 0$ is valid. However, in the case of a fluid bed process, the evaporation rate dominates over rewetting, and thus the evaporation term needs to be taken into account [18].

Consolidation is a negative growth process which represents the compacting of granules due to the escape of air from the pores. It can modeled as an empirical exponential decay relation [94], given by

$$\frac{d\varepsilon}{dt} = -c(\varepsilon - \varepsilon_{\min}), \tag{30}$$

$$\frac{dg}{dt} = \frac{c(s + l + g)(1 - \varepsilon_{\min})}{s} \times \left[l - \frac{\varepsilon_{\min} s}{1 - \varepsilon_{\min}} + g \right] \tag{31}$$

where the porosity ε is

$$\varepsilon = \frac{l+g}{s+l+g} \tag{32}$$

Here $\varepsilon_{\min}$ is the minimum porosity of the granules and c is the compaction rate constant.

The above equations adequately represent the growth terms that are observed on the left side of the PBM (Eq. (26)). The source terms on the right side of Eq. (26) comprise of nucleation, agglomeration, and breakage. Aggregation of particles is one of the dominant mechanisms that enables size enlargement during granulation. The aggregation of particles is obtained as a combination of the formation and depletion terms as shown in Eqs. (34) and (35):

$$\Re_{\text{agg}}(s,l,g) = \Re_{\text{agg}}^{\text{form}} - \Re_{\text{agg}}^{\text{dep}} \tag{33}$$

such that

$$\Re_{\text{agg}}^{\text{form}} = \frac{1}{2}\int_{s_{\min}}^{s_{max}}\int_{0}^{l_{\max}}\int_{0}^{g_{\max}} \beta(s', s-s', l', l-l', g', g-g') \times F(s', l', g', t)F(s-s', l-l', g-g', t)ds'dl'dg' \tag{34}$$

$$\Re_{\text{agg}}^{\text{dep}} = F(s,l,g,t)\int_{s_{\min}}^{s_{max}}\int_{0}^{l_{\max}}\int_{0}^{g_{\max}} \beta(s', s-s', l', l-l', g', g-g') \times F(s', l', g', t)ds'dl'dg' \tag{35}$$

where $s_{\min}$ is the minimum solid volume of the particles present in the system and $\beta(s', s-s', l', l-l', g', g-g')$ is the size-dependent aggregation kernel that describes rate expression for aggregation of two granules of internal coordinates (s', l', g') and $(s-s', l-l', g-g')$. The formation term accounts for the new particles that are formed after coalescence, whereas the depletion term accounts for the particles that are lost when the smaller particles coalesce and aggregate. The PBM has been validated against experimental data by calibrating the model using parameter estimation techniques. Figure 7 shows agreement between the experimental and predicted results for the mean diameter and porosity of the product class for the granules. It can be seen that the model is able to

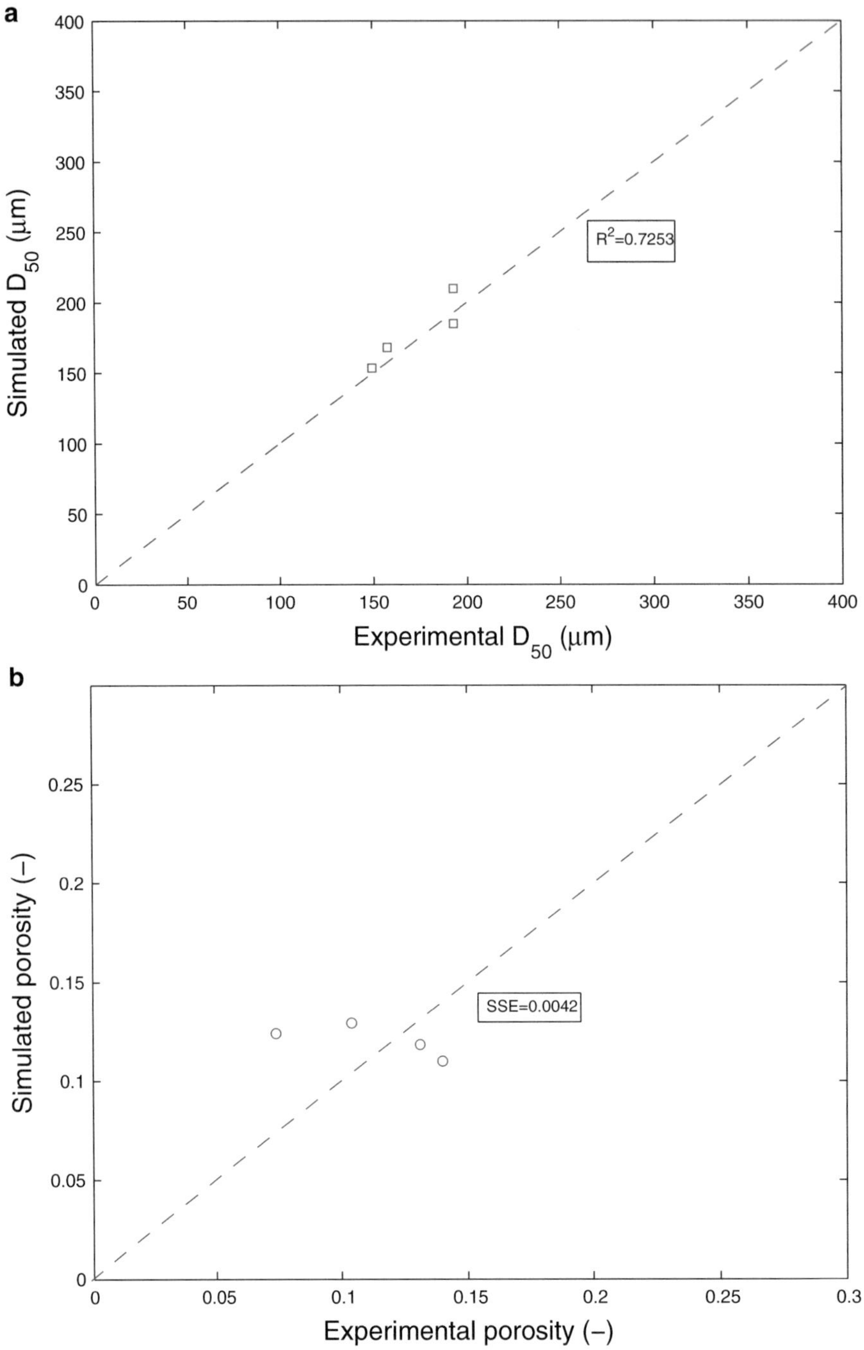

Fig. 7 Estimations and predictions for a high shear granulator using PBM. (**a**) Experimental and estimated simulated D_{50}, (**b**) experimental and estimated simulated porosity, (**c**) experimental and predicted D_{50}, and (**d**) experimental and predicted porosity

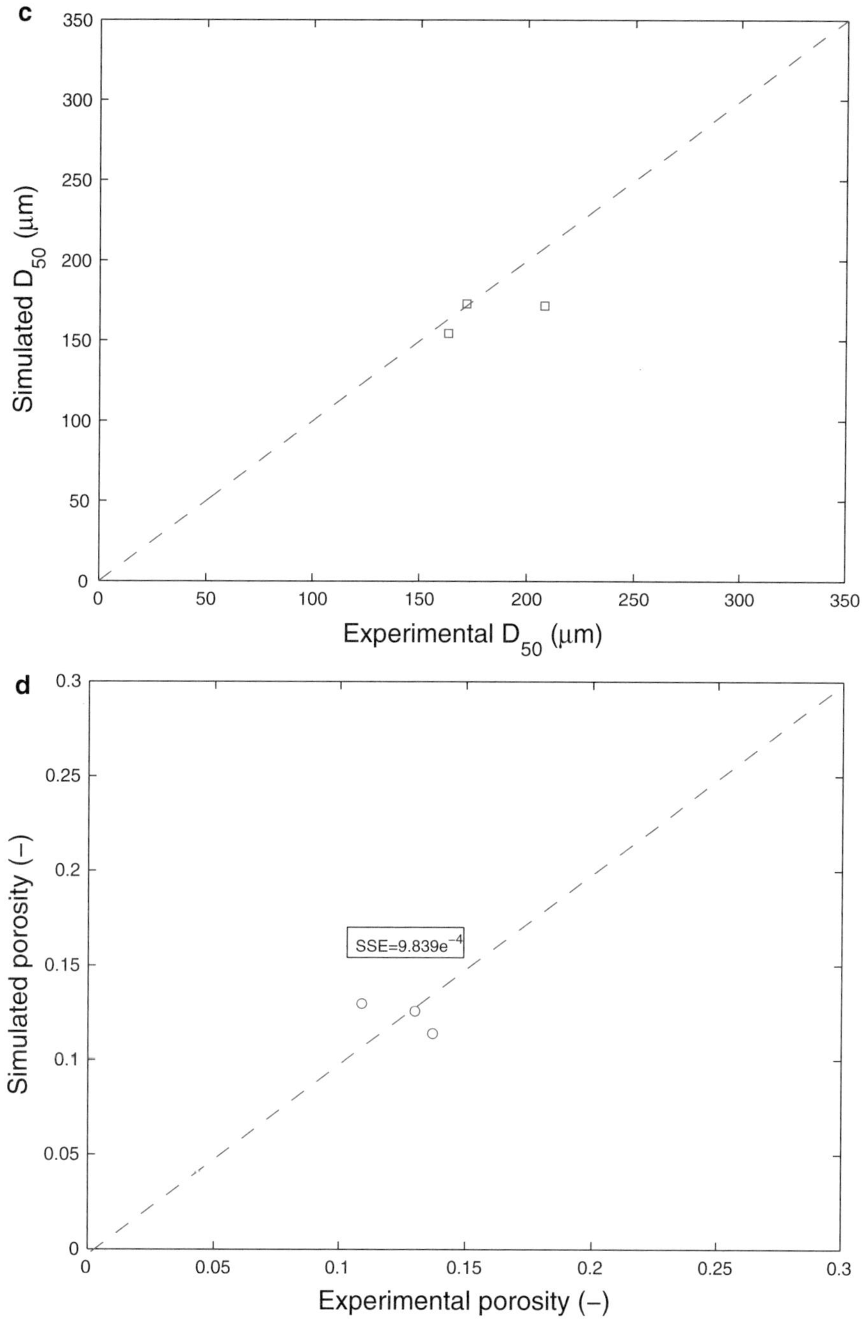

Fig. 7 (continued)

capture the experimental trends and also make close predictions. The 45^0 dotted line depicts the condition where the simulated values equal the experimental results (perfect agreement). The proximity of the points to the dotted 45^0 line shows the accuracy in the approximation. This suggests the applicability towards QbD for improved process operation.

The appropriate selection of the aggregation kernel is also a crucial task in accurate modeling of a granulating system. Many empirical kernels have been observed in literature, but a mechanistic kernel is most desirable as it takes into account the various material and system properties and can hence be extrapolated for different systems. An empirical kernel requires more number of parameters that have to be tuned in order to tailor the kernel to a particular system. This makes an empirical kernel extremely system specific and thus cannot be generalized. A list of the various aggregation kernels available in literature has been shown in a tabular form by Cameron et al. [12]. The approach based on [26] and [50, 51] was implemented by Immanuel and Doyle [37] in order to obtain a mechanistic kernel, which takes into account various system parameters, such as particle diameter and temperature, resulting in a more generalized kernel which requires fewer estimated parameters. A modified physical-based kernel has also been proposed recently, which considers variable binder thickness for obtaining the aggregation kernel [52]. A novel semi-mechanistic kernel has been recently proposed that takes into account the influence of various measurable operating parameters on the final granule properties and yet is computationally inexpensive [21]. This model can effectively capture the steady and induction growth behaviors that have been observed experimentally. Figure 8 depicts the ability of the semi-mechanistic kernel to be able to capture the various growth behaviors. For highly porous starting particles, the growth in particle size is not observed immediately after the onset of liquid addition. Growth is brought about only after there is sufficient consolidation to squeeze out liquid which can then aid with agglomeration. Therefore, there is a delay with the growth of particle size. This scenario is called induction growth. The granulation of less porous particles, however, starts immediately after the addition of liquid. This is known as steady growth. Figure 8 not only captures the influence of crucial operating parameters, e.g. viscosity (higher viscosity suggests higher viscous forces thus forming bigger granules), but is also able to demarcate the existence of induction vs steady growth. As expected intuitively, the growth of primary particles is aided on increasing the viscosity of the binder. The highly viscous binder is more efficient with dissipating the kinetic energy through viscous forces.

Breakage is the disintegration of a particle into two or more fragments and is mainly governed by attrition and impact. It plays a crucial role in controlling the final granule size distribution in high shear granulators. Many published breakage kernels have been presented in literature, one of which defines the breakage kernel in terms of empirical parameters and shear rate [64]. Mechanistic kernels have also been proposed in literature, where the kernel has

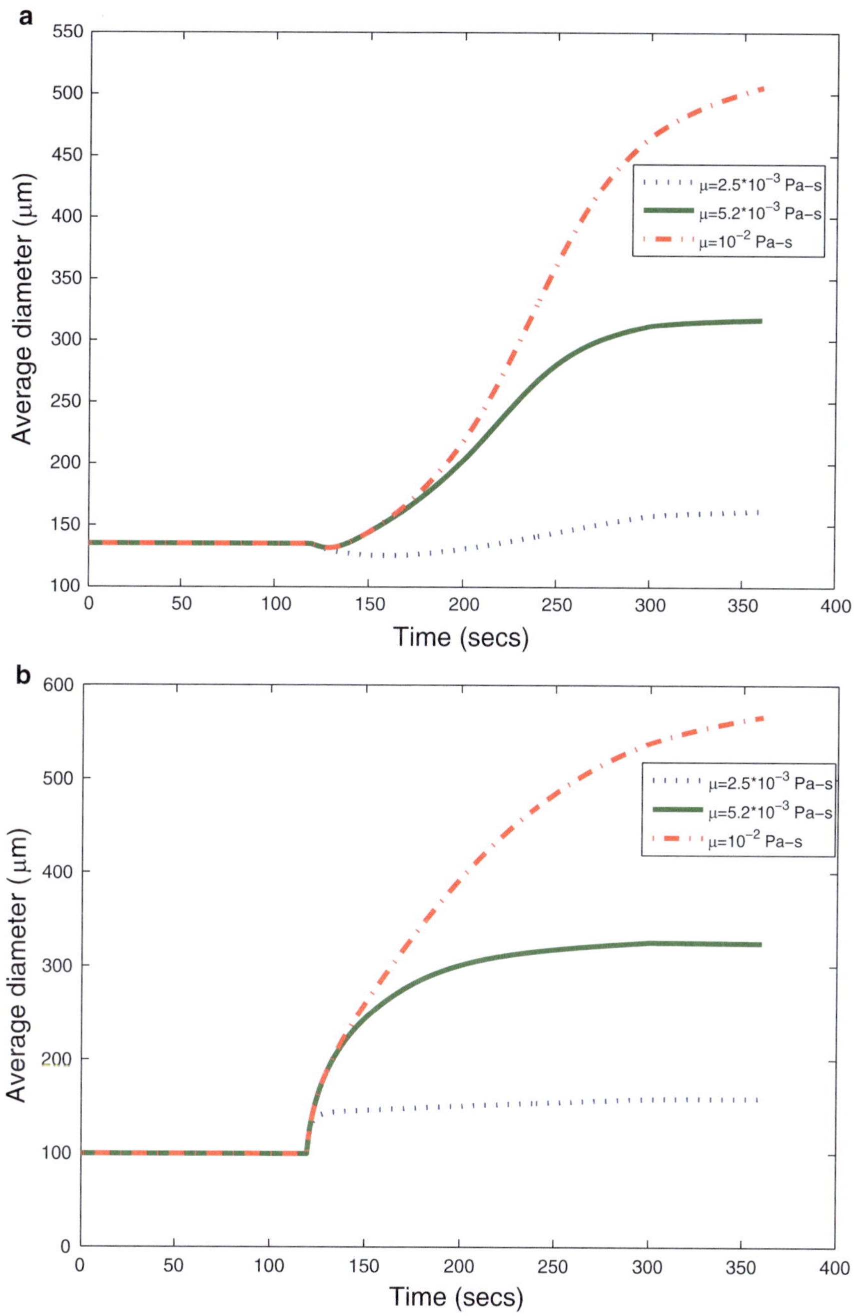

Fig. 8 Effect of viscosity from model results revealing steady and induction growth behavior for low and high porous particles respectively. (**a**) Induction growth for highly porous particles and (**b**) steady growth for less porous particles

been expressed to be proportional to the ratio of the external stress to the intrinsic strength [73].

Several models have been developed in order to quantify the breakage distribution function [65]. Using this information, the breakage term can be broken into its corresponding birth and death terms as

$$\mathfrak{R}_{\text{break}}(s,l,g) = \mathfrak{R}_{\text{break}}^{\text{form}} - \mathfrak{R}_{\text{break}}^{\text{dep}}, \tag{36}$$

such that the birth and death terms can be explained using Eqs. (37) and (38):

$$\mathfrak{R}_{\text{break}}^{\text{form}} = \int_0^{s_{\max}} \int_0^{l_{\max}} \int_0^{g_{\max}} K_{\text{break}}(s,l,g)\, b(s', s-s', l', l-l', g', g-g') \times F(s', l', g', t)\, ds'\, dl'\, dg' \tag{37}$$

$$\mathfrak{R}_{\text{break}}^{\text{dep}} = K_{\text{break}}(s,l,g) F(s,l,g,t). \tag{38}$$

A more mechanistic approach towards obtaining the breakage kernel has also been proposed by Ramachandran et al. [73] where the K_{break} has been represented as a ratio of the external stress to the internal stress. These stresses have been further obtained by applying the physics of the system combined with the information of the various microscopic properties. Breakage has also been studied using a non-linear effect on the mechanism for milling processes by Bilgili and Scarlett [6].

The effect of the various underlying mechanisms governing granulation has been presented in [15]. The effect of the primary particles and the binder addition mode has also been studied.

A continuous granulation operation has also been studied recently in the work by Barrasso et al. [4]. The continuous granulator has been subdivided into three distinct zones demarcating the premixing zone, the spray zone, and the wet massing zone. The multidimensional PBM for the continuous process is expressed as a function of the internal and the external coordinates in order to address the inhomogeneities based on the spatial positions (as shown in Eq. (39)):

$$
\begin{aligned}
&\frac{\partial}{\partial t}F(s_1,s_2,l,g,x,y,t)+\frac{\partial}{\partial s_1}\left[F(s_1,s_2,l,g,x,y,t)\frac{ds_1}{dt}\right]\\
&\quad+\frac{\partial}{\partial s_2}\left[F(s_1,s_2,l,g,x,y,t)\frac{ds_2}{dt}\right]+\frac{\partial}{\partial l}\left[F(s_1,s_2,l,g,x,y,t)\frac{dl}{dt}\right]\\
&\quad+\frac{\partial}{\partial g}\left[F(s_1,s_2,l,g,x,y,t)\frac{dg}{dt}\right]\frac{\partial}{\partial x}\left[F(s_1,s_2,l,g,x,y,t)\frac{dx}{dt}\right]\\
&\quad+\frac{\partial}{\partial y}\left[F(s_1,s_2,l,g,x,y,t)\frac{dy}{dt}\right]\\
&=\dot{F}_{\text{in}}(s_1,s_2,l,g,x,y,t)+\dot{F}_{\text{out}}(s_1,s_2,l,g,x,y,t)\\
&\quad+\mathfrak{R}_{\text{nuc}}+\mathfrak{R}_{\text{agg}}+\mathfrak{R}_{\text{break}}
\end{aligned}
\tag{39}
$$

The modeling results could very well capture the experimental trends that were observed in practice (see Fig. 9). A detailed parametric study involving the study of the various crucial operating parameters has also been studied [4]. Findings suggest that the residence time and liquid-to-solid ratio are the most critical parameters in determining the product size distribution in twin screw granulation. The residence time is the result of design parameters, such as screw length and configuration, as well as process parameters, such as screw speed and throughput. The model calibration for the granulation process requires experimental data consisting of the particle size distribution and porosity in typical cases. The tight quality criteria in the pharmaceutical industry requires the particle size to lie within a certain range, such that the dissolution and bioavailability of the drug can be controlled. The granule size can be measured using various equipments measuring the size or using the sieve analysis. Sieve analysis is a more crude method and is generally associated with a relatively high measurement error [63].

2.4 Milling

Milling is a particle size reduction process often used in pharmaceutical manufacturing of solid oral dosage forms. Used for delumping in the direct compaction manufacturing route, milling processes have additional purposes in wet and dry granulation manufacturing routes. In dry granulation, a roller compactor is used to form compacted ribbons from fine powder. Milling is then used to produce granules from these ribbons. In manufacturing routes using wet granulation, milling operations can be used to reduce oversize particles following granulation and increase amount of the product within the desired size class.

The conical screen mill, which is most widely used for pharmaceutical applications, applies strong shear forces on particles that are trapped between the impeller and the screen, resulting in particle attrition [3]. Small fragment particles are able to pass through the

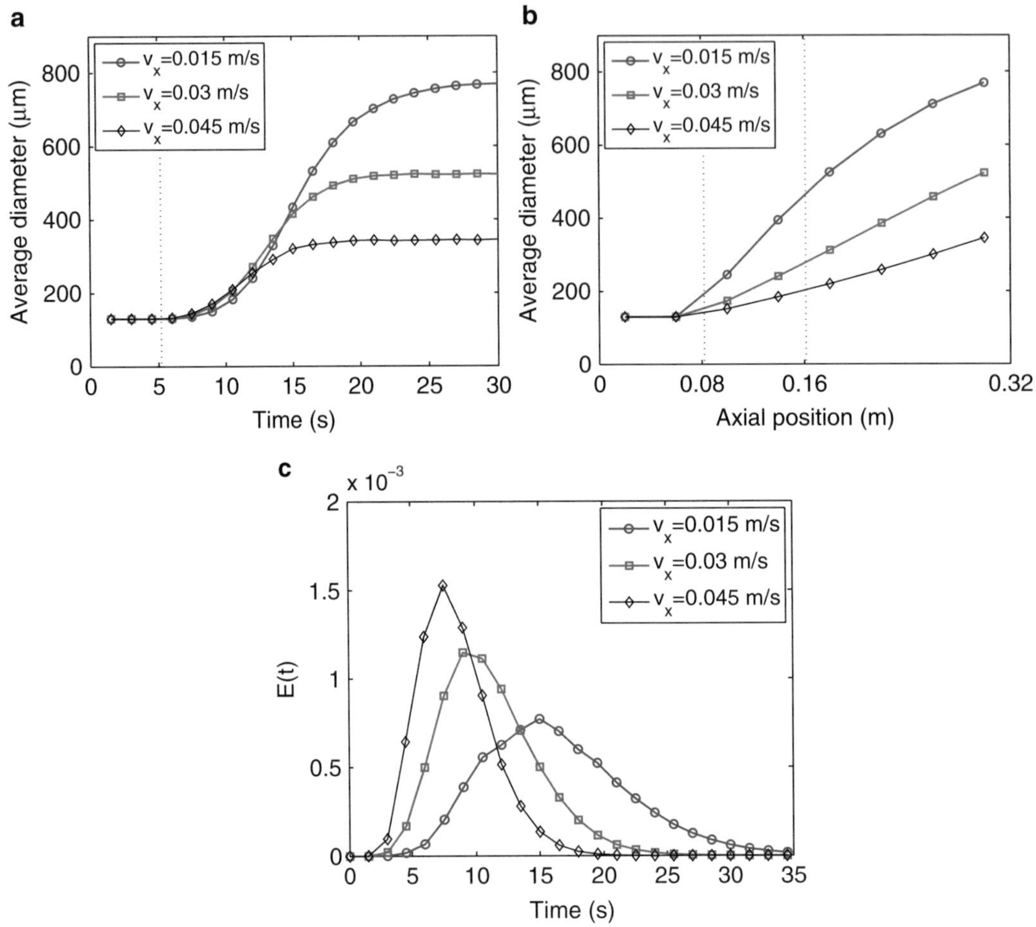

Fig. 9 Average diameter at the outlet vs. time for varying (**a**) axial velocity, (**b**) average diameter vs. axial position for varying axial velocities, where *vertical lines* represent the axial bounds of the spray zone. (**c**) Residence time distributions for varying axial velocities

screen, while larger fragments are contained and experience additional breakage. In all milling processes, the primary mechanism is particle breakage. Particles are subjected to shear or impact forces, which break larger particles into fragments. The breakage rate, distribution of fragment particles, and screen classification determine the product size distribution. These factors can depend on process parameters, such as impeller speed and screen size, as well as material properties, such as ribbon strength and particle size. PBMs can be employed to understand these effects.

Typically, milling processes have been modeled using 1-D PBMs, accounting for variations in particle size [3, 6, 23, 78]. A general 1-D PBM for milling is presented in Eq. (40), which describes the rate of change in the number of particles of each size class in the system.

$$\frac{\partial F(u,t)}{\partial t} = \mathfrak{R}_{\text{break}}(u,t) = \mathfrak{R}_{\text{break, form}}(u,t) - \mathfrak{R}_{\text{break, dep}}(u,t) \tag{40}$$

Here, $F(u, t)$ represents the number of particles in the system of size u at time t. $\mathfrak{R}_{\text{break}}$ is the breakage term, accounting for the formation of particles due to the breakage of larger particles, $\mathfrak{R}_{\text{break,form}}$, and the depletion of particles as they break into smaller fragments, $\mathfrak{R}_{\text{break,dep}}$. Expressions for the formation and depletion rates are given in Eqs. (41) and (42):

$$\mathfrak{R}_{\text{break, dep}}(u,t) = S(u)F(u,t) \tag{41}$$

$$\mathfrak{R}_{\text{break, form}}(u,t) = \int_{u}^{\infty} S(v)F(v)b(u,v)dv \tag{42}$$

Here, $S(u)$ is the breakage rate kernel for a particle of size u, and $b(u, v)$ is the fragment distribution, describing the probability that a particle of size u will result in a fragment of size v. While the depletion rate depends only on the number of particles and breakage kernel, the formation rate is more complex. The rate of fragment formation depends on the breakage rates and numbers of particles of all size classes larger than the size class in consideration, resulting in an integral term. Additionally, the breakage distribution function is necessary to describe the sizes of the fragments.

Some studies use a constant rate kernel [78], while various size-dependent rate kernels are also found in the literature [64, 65, 95]. The breakage rate kernel depends on the material properties of the formulation, as well as the process conditions. While these kernels are empirical, a mechanistic breakage kernel for granulation was developed by Ramachandran et al. [73] and can be applied to milling processes. Similarly, various breakage distribution functions have been proposed, including the Hill–Ng distribution [25], a bimodal lognormal distribution [78], and a lognormal distribution [3]. Selection of an appropriate breakage rate kernel and fragment distribution depends on the desired complexity of the model and the ability to describe experimental data.

For a mill operated in batch mode, the mill is treated as a closed system, and any material that passes through the screen is considered a part of the system. The breakage rate $S(u)$ is used to characterize the screen, and small particles are represented with breakage rates of zero as they have passed through the screen and can no longer break. The size cut-off value is often extracted from experimental data [78].

The breakage rate kernel, fragment distribution, and screen model contain parameters that are unknown and must be evaluated empirically. Parameter estimation has been employed to determine unknown parameters by Reynolds [78], Capece et al. [13], and Barrasso et al. [3].

As a case study, Barrasso et al. [3] developed a 1-D PBM for a continuous conical screen mill, accounting for the effects of impeller speed and screen size. Experimental data was collected from a conical screen mill (Quadro 197-S) using pure microcrystalline cellulose (Avicel PH 200). A full-factorial 2 × 3 × 2 experiment was performed to measure the effects of screen size (3 levels), impeller speed (2 levels), and ribbon density (2 levels) on the product size distribution and mass hold-up.

The PBM incorporated inlet and outlet flow rates, $\dot{F}_{\text{in}}$ and $\dot{F}_{\text{out}}$, to represent the feed and product streams of the continuously operated mill, as shown in Eq. (43) [3]:

$$\frac{\partial F(u,t)}{\partial t} = \mathfrak{R}_{\text{break}}(u,t) + \dot{F}_{\text{in}}(u,t) - \dot{F}_{\text{out}}(u,t) \tag{43}$$

Using this approach, $S(u)$ only describes the breakage rate but does not account for the screen. An expression for $\dot{F}_{\text{out}}$ was formulated to describe the likelihood that a particle will leave the mill, shown in Eq. (44):

$$\dot{F}_{\text{out}} = (\mathfrak{R}_{\text{break,form}}(u) - \mathfrak{R}_{\text{break,dep}}(u) + \dot{F}_{\text{in}}(u))(1 - f_d(u)), \tag{44}$$

where f_d is a classification function given by Eq. (45) [3]:

$$f_d(u) = \begin{cases} 0 & \text{for } d(u) \le (1-\delta) d_{\text{screen}} \\ \dfrac{d(u) - (1-\delta) d_{\text{screen}}}{\delta\, d_{\text{screen}}} & \text{for } (1-\delta) d_{\text{screen}} \le d(u) \le d_{\text{screen}} \\ 1 & \text{for } d(u) > d_{\text{screen}} \end{cases} \tag{45}$$

In this equation, $d(u)$ represents the diameter of a particle in size class u, d_{screen} is the screen aperture, and $\delta\, d_{\text{screen}}$ is the cut-off size at which particles are large enough to be held up by the screen, assumed to be proportional to the screen size.

An empirical shear rate- and size-dependent breakage kernel was adapted from Pandya and Spielman [64] to account for the effect of impeller speed, v_{imp} on the breakage rate, shown in Eq. (46):

$$S(u) = P_1 v_{\text{imp}} u^{P_2} \tag{46}$$

In this equation, P_1 is a rate coefficient and P_2 describes the size-dependence of the breakage rate. Both parameters must be evaluated experimentally.

Finally, a log-normal fragment size distribution was assumed, given by Eq. (47):

$$b(u, v) = \frac{C(v)}{u\sigma} \exp\left[-\frac{(\log u - \mu(v))^2}{2\sigma^2}\right], \tag{47}$$

where $\mu(v)$ and σ describe the mean and variance of the fragment size distribution formed from a particle of size v, and $C(v)$ is a normalization constant [3]. The mean of the fragment distribution was defined to be proportional to the log of the parent particle size.

Using this data, parameter estimation was performed for each ribbon density to determine five unknown model parameters. For each ribbon density, four of the six experiments were used in parameter estimation. The calibrated model was then used to simulate the remaining two experiments (at the intermediate screen size), and the resulting particle size distributions were compared. The simulated and measured particle size distributions for the low density ribbon showed strong agreement, as shown in Fig. 10 [3]. Once validated, a PBM can be used as a tool to predict the CQAs based on process parameters, demonstrating a model-based approach to QbD.

3 Numerical Techniques to Solve Population Balance Models

The solution of a population balance equation can be obtained using various methods, such as direct discretization, Monte Carlo and the method of moments. Direct discretization is a straightforward approach to obtain an accurate estimate of the particle size distribution, but a disadvantage associated with this methods is the inaccuracy with tracking the moments [41]. Monte Carlo methods are more stochastic in nature and can be used to obtain the solution of multi-component or poly-disperse population balance equation, provided that the probability of dynamic behavior in the system obeys the balance principle for system details, the time step for each successful event can be calculated accurately and all the events occurring in the system are mutually dependant [29, 59]. Various other techniques, such as the method of classes [69], the method of characteristics [28], the method of moments [44, 56], and high resolution algorithms [34] have been used to solve PBMs. Direct discretization techniques, such as the finite difference method, finite element method, and finite volume method, are quite popular for solving PBMs [36, 55]. A more systematic approach known as the hierarchical two tier algorithm was proposed by Immanuel and

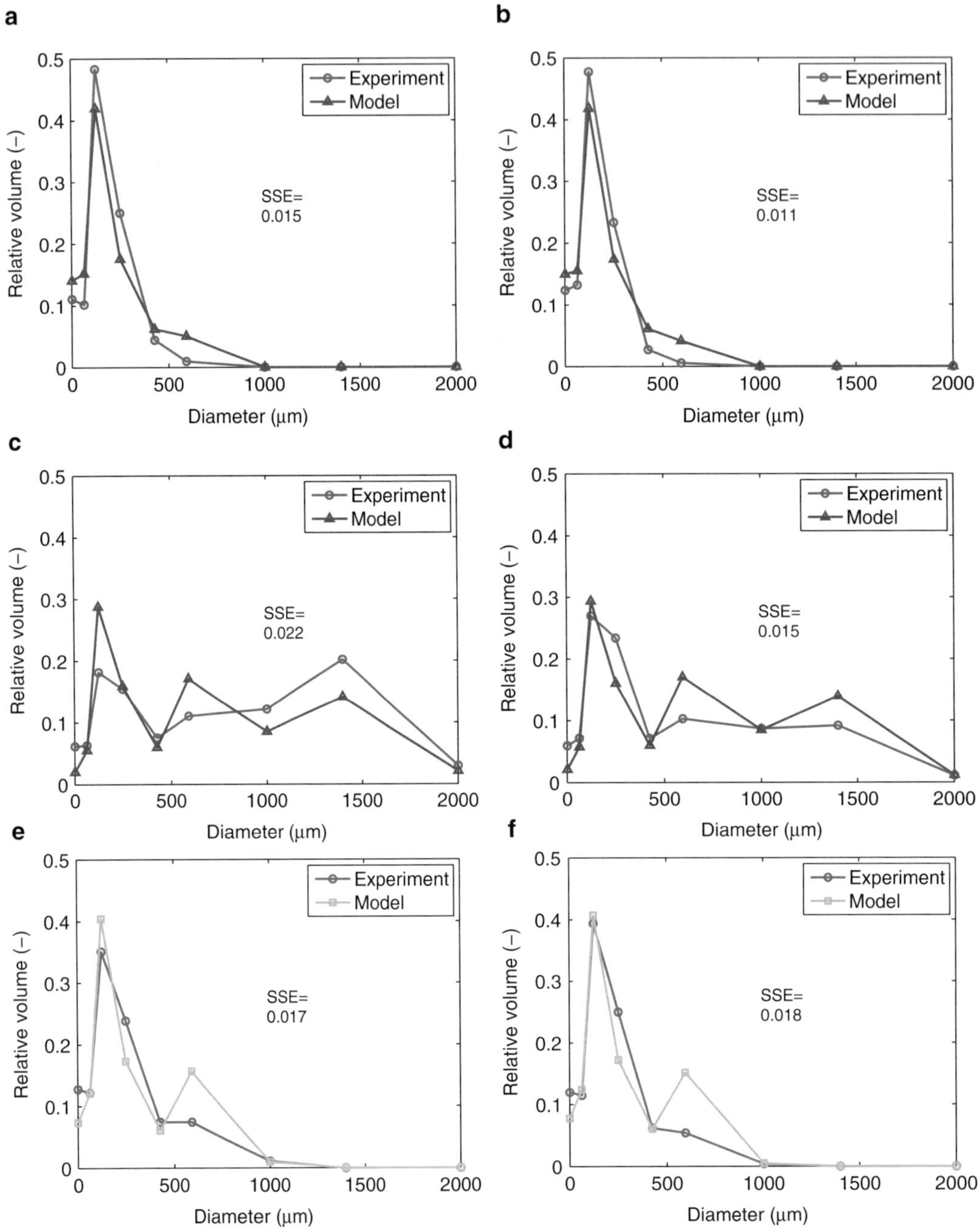

Fig. 10 Experimental and simulated particle size distributions from low density ribbons using (**a**–**d**) parameter estimation and (**e**–**f**) predictive modeling, by screen aperture and impeller speed [3]. (**a**) 991 μm, 3350 RPM, (**b**) 991 μm, 4923 RPM, (**c**) 3175 μm, 3350 RPM, (**d**) 3175 μm, 4923 RPM, (**e**) 1575 μm, 3350 RPM, and (**f**) 1575 μm, 4923 RPM

Doyle III [36]. The partial differential equation can be discretized (with respect to spatial/internal coordinates) to obtain a set of ODEs (as shown in Eq. (48) for the 3-D PBM for granulation), which can then be solved using a standard Runge Kutta or Euler technique.

$$\begin{aligned}\frac{dF'_{i,j,k}}{dt} &+ \frac{F'_{i,j,k}}{\Delta s_i}\frac{ds}{dt}\bigg|_{s_i} - \frac{F'_{i,j,k+1}}{\Delta s_{i+1}}\frac{ds}{dt}\bigg|_{s_{i+1}} + \frac{F'_{i,j,k}}{\Delta l_j}\frac{dl}{dt}\bigg|_{l_j}\\ &- \frac{F'_{i,j,j+1}}{\Delta l_{j+1}}\frac{dl}{dt}\bigg|_{l_{j+1}} + \frac{F'_{i,j,k}}{\Delta g_k}\frac{dg}{dt}\bigg|_{g_k} - \frac{F'_{i,j,k+1}}{\Delta g_{k+1}}\frac{dg}{dt}\bigg|_{g_{k+1}}\\ &= \Re_{\text{agg}}(s_i, l_j, g_k) + \Re_{\text{break}}(s_i, l_j, g_k)\end{aligned} \tag{48}$$

Here $F'_{i,j,k} = \int_{s_i}^{s_{i+1}}\int_{l_j}^{l_{j+1}}\int_{g_k}^{g_{k+1}} F(s,l,g)\ ds\ dl\ dg$, s_i is the value of the solid volume at the upper end of the *i*th bin along the solid volume axis, l_j is the value of the liquid volume at the upper end of the *j*th bin along the liquid volume axis, and s_k, l_k, and g_k are the values of the solid, liquid, and gas volume at the upper end of the *i*th, *j*th, and *k*th bins along the solid, liquid, and gas volume axes, respectively. Δs_i, Δl_j and Δg_k are the sizes of the *i*th, *j*th and *k*th bins.

The solution to PBMs is highly computationally expensive. It is thus desirable to develop approaches by which the computational complexity can be alleviated. The discretization of the PBM can be performed using a linear or a nonlinear grid. A linear grid requires a large number of bins to span a certain size range. However a nonlinear grid can span the same range using fewer bins, thus reducing the computational overhead [17]. Various algorithms have been developed to implement a nonlinear grid for discretization [14, 40, 41, 49]. It is most accurate to employ multi-dimensional PBMs to represent the various particulate processes [38]. However, sometimes there are difficulties in solving such system of equations due to the "curse of dimensionality." It has been pre-established that with an increase in the dimensionality of the PBM, the computation for the aggregation term increases polynomially [67]. The computational expense associated with calculating the aggregation term has been split or redistributed and the overall solution technique has been sped up by parallelizing the code [67, 68].

Another approach for alleviating the computational load for the solution to PBMs is the development of reduced order models. A lower dimensional model was developed by Barrasso and Ramachandran [2] which involved lumping the model in terms of certain independent coordinates in order to reduce the computational expense. The results from the reduced order model suggest a reasonable approximation of the original high dimensional model. As expected, the accuracy of the results reduces as the number of

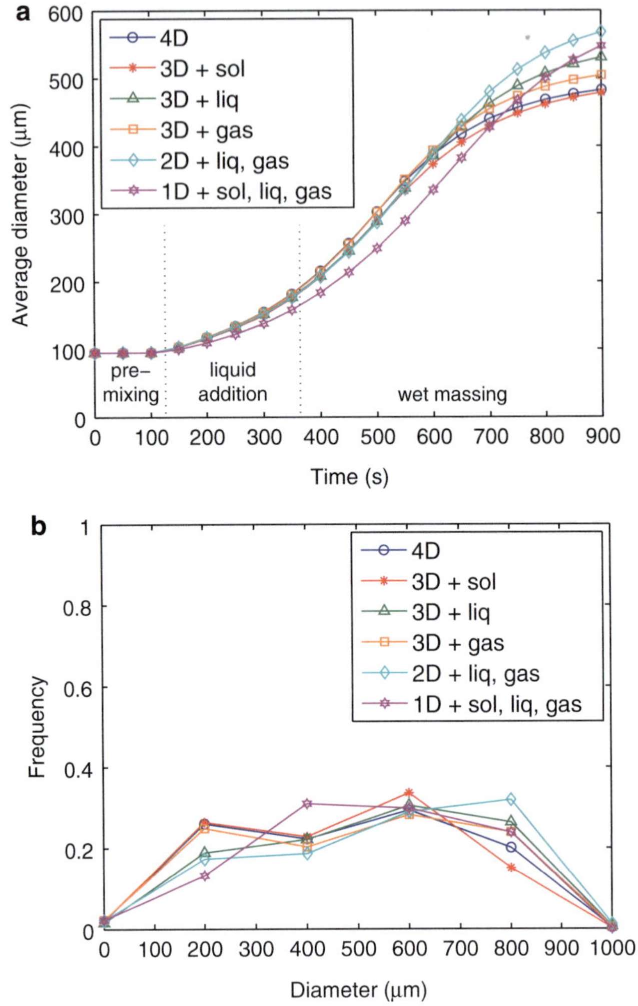

Fig. 11 Comparison of output properties for particle size based on different lumped models. (**a**) Average diameter over time and (**b**) normalized particle size distribution at final time

lumped variables is increased (see Fig. 11). It is thus advisable to choose an appropriate framework with a certain extent of lumping which balances the speed and accuracy of the simulations. The ability to discretize a differential equation using a larger number of finer grids reduces the chances of introducing a discretization error (which is a function of the step size for the discretization) into the solution. The appearance of such an instance can be minimized by using a large number of bins for the discretization and also keeping the associated computational time at a reasonable value. For this purpose, the tensor-decomposition approach has been used (implemented in MATLAB by Oseledets et al.) which compresses large volumes of data in the form of a tensor transformation thus enabling significant savings in the computational time and memory for the simulations [62]. The aggregation term has been calculated as a convolution with the help of fast fourier transforms (FFT). The breakage term and the associated breakage distribution is calculated by separation of variables and reformulating the summation to calculate the equivalent integral terms. The detailed approach for performing the tensor decomposition on the pre-existing high dimensional model has been described in [19].

4 Parameter Estimation

Effective parameter estimation is very crucial for model calibration in order to utilize the mathematical model for predictive purposes. Estimation of empirical parameters requires the need for matching the model outcome with the experimental data. The required experimental data for model calibration might vary for every unit operation. Typically for crystallization processes, the experimental data utilized are the solute concentration in the mother liquor, crystal size [5], and shape. These quantities can be used to fit the model output for calibration purposes. In order to study mixing, the typical data measured in the experiments are the outlet concentration of the powder over time. This information can be utilized to derive various quantities such as RSD and RTD [85]. For studying the agglomeration associated with the mixing process, particle size data can be utilized. In case of granulation processes, the particle size and porosity data are typically considered for parameter estimation purposes [20]. Particle size data also plays a crucial role in the model calibration of milling processes [3].

Parameter estimation is performed by implementing an optimization algorithm that can minimize the mismatch between the model outcome and the experimental data. The objective function ϕ can be written as

$$\phi = (CQA_{\text{simulated}} - CQA_{\text{experimental}})^2 \tag{49}$$

Here, CQA is utilized for the simulated and the experimental results. The CQA varies for each unit operation. The objective function/error can be minimized using various optimization algorithms, such as Nelder–Mead algorithm, various gradient based algorithms, or meta-heuristic techniques (such as genetic algorithm, particle swarm algorithm). With the availability of extensive data, and based on the needs, multiple CQAs can be simultaneously used to fit the model. This can be implemented using a multi-objective optimization algorithm. There are various means by which multiple objective functions can be simultaneously taken into consideration. Using multiple objective functions, there is a need to obtain the pareto optimal solution. The multiple objective functions can be tackled in various ways—sum/weighted sum of objective functions, min-max method, ε -constraint method. The optimization algorithm efficiently minimizes the error between the simulated and the experimental results thus fitting the model to the system under consideration. The calibrated model can then be utilized for making predictions of the process outcome.

5 Flowsheet Applications of PBM in Pharmaceutical Manufacturing

Detailed process modeling is highly beneficial towards improved process operation for pharmaceutical drug manufacturing. The formulation of a PBM requires an in-depth understanding of the process physics and associated risks. Thus, development of a first principles-based PBM leads to compliance with the guidelines of QbD as proposed by the FDA. Using effective parameter estimation techniques to calibrate the model (as discussed in the previous sections) contributes to the predictive ability of the PBM framework. PBMs can also be utilized for performing control and optimization of various particulate processes. Some of the works presenting a control strategy for improved process operation include [54, 72, 93].

In recent years, a large thrust is observed, pushing the drug manufacturing mode from batch to continuous. This would not only involve modernizing the overall process with more sophisticated equipment/control strategies but also the need for modifying the mathematical models for improved process understanding [45]. Some benefits of adopting the continuous manufacturing route include the use of the same equipment for the production of variable quantities of drug, thus minimizing the need for scale-up studies and reducing the time-to-market. This also enables the entire setup to fit in a much smaller space. Also, in a continuous setting, the human factor is significantly reduced through automation of operation and thus overheads incurred due to labor can be minimized. Drug products can be continuously manufactured using one of the three primary routes, as selected based on the material being handled. These routes are, namely, direct compaction, roller compaction, and wet granulation [8]. Figure 12 shows the continuous mode for running the upstream and downstream processes for drug manufacturing. Within this flowsheet, various unit operations can be addressed using a PBM are crystallization, granulation, milling, mixing, and coating. In the previous sections a brief description has been provided on the extension of the PBMs for batch operation to continuous operation. Boukouvala et al. [9, 10] have performed detailed studies on simulating and optimizing the downstream pharmaceutical drug manufacturing in continuous mode. Sen et al. [84, 87] have conducted studies on developing and analyzing the upstream purification process in the continuous mode.

The flowsheet modeling approach provides a more accurate representation of the continuous manufacturing mode due its consideration of the unit operations in an integrated framework. All the unit operations within the flowsheet framework are not represented using population balances, however several complex unit operations are described using population balance models. The properties of the solid/mixture change at the exit of every unit operation is accounted into the flowsheet model and can successfully describe/explain the evolution of powder properties over the continuous manufacturing framework.

Figure 12 also shows the critical quality attributes (CQA) that are of concern at the exit of each unit operation. It can be seen that

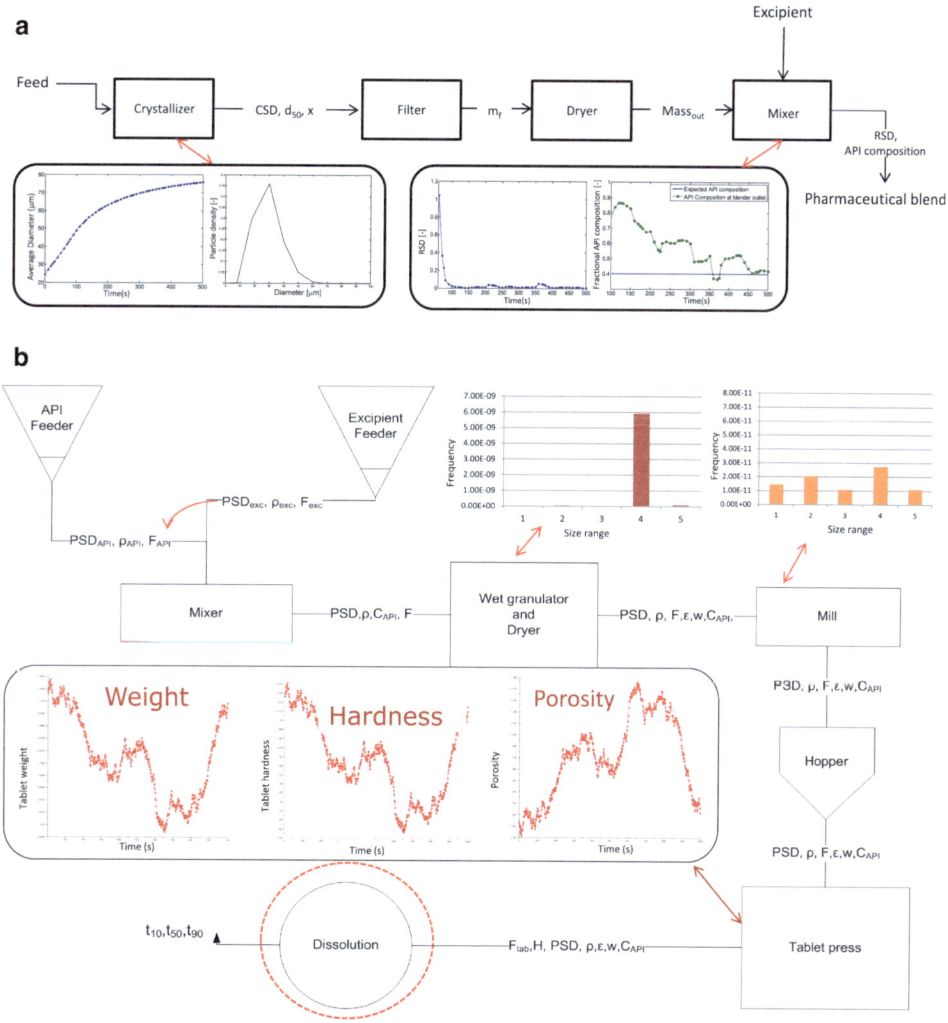

Fig. 12 Flowsheet modeling of continuous downstream pharmaceutical drug manufacturing. (**a**) Upstream tablet manufacturing and (**b**) downstream tablet manufacturing using wet granulation route

using the corresponding model, such exit properties can be tracked. Boukouvala et al. [9] have also performed a sensitivity analysis for the flowsheet models and have studied how noise gets propagated through the multiple unit operations present in the flowsheet. This enables the identification of noise propagation through the continuous operation and can provide the upper bounds to the extent of upset that would keep the system within the acceptable range of operation. Sen et al. [87] performed optimization studies coupled with a multiscale model on the upstream pharmaceutical manufacturing process in order to maximize the efficiency of the overall operation. The crucial parameters affecting each unit operation were identified, which were then fed into the optimization algorithm with the flexibility to be manipulated such that the overall flowsheet could be optimized. Various PBMs were also a part of the flowsheet, which enabled to alleviate the inefficiencies associated with the continuous operation mode. With the development of sophisticated and mechanistic models that can effectively capture the trends in continuous manufacturing operation, the overall understanding of the continuous operation can be enriched and the operation of the process can be improved.

6 Conclusions

There exists a significant amount of inefficient operation in the pharmaceutical industry which suggests the need for a model-based system approach in order to alleviate the operation of the process and reduce wastes. Using a model based approach, the number of experimental trials can be significantly reduced, thus accelerating the profits made by the industry. Population balance models are highly useful for modeling the processes relevant to the pharmaceutical industry, due to its ability to effectively capture the dynamics of a discrete particulate process. Population balance models can be used to describe the process involving solid handling and can be further used to make effective predictions. Parameter estimation techniques enable the quantification of the empirical constants within the model thus extending the utility of the model to make predictions. A brief overview of the parameter estimation techniques is presented in this chapter and provides guidance on the applicability of this framework for predictive purposes. This aligns very closely with the principles of QbD/Process Analytical Tools (PAT) that has been recently proposed by the US FDA.

In order to effectively use this framework for improving the process operation, development of numerical techniques for solution of PBMs is essential. Various numerical techniques have surfaced over the past decade that enable effective solution to PBMs. Some of those techniques have also been briefly discussed in Sect. 3. The "curse of dimensionality" can be overcome using reduced order models as discussed above. Since pharmaceutical drug manufacturing mostly involves the handling of powder and other particulate matter (such as crystals), employing PBMs for modeling such processes is an effective approach. With the advent of a paradigm shift in the mode of process operation for drug manufacturing, the operation of the continuous manufacturing process requires the study of the integrated process using a flowsheet modeling approach. Several unit operations within the flowsheet model are represented using PBMs. From the discussions presented above, it can be clearly inferred that PBMs play a vital role in modeling pharmaceutical processes, primarily due to the inherent discrete nature of PBMs.

References

1. Abbas A, Romagnoli JA (2007) Multiscale modeling, simulation and validation of batch cooling crystallization. Sep Purif Technol 53 (2):153–163
2. Barrasso D, Ramachandran R (2012) A comparison of model order reduction techniques for a four-dimensional population balance model describing multi-component wet granulation processes. Chem Eng Sci 80:380–392
3. Barrasso D, Oka S, Muliadi A, Litster JD, Wassgren C, Ramachandran R (2013) Population balance model validation and prediction of cqas for continuous milling processes: toward QbD in pharmaceutical drug product manufacturing. J Pharm Innov 8:147–162
4. Barrasso D, Walia S, Ramachandran R (2013) Multi-component population balance modeling of continuous granulation processes: a parametric study and comparison with experimental trends. Powder Technol 241:85–97
5. Besenhard MO, Chaudhury A, Vetter T, Ramachandran R, Khinast J (2015) Evaluation of parameter estimation methods for crystallization processes modeled via population balance equations. Chem Eng Res Des 94: 275–289
6. Bilgili E, Scarlett B (2005) Population balance modeling of non-linear effects in milling processes. Powder Technol 153(1):59–71
7. Borissova A (2009) General systems modeling of multi-phase batch crystallization from solution. Chem Eng Process Process Intensif 48 (1):268–278
8. Boukouvala F, Ramachandran R, Muzzio FJ, Ierapetritou M (2011) Computer-aided design of an integrated pharmaceutical process. Comput Aided Chem Eng 6:100–105
9. Boukouvala F, Niotis V, Ramachandran R, Muzzio FJ, Ierapetritou MG (2012) An integrated approach for dynamic flowsheet modeling and sensitivity analysis of a continuous tablet manufacturing process. Comput Chem Eng 42:30–47
10. Boukouvala F, Chaudhury A, Sen M, Zhou R, Mioduszewski L, Ierapetritou M, Ramachandran R (2013) Computer-aided flowsheet simulation of a pharmaceutical tablet manufacturing process incorporating wet granulation. J Pharm Innov 8(1):11–27
11. Braatz RD (2002) Advanced control of crystallization processes. Annu Rev Control 26 (1):87–99
12. Cameron I, Wang F, Immanuel C, Stepanek F (2005) Process systems modelling and applications in granulation: a review. Chem Eng Sci 60 (14):3723–3750
13. Capece M, Bilgili E, Dave R (2011) Identification of the breakage rate and distribution parameters in a non-linear population balance model for batch milling. Powder Technol 208 (1):195–204
14. Chakraborty J, Kumar S (2007) A new framework for solution of multidimensional population balance equations. Chem Eng Sci 62 (15):4112–4125
15. Chaudhury A, Ramachandran R (2013) Integrated population balance model development and validation of a granulation process. Part Sci Technol 31(4):407–418
16. Chaudhury A, Tabora JE, Murugesan S, Remy B, Ramachandran R (2011) Application of a 2-d population balance model to a pharmaceutical crystallization process. In: Proceedings of 2011 AIChe annual meeting, Minneapolis, October 2011
17. Chaudhury A, Kapadia A, Prakash AV, Barrasso D, Ramachandran R (2013) An extended cell-average technique for a multi-dimensional

population balance of granulation describing aggregation and breakage. Adv Powder Technol 24(6):962–971
18. Chaudhury A, Niziolek A, Ramachandran R (2013) Multi-dimensional mechanistic modeling of fluid bed granulation processes: an integrated approach. Adv Powder Technol 24 (1):113–131
19. Chaudhury A, Oseledets IV, Ramachandran R (2013) A computationally efficient technique for the solution of multi-dimensional population balance models of granulation via tensor decomposition. Comput Chem Eng 61:234–244
20. Chaudhury A, Barrasso D, Pandey P, Wu H, Ramachandran R (2014) Population balance model development, validation, and prediction of cqas of a high-shear wet granulation process: towards qbd in drug product pharmaceutical manufacturing. J Pharm Innov 9(1):53–64
21. Chaudhury A, Wu H, Khan M, Ramachandran R (2014) A mechanistic population balance model for granulation processes: effect of process and formulation parameters. Chem Eng Sci 107:76–92
22. Costa CBB, Maciel MRW, Filho RM (2007) Considerations on the crystallization modeling: population balance solution. Comput Chem Eng 31(3):206–218
23. Datta A, Rajamani RK (2002) A direct approach of modeling batch grinding in ball mills using population balance principles and impact energy distribution. Int J Miner Process 64(4):181–200
24. Denis C, Hemati M, Chulia D, Lanne JY, Buisson B, Daste G, Elbaz F (2003) A model of surface renewal with application to the coating of pharmaceutical tablets in rotary drums. Powder Technol 130(1–3):174–180
25. Diemer RB Jr, Spahr DE, Olson JH, Magan RV (2005) Interpretation of size reduction data via moment models. Powder Technol 156(2–3):83–94
26. Ennis BJ, Tardos G, Pfeffer R (1991) A microlevel-based characterization of granulation phenomena. Powder Technol 65 (1–3):257–272
27. Fazli NA, Samad A, Singh R, Sin G, Gernaey KV, Gani R (2011) A generic multi-dimensional model-based system for batch cooling crystallization processes. Comput Chem Eng 35(5):828–843
28. Févotte F, Févotte G (2010) A method of characteristics for solving population balance equations (pbe) describing the adsorption of impurities during crystallization processes. Chem Eng Sci 65(10):3191–3198
29. Fichthom KA, Weinberg WH (1991) Theoretical foundations of dynamical monte carlo simulations. J Chem Phys 95(2):1090–1096
30. Freireich B, Li J, Litster J, Wassgren C (2011) Incorporating particle flow information from discrete element simulations in population balance models of mixer-coaters. Chem Eng Sci 66(16):3592–3604
31. Gernaey KV, Gani R (2010) A model-based systems approach to pharmaceutical product-process design and analysis. Chem Eng Sci 65:5757–5769
32. Gerogiorgis DI, Barton PI (2009) Steady-state optimization of a continuous pharmaceutical process. In: 10th International symposium on process systems engineering: Part A, computer aided chemical engineering, vol 27, pp 927–932
33. Goldhirsch I (2008) Introduction to granular temperature. Powder Technol 182:130–136
34. Gunawan R, Fusman I, Braatz RD (2004) High resolution algorithms for multidimensional population balance equations. AIChE J 50(11):2738–2749
35. Hounslow MJ, Ryall RL, Marshall VR (1988) A discretized population balance for nucleation, growth, and aggregation. AIChE J 34 (11):1821–1832
36. Immanuel CD, Doyle III FJ (2003) Computationally efficient solution of population balance models incorporating nucleation, growth and coagulation: application to emulsion polymerization. Chem Eng Sci 58(16):3681–3698
37. Immanuel CD, Doyle III FJ (2005) Solution technique for a multi-dimensional population balance model describing granulation processes. Powder Technol 156(2–3):213–225
38. Iveson SM (2002) Limitations of one-dimensional population balance models of wet granulation processes. Powder Technol 124:219–229
39. Klatt KU, Marquardt W (2009) Perspectives of process systems engineering-personal views from academia and industry. Comput Chem Eng 33:536–550
40. Kumar J, Peglow M, Warnecke G, Heinrich S, Mörl L (2006) Improved accuracy and convergence of discretized population balance for aggregation: the cell average technique. Chem Eng Sci 61(10):3327–3342
41. Kumar S, Ramkrishna D (1996) On the solution of population balance equations by discretization–i. A fixed pivot technique. Chem Eng Sci 51(8):1311–1332
42. Kumar R, Wassgren C (2013) Angular circulation speed of tablets in a vibratory tablet

coating pan. AAPS PharmSciTech 14(1): 339–351

43. Laloue N, Couenne F, Gorrec YL, Kohl M, Tanguy D, Tayakout-Fayolle M (2007) Dynamic modeling of a batch crystallization process: a stochastic approach for agglomeration and attrition process. Chem Eng Sci 62 (23):6604–6614
44. Lee K (1983) Change of particle size distribution during brownian coagulation. J Colloid Interface Sci 92(2):315–325
45. Leuenberger H (2001) New trends in the production of pharmaceutical granules: batch versus continuous processing. Eur J Pharm Biopharm 52:289–298
46. Li J, Freireich B, Wassgren C, Litster JD (2012) A general compartment-based population balance model for particle coating and layered granulation. AIChE J 58(5): 1397–1408
47. Lindenberg C, Krättli M, Cornel J, Mazzotti M (2009) Design and optimization of a combined cooling/antisolvent crystallization process. Cryst Growth Des 9(2):1124–1136
48. Litster J, Ennis B (2004) The science and engineering of granulation processes. Kluwer Academic Publishers, Dordrecht
49. Litster JD, Smit DJ, Hounslow MJ (1995) Adjustable discretized population balance for growth and aggregation. AIChE J 41(3): 591–603
50. Liu LX, Litster JD (2002) Population balance modelling of granulation with a physically based coalescence kernel. Chem Eng Sci 57 (12):2183–2191
51. Liu LX, Litster JD, Iveson SM, Ennis BJ (2000) Coalescence of deformable granules in wet granulation processes. AIChE J 46(3): 529–539
52. Liu L, Robinson D, Addai-Mensah J (2012) Population balance based modelling of nickel laterite agglomeration behaviour. Powder Technol 223:92–97
53. Long C, Gantt J, Gatzke E (2005) Batch granulation control using a simplified population balance and nonlinear model predictive control. In: Proceedings of the 2005 American control conference, 2005, vol 2, pp 949–954
54. Ma DL, Tafti DK, Braatz RD (2002) Optimal control and simulation of multidimensional crystallization processes. Comput Chem Eng 26(7–8):1103–1116
55. Marchal P, David R, Klein J, Villermaux J (1988) Crystallization and precipitation engineering—i. An efficient method for solving population balance in crystallization with agglomeration. Chem Eng Sci 43(1):59–67
56. Marchisio DL, Soos M, Sefcik J, Morbidelli M, Barresi AA, Baldi G (2006) Effect of fluid dynamics on particle size distribution in particulate processes. Chem Eng Technol 29 (2):191–199
57. Maronga S, Wnukowski P (1997) Establishing temperature and humidity profiles in fluidized bed particulate coating. Powder Technol 94 (2):181–185
58. Maronga S, Wnukowski P (1997) Modelling of the three-domain fluidized-bed particulate coating process. Chem Eng Sci 52 (17):2915–2925
59. Marshall CL Jr, Rajniak P, Matsoukas T (2011) Numerical simulations of two-component granulation: comparison of three methods. Chem Eng Res Des 89(5):545–552
60. Miki H, Terashima T, Asakuma Y, Maeda K, Fukui K (2005) Inclusion of mother liquor inside kdp crystals in a continuous msmpr crystallizer. Sep Purif Technol 43(1):71–76
61. Nakamura H, Abe E, Yamada N (1998) Coating mass distributions of seed particles in a tumbling fluidized bed coater part ii. A monte carlo simulation of particle coating. Powder Technol 99(2):140–146
62. Oseledets IV, Tyrtyshnikov EE (2009) Breaking the curse of dimensionality, or how to use SVD in many dimensions. SIAM J Sci Comput 31(5):3744–3759
63. Pandey P, Tao J, Chaudhury A, Ramachandran R, Gao JZ, Bindra DS (2013) A combined experimental and modeling approach to study the effects of high-shear wet granulation process parameters on granule characteristics. Pharm Dev Technol 18(1):210–224
64. Pandya J, Spielman L (1983) Floc breakage in agitated suspensions: effect of agitation rate. Chem Eng Sci 38(12):1983–1992
65. Pinto MA, Immanuel CD, Doyle III FJ (2007) A feasible solution technique for higher-dimensional population balance models. Comput Chem Eng 31(10):1242–1256
66. Portillo PM, Muzzio FJ, Ierapetritou MG (2008) Using compartment modeling to investigate mixing behavior of a continuous mixer. J Pharm Innov 3:161–174
67. Prakash AV, Chaudhury A, Barrasso D, Ramachandran R (2013) Simulation of population balance model-based particulate processes via parallel and distributed computing. Chem Eng Res Des 91(7):1259–1271
68. Prakash AV, Chaudhury A, Ramachandran R (2013) Parallel simulation of population balance model-based particulate processes using multicore CPUs and GPUs. Model Simul Eng. doi:10.1155/2013/475478

69. Puel F, Févotte G, Klein J (2003) Simulation and analysis of industrial crystallization processes through multidimensional population balance equations. Part 1: a resolution algorithm based on the method of classes. Chem Eng Sci 58(16):3715–3727
70. Qamar S, Galan K, Elsner MP, Hussain I, Seidel-Morgenstern A (2013) Theoretical investigation of simultaneous continuous preferential crystallization in a coupled mode. Chem Eng Sci 98:25–39
71. Ragnarsson G, Johansson MO (1988) Coated drug cores in multiple unit preparations influence of particle size. Drug Dev Ind Pharm 14 (15–17):2285–2297
72. Ramachandran R, Chaudhury A (2012) Model-based design and control of a continuous drum granulation process. Chem Eng Res Des 90(8):1063–1073
73. Ramachandran R, Immanuel CD, Stepanek F, Litster JD, Doyle III FJ (2009) A mechanistic model for breakage in population balances of granulation: theoretical kernel development and experimental validation. Chem Eng Res Des 87(4):598–614
74. Ramkrishna D (2000) Population balances. Academic, San Diego
75. Reklaitis GV, Khinast J, Muzzio F (2010) Pharmaceutical engineering science - new approaches to pharmaceutical development and manufacturing. Chem Eng Sci 65:iv–vii
76. Remy B (2010) Granular flow, segregation and agglomeration in bladed mixers. PhD thesis, Rutgers, The State University of New Jersey
77. Remy B, Khinast J, Glasser B (2009) Discrete element simulation of free flowing grains in a four-bladed mixer. AIChE J 55:2035–2058
78. Reynolds GK (2010) Modelling of pharmaceutical granule size reduction in a conical screen mill. Chem Eng J 164(2–3):383–392
79. Ronsse F, Pieters J, Dewettinck K (2007) Combined population balance and thermodynamic modelling of the batch top-spray fluidised bed coating process. Part I—model development and validation. J Food Eng 78 (1):296–307
80. Salman AD, Hounslow MJ, Seville JPK (2007) Granulation. Elsevier, Amsterdam
81. Sastry KV (1975) Similarity size distribution of agglomerates during their growth by coalescence in granulation or green pelletization. Int J Miner Process 2(2):187–203
82. Sen M, Ramachandran R (2013) A multi-dimensional population balance model approach to continuous powder mixing processes. Adv Powder Technol 24:51–59
83. Sen M, Singh R, Vanarase A, John J, Ramachandran R (2012) Multi-dimensional population balance modeling and experimental validation of continuous powder mixing processes. Chem Eng Sci 80:349–360
84. Sen M, Chaudhury A, Singh R, John J, Ramachandran R (2013) Multi-scale flowsheet simulation of an integrated continuous purification-downstream pharmaceutical manufacturing process. Int J Pharm 445(1–2):29–38
85. Sen M, Chaudhury A, Singh R, Ramachandran R (2014) Two-dimensional population balance model development and validation of pharmaceutical crystallization processes. American Journal of Modern Chemical Engineering 1: 13–29
86. Sen M, Dubey A, Singh R, Ramachandran R (2013) Mathematical development and comparison of a hybrid pbm-dem description of a continuous powder mixing process. J Powder Technol. dx.doi.org/10.1155/2013/843784
87. Sen M, Rogers A, Singh R, Chaudhury A, John J, Ierapetritou MG, Ramachandran R (2013) Flowsheet optimization of an integrated continuous purification-processing pharmaceutical manufacturing operation. Chem Eng Sci 102:56–66
88. Sherony DF (1981) A model of surface renewal with application to fluid bed coating of particles. Chem Eng Sci 36(5):845–848
89. Stephanopoulos G, Reklaitis GV (2011) Process systems engineering: from solvay to modern bio- and nanotechnology: a history of development, successes and prospects for the future. Chem Eng Sci 66(19):4272–4306
90. Tan HS, Goldschmidt MJV, Boerefijn R, Hounslow MJ, Salman AD, Kuipers JAM (2004) Building population balance model for fluidized bed melt granulation: lessons from kinetic theory of granular flow. Powder Technol 30:103–109
91. Vanarase AU (2011) Design modeling and real time monitoring of continuous powder mixing processes. PhD thesis, Rutgers, The State University of New Jersey
92. Vanarase A, Muzzio F (2011) Effect of operating conditions and design parameters in a continuous powder mixer. Powder Technol 208(1):26–36
93. Vanderroost M, Ronsse F, Dewettinck K, Pieters J (2008) Population balance model for quality control of coating processes in fluidised beds. In: Proceedings of XVI international conference on bioencapsulation, p 23-1
94. Verkoeijen D, Pouw GA, Meesters GMH, Scarlett B (2002) Population balances for

particulate processes–a volume approach. Chem Eng Sci 57(12):2287–2303

95. Vogel L, Peukert W (2003) Breakage behaviour of different materials—construction of a mastercurve for the breakage probability. Powder Technol 129(1–3):101–110
96. Wang F, Ge X, Balliu N, Cameron I (2006) Optimal control and operation of drum granulation processes. Chem Eng Sci 61(1):257–267
97. Wauters P (2001) Modelling and mechanisms of granulation. PhD thesis, Delft University of Technology, Delft
98. Wnukowski P, Setterwall F (1989) The coating of particles in a fluidized bed (residence time distribution in a system of two coupled perfect mixers). Chem Eng Sci 44(3):493–505

Chapter 3

A Quantitative Approach to Understand Raw Material Variability

Sara Koynov and Fernando J. Muzzio

Abstract

The properties of raw materials can affect the properties and quality of intermediates and the final product. The flow behavior of powders, the form of the raw materials for solid oral drug product manufacturing processes, is poorly understood. Thus, the prediction of manufacturability and process performance is difficult. This is further complicated with variability in raw material flow properties. This chapter discusses seven properties typically characterized and the techniques used to measure them. The state of the art in applying this information to define a formulation of manufacturing process is also discussed.

Key words Powder, Flowability, Characterization

1 Introduction

The current state of the art in characterizing raw material variability is to use a set of established techniques to measure a set of properties. The results are then used to inform formulation and processing decisions. This chapter discusses the properties typically characterized and the techniques used to measure them. The state of the art in applying this information to define a formulation of manufacturing process is also discussed.

It is a truism that the properties of raw materials can affect the properties and quality of the final product [1]. Therefore, understanding the properties of the raw materials is critical. Further, raw materials often have variability in their properties, and so it is prudent to monitor raw material properties and develop effective acceptance criteria in order to avoid easily preventable product/process failures.

The raw materials of interest here are powders and granular materials. Powders are complex and their behavior is poorly understood [2]. The flowability, or flow-related behavior, of a powder is a multidimensional characteristic. There are no general constitutive

Marianthi G. Ierapetritou and Rohit Ramachandran (eds.), *Process Simulation and Data Modeling in Solid Oral Drug Development and Manufacture*, Methods in Pharmacology and Toxicology,
DOI 10.1007/978-1-4939-2996-2_3, © Springer Science+Business Media New York 2016

equations or unifying framework to describe powder dynamics. Instead, experimental characterization techniques and empirical correlations are used. There are several characterization techniques available, each measuring a slightly different aspect of flow behavior. These techniques also test powder in a variety of states: from fully aerated to varying degrees of consolidation, from confined to unconfined, from incipient slipping to steady state flow. It is also frequently the case that a single characterization parameter is not sufficient to fully describe the flow behavior; the information from multiple tests is often required. Although there are several techniques, the tests can be classified as the measurement of one or more of the following properties: cohesion, bulk density, permeability, friability, hydrophobicity, electrostatics, and segregation tendency. Usually, rather than characterized directly, the effects of particle size and particle shape are incorporated into the cohesion and bulk density properties, respectively. As the particle size decreases, there is an accompanying increase in cohesion due to van der Waals forces. As a result, the effect of altering the particle size on the flowability of a powder can be ascertained by measuring the cohesion of the material, similarly to the way in which a change in the molecular weight of a polymer is often described in terms of the changes in viscosity of polymer solutions. The mechanism by which the particle shape affects the flowability of a material is through the mechanical interlocking between the particles. As compared to freely flowing spherical particles, particles with a large degree of interlocking can exhibit hindered flow.

It is important to understand raw material properties because of their great impact on the success of the manufacturing process. Variation in raw material properties will in turn become variation in blend properties eventually becoming variation in final product quality. Raw material properties should not be characterized in vacuum, but accompanying understanding of the manufacturing process. Identifying the relationship between raw material properties and process performance is critical to determining which raw material properties should be measured and the magnitude of variation in those properties will cause processing issues downstream.

Several studies have been performed in an effort to elucidate the relationship between raw material properties and process performance. These studies typically seek one of two goals: improving powder flow or understanding individual unit operations. Powder flow is frequently improved by altering the surface properties of the particles. Flow additives such as magnesium stearate or colloidal silica will often dry-coat a fraction of the outer surface of poorly flowing particles. Magnesium stearate decreases the attraction force between particles either by rendering the surface hydrophobic, thus decreasing the formation of capillary bridges, or by physically separating particles, thus reducing the van der Waals forces [3–5].

Colloidal silica reduces the electrostatic properties of the particles [6–9]. In addition, unit operations have been individually investigated including hopper flow, spheronization, mixing, tabletting, and capsule filling [10–15].

This chapter discusses seven raw material properties. For each property, the definition, physical meaning, and available techniques to measure them will be given.

2 Material Properties

2.1 Cohesion

The cohesion of a material is indicative of how well a material flows or how "sticky" it is. Particles tend to stick together when the interparticle forces exceeds the gravitational force pulling them apart. This ratio is described for dense flows using the granular bond number [16].

$$\mathrm{Bo_g} = \frac{n^{\mathrm{c}}}{mg}$$

Where n^{c} is the maximum attractive force and mg is the gravitational force exerted on a particle. Sources of the attractive interparticle forces include van der Waals forces, capillary forces from liquid bridges, and electrostatics originating from unlike charges or dipolar moments.

Cohesion can be measured, among other methods, by using avalanche testers (such as the Gravitational Displacement Rheometer [17, 18]) or shear cells [19]. During avalanche testing, cohesion is determined based on the size of the avalanches that are formed during tumbling flow; that is, the scale of the self-assembled chunks formed within the powder bed. The shear cell measures the degree by which arches formed within the powder bed strengthen (more, stronger arches tend to form in cohesive powders) in response to increase applied normal stress. The shear cell also measures cohesion as the shear stress required to move particles against one another under zero normal stress. The resulting flow index or flow function metrics, in addition to the cohesion parameter, can be used to rank the flowability of materials and to predict the behavior of a new material in a manufacturing process based on prior experience with other materials.

2.1.1 Avalanche Testers

Commercially available avalanche testers, including the AeroFlow, Revolution Powder Analyzer, and the GDR, have been used for industrial applications in addition to academic studies [20–22]. The primary difference between the three technologies is the length of the cylinder used; the cylinder used in the GDR is much longer to reduce the effect of friction with the side walls. The GDR method is described in detail here. A comprehensive discussion comparing the GDR to other devices exists in the literature [23].

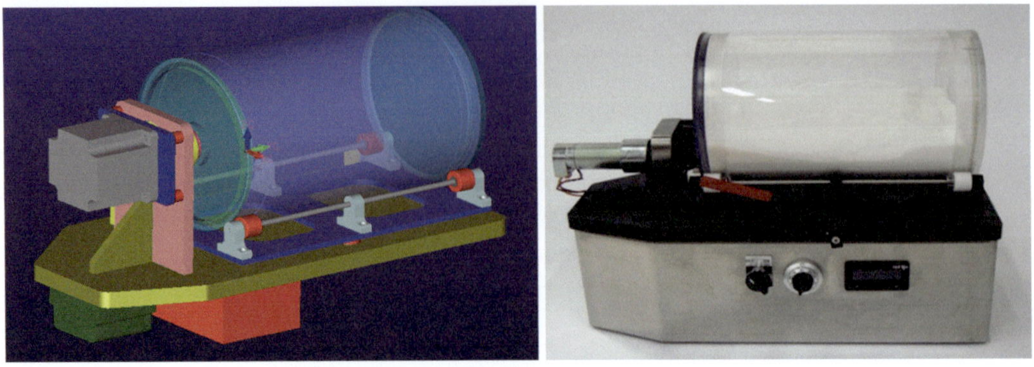

Fig. 1 Schematic (*left*) and image (*right*) of the GDR experimental setup

The GDR, a technology developed at Rutgers based on the method described by Davies et al. [24], is comprised of a rotating cylinder placed on a table supported by a hinge and a load cell. The experimental setup is presented in Fig. 1. To test the cohesive nature of a material, a characteristic known as the flow index is measured. An acrylic cylinder is given sufficient length to eliminate the wall effects on the powder flow. As the cylinder rotates, the powder flows in avalanches. As the avalanches fall, their force is measured by the load cell. Using these measurements, the sizes of the avalanches are determined. The sizes of the avalanches are correlated to the cohesiveness of the powder; a more cohesive powder will flow in large avalanches.

The load cell signal is recorded for rotation rates of 5, 10, 15, and 20 rpm. The load cell signal is filtered and normalized, in order to account for the material density. The standard deviation of the signals is then calculated for each rotation rate. The average of the standard deviations at 5, 10, 15, and 20 rpm is then reported as the flow index. Since the variation in the load cell signal increases with larger avalanches, lower flow index values indicate better flow.

2.1.2 Shear Cell

The shear cell testing methodology was originally developed by A. W. Jenike for the specific application of sizing hoppers and silos from the principals of solid state mechanics [25]. They are now commonly used to rank granular materials according to their flowability, i.e., how cohesive the material is. From a series of shear tests, the angle of internal friction, the angle of wall friction, the slope of the hopper walls, and other design parameters can be extracted. This methodology has since been more generally applied in the field of powder characterization.

This characterization technique has become so prevalent that international standards detailing the procedure have been defined [26]. Because the shear cell is commonly used, many studies and much research have been done involving the shear cell. Much of this work has been focused on determining if the shear cell

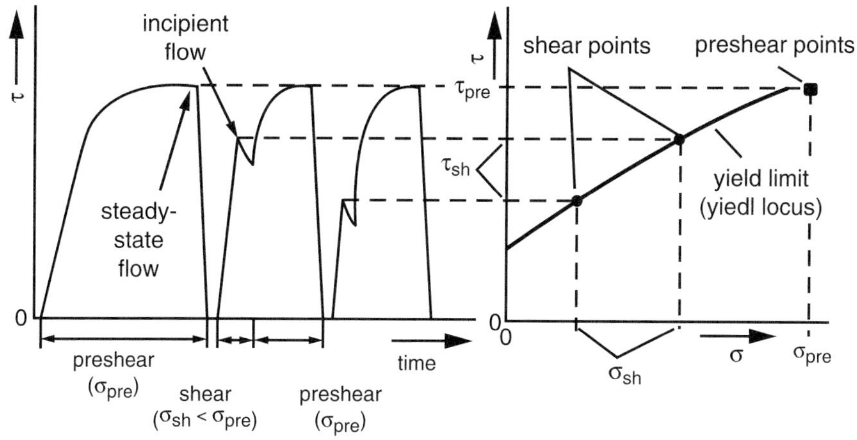

Fig. 2 The yield locus is derived from shear stress data. Reprinted from ref. [62] with permission

measurements are applicable for specific situations [22]. In addition, changes in various aspects of the procedure (ones not dictated by the standards) have been studied [27]. Lastly, some studies have been conducted that compare the measurements of a particular shear cell parameter between different types of shear testers [28].

While original shear cells used translational motion to measure "incipient failure", more modern versions of the technique use rotational motion to impart large strain and to achieve "steady state" mechanical behavior. The standard procedure involves three steps: pre-compaction of the powder bed, pre-shearing of the powder bed until steady state flow is achieved (powder stresses are equal in all directions and the bulk density is constant), and shearing until the powder yields. The pre-shearing/shearing process is repeated 4–5 times using normal stresses 20–80 % of the consolidation stress. The result of the above process is a series of measurements of corresponding normal and shear stresses and is called the "yield locus". Figure 2 depicts how the yield locus is derived from the raw data. The yield locus is fit with a best-fit line that is extrapolated to the *y*-axis. This corresponds to the shear stress at zero normal stress, or cohesion.

In addition, Mohr circle analysis is performed on the yield locus. Mohr circle analysis is a geometric representation of a coordinate transformation to identify the principal stresses. Two circles are used. The first goes through the origin and is tangent to the best-fit line through the yield locus. This circle represents the conditions present at the free surface of an arch (as is present in hopper flow). The second circle is tangent to the yield locus and passes through the pre-shear (steady state flow) point. The principal stresses extracted from this analysis are called the unconfined yield stress and the compacting stress (major principal stress), i.e., the yield stress of the powder in the free surface of the arch (as in

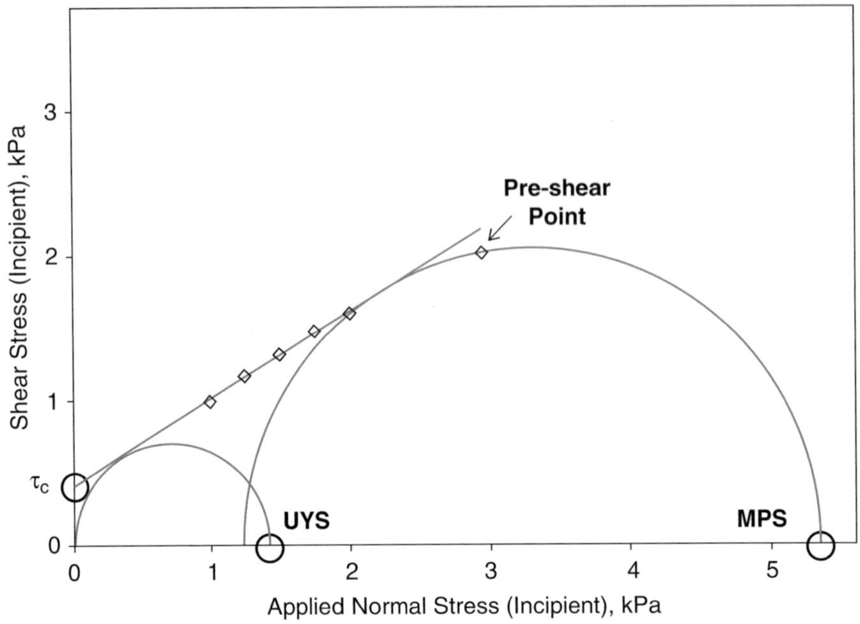

Fig. 3 Yield locus with best-fit line and Mohr circles is shown. τ_c is the cohesion, UYS is the unconfined yield stress, and MPS is the compacting stress

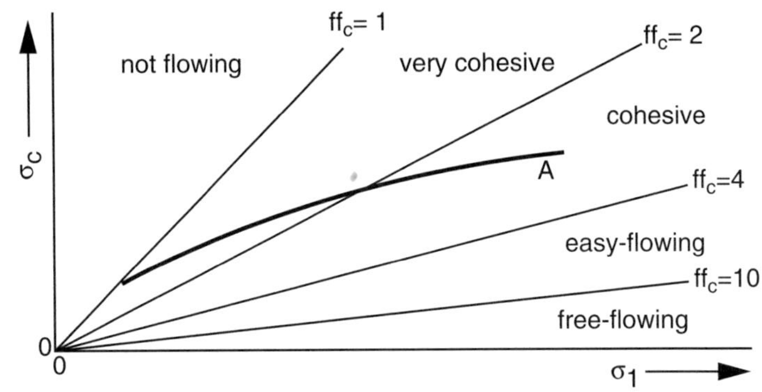

Fig. 4 A typical flow function (*left*) and the categorization of the material (*right*). Reprinted from ref. [62] with permission

hopper flow) and the compacting stress in the powder bed, respectively as shown in Fig. 3. The yield locus is measured at several consolidation stresses and the unconfined yield stress and compacting stress is extracted from each yield locus. The unconfined yield stress–compacting stress pairs are plotted from each yield locus to give the flow function, *see* Fig. 4. The slope of the flow function indicates how well a powder flows; the steeper the slope, the more cohesive and poorly flowing the powder.

2.2 Bulk Density

The bulk density of a material is indicative of a material's flowability and the degree to which the material may expand or consolidate under various conditions occurring during manufacturing [29, 30]. The bulk density, defined as the ratio of the mass of powder sample to the volume of that powder sample, takes into account both the particle density as well as the packing of the powder bed [31]. As this measurement is highly dependent on the consolidation, or packing, state of the material, the aerated bulk density measurement is traditionally paired with the tapped bulk density (the density of the bulk with a high degree of packing resulting from extensive settling). The tapped bulk density is measured according to Standard Test Method ASTM D6393-08 "Bulk Solids Characterization by Carr Indices" using instruments such as the Tap Density Volumeter (Quantachrome) and Hosokawa Micron Powder Tester (Hosokawa, Japan).

The compressibility of a granular material compares the aerated bulk density to the tapped bulk density. There are several acceptable ways of expressing the relationship between the aerated and tapped bulk density including the Carr index, C [32], and Hausner ratio, HR [33].

$$C = \frac{V_0 - V_N}{V_0}$$

$$\mathrm{HR} = \frac{\rho_N}{\rho_0} = \frac{V_0}{V_N}$$

where ρ_0, V_0, ρ_N, and V_N, are the aerated bulk density and powder volume, and the tapped bulk density and powder volume after N taps, respectively.

More recently, the bulk density has been measured over a range of packing states by varying the consolidation stress applied to the powder bed, such as with the Freeman Technology FT4 [34]. Alternatively, the difference in density between the tapped state and the dilated state can be measured using the dilation method from the Gravitational Displacement Rheometer [35].

2.2.1 FT4: Compressibility

The compressibility test is part of the Freeman Technology FT4 Powder Rheometer suite. This test measures bulk density over a range of packing states comparing the results to an initial, conditioned bulk density. A schematic of the test and sample raw data that can be obtained are shown in Fig. 5. The test begins by conditioning the powder by passing a helical blade through the bed in order to erase the history of the powder and create a uniform, reproducible packing state. A normal force is then slowly applied using a vented piston (allowing air to escape from the power bed). The normal force ranges from 0.5 to 15 kPa and is applied over ten intervals, holding each load for 60 s. The change in volume due to

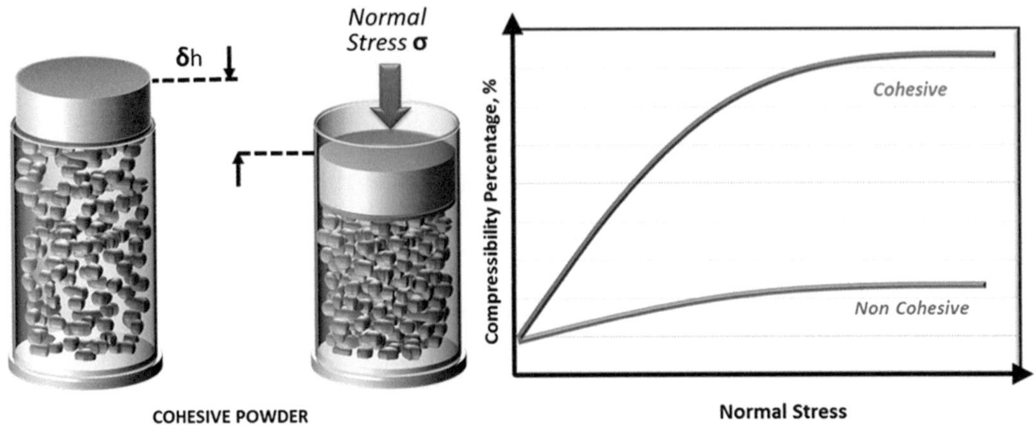

Fig. 5 Courtesy of Freeman Technology, Inc (reprinted with permission). Schematic of compressibility test (*left*) and sample raw data obtained from compressibility test (*right*)

the compression is measured at each interval. The compressibility index (I_c), is calculated as a ratio of the initial and final densities, analogous to the Hausner ratio. The compressibility (Comp %) is calculated as the percent change in volume after compression, analogous to the Carr index.

$$\text{Comp }\% = 100 \times \frac{V_C - V_P}{V_C}$$

$$I_C = \frac{\rho_P}{\rho_C}$$

where ρ_C, V_C, ρ_P, and V_P are the conditioned bulk density and powder volume, and bulk density and powder volume after compression with piston, respectively.

2.2.2 GDR: Dilation

Using the GDR experimental setup described in Sect. 2.1.1 and illustrated in Fig. 1, a characteristic known as the dilation can be measured. Here, the powder is initially at or close to an asymptotic tapped density state, usually obtained by tapping the cylinder 500 times using a tap density tester. The dilation is a measure of the expansion of the powder bed upon rotation of the cylinder [17, 35].

First, the material is loaded into the cylinder and the cylinder is tapped to consolidate the powder. The material is loaded into the cylinder such that after tapping the powder bed volume is about 40 % of the cylinder volume. The cylinder is then placed on top of the load cell in front of an IR camera. The camera begins recording images of the tapped powder bed and continues as the cylinder rotates at 15 rpm. The change in volume of the powder bed upon rotation is determined by image analysis. The change in volume is reported as the dilation, in percent. The dilation is analogous to the Carr index, comparing the tapped bulk density to the dilated bulk density [36].

$$\text{Dilation }\% = \frac{V_{\text{new}} - V_{\text{initial}}}{V_{\text{initial}}}$$

where V_{new} and V_{initial} are the dilated and tapped bulk densities, respectively.

One advantage of measurements such as the Carr Indx, the Hausner ratio, or the dilation, which normalize the measurement using a ratio of densities (or volumes), is that they yield a measurement that is independent of the absolute density, thus facilitating comparison between powders that are composed of intrinsically heavier or lighter materials.

2.3 FT4: Permeability

Permeability is a measure of how well air passes through a powder bed. In order for powder to flow smoothly, air must be able to travel upstream; otherwise vacuum is created. This is particularly critical at points during a manufacturing process where the powder is transitioning from a static (or nearly static) to a dynamic state, e.g., flow from a hopper, compaction in a tablet die, the initial aeration in a fluidized bed, or when the powder must fill a small cavity, such as in tabletting. The permeability of a granular material is influenced by particle properties such as size, shape, surface texture, and stiffness as well as bulk properties such as cohesion and bulk density.

Permeability is typically measured as the pressure drop across a powder bed over a range of consolidation states, as achieved by the FT4 from Freeman Technology [37]. A schematic of the test and sample raw data that can be obtained are shown below in Fig. 6. The powder sample is loaded into the test vessel and then is conditioned using a helical blade. The conditioning step creates a uniform, reproducible packing state. An upward air velocity of 2 mm/s is added to the bottom of the powder bed. A normal force is then applied to the powder bed with a vented piston (to allow the removal of air from the system) over a range of 0.5–15 kPa. At each normal force, the pressure drop across the bed is measured.

A low, flat pressure drop profile is indicative of a highly permeable material. A high, curved pressure drop profile is indicative of a material with low permeability.

2.4 Friability

Friability is an indication of a material's propensity for attrition and sensitivity to shear. The ability to identify powders that are potentially sensitive to processing can lead to improvements in process design. For materials that are prone to attrition or sensitive to high levels of shear, certain unit operations can be detrimental to the

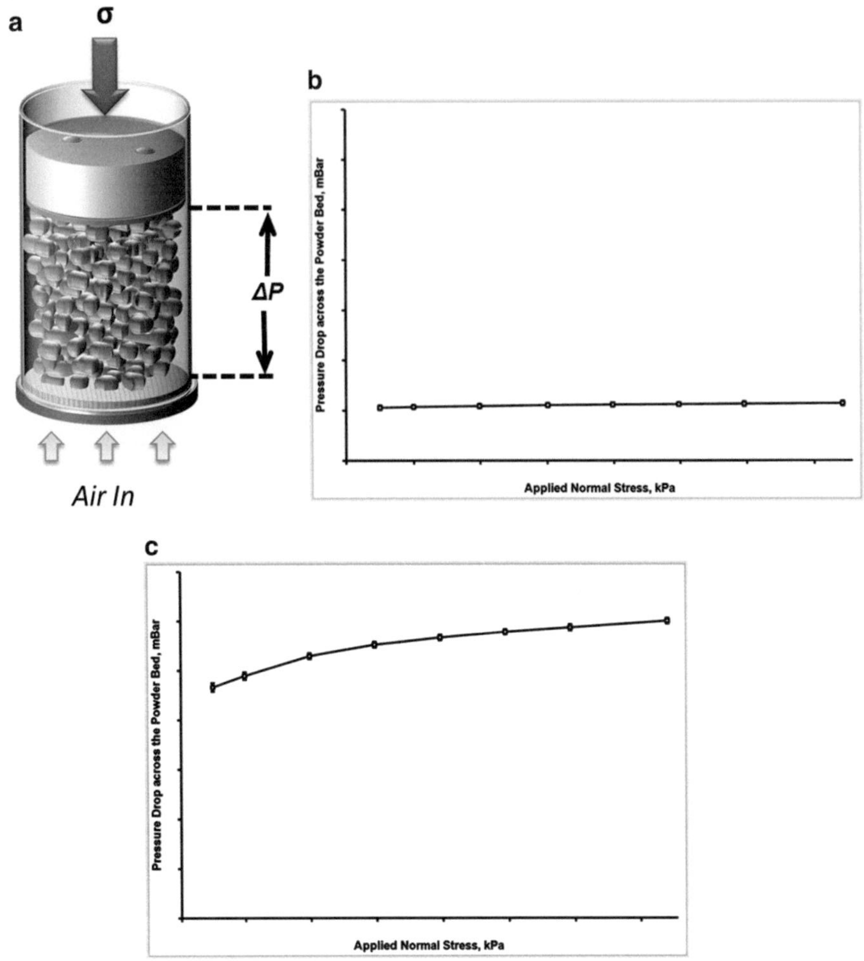

Fig. 6 Courtesy of Freeman Technology, Inc (reprinted with permission). Schematic of permeability test (**a**) and sample raw data obtained (**b** and **c**)

final product quality. Therefore, unit operations that create high shear environments, such as milling, are not recommended for these materials. Instead, it is recommended that friable materials be diverted around such unit operations so that non-friable materials can still be processed in high shear environments. In addition, variability in the raw material properties of friable materials could be magnified during processing.

The friability of a granular material is determined by measuring changes in particle properties after exposure to a simulated processing environment. These changes can be in mass [38] or in the amount of energy required to pass a blade through the powder bed [39].

2.4.1 Friability: Lost Mass

The attrition of granular materials with larger particle size, such as the product of a granulation process or a finished tablet, manifests as the breaking off of pieces of the particles [40]. The extent to which this occurs, known as the friability, is measured by tracking the change in mass of the particles as a result of exposure to stress. Instruments such as the Friabilator [41] and the Friabimat [42] simulate the stresses particles experience during processing and handling. Each of these instruments follows the same basic procedure. The sample of material to be tested is initially weighed, m_i. The sample is then loaded into the testing apparatus, the drum of the Friabilator or bottle of the Friabimat shown in Fig. 7. The sample is then agitated by the rotation of the drum or the oscillation of the bottle through an arc. After agitation, the sample is weighed, m_f. The friability is calculated as the percent mass lost during testing.

$$\%\ \text{mass lost} = 100 \times \left(\frac{m_i - m_f}{m_i}\right)$$

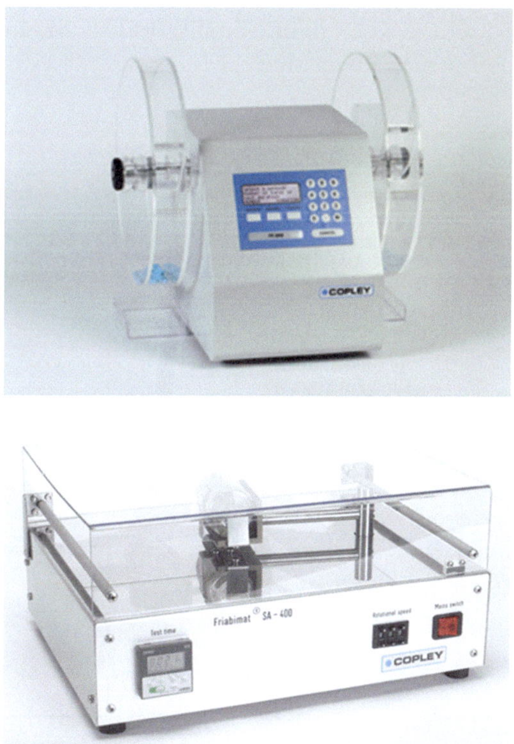

Fig. 7 Courtesy of Copley Scientific (reprinted with permission). A Friabilator and Friabimat SA-400 used for friability testing

2.4.2 FT4: Dynamic Test

The friability of powders and granular materials, including those with smaller particle size, can be measured as a function of the energy required to move a helical blade through a powder bed, known as the flow energy, *E*. In other words, the energy required to make powder flow around the blade. The Freeman Technology FT4 dynamic test procedure measures this energy over time and shear rate (variable blade tip speed) [43]. The powder sample is loaded into the test vessel and then is conditioned using a helical blade. The conditioning step creates a uniform, reproducible packing state. The helical blade is then used during the testing cycle where the flow energy, the energy consumed to make the powder flow, is measured on the downward and upward traverse. The conditioning step followed by a testing cycle is repeated seven times with identical conditions then four additional repetitions are executed with variable blade tip speed [44]. A friable material may exhibit a change in the flow energy over time and over a variable shear rate due to attrition, agglomeration, or segregation. In addition, some materials that are cohesive or highly compressible can become caked or agglomerated during flow. This behavior can also manifest as a change in the flow energy. Figure 8 shows a schematic of the helical blade pattern as well as sample data obtained from the dynamic test procedure.

The stability index, SI, indicates the change in flow energy over time.

$$\mathrm{SI} = \frac{E_7}{E_1}$$

where E_1 is the flow energy required on the first downward traverse blade pass and E_7 is the flow energy required on the seventh

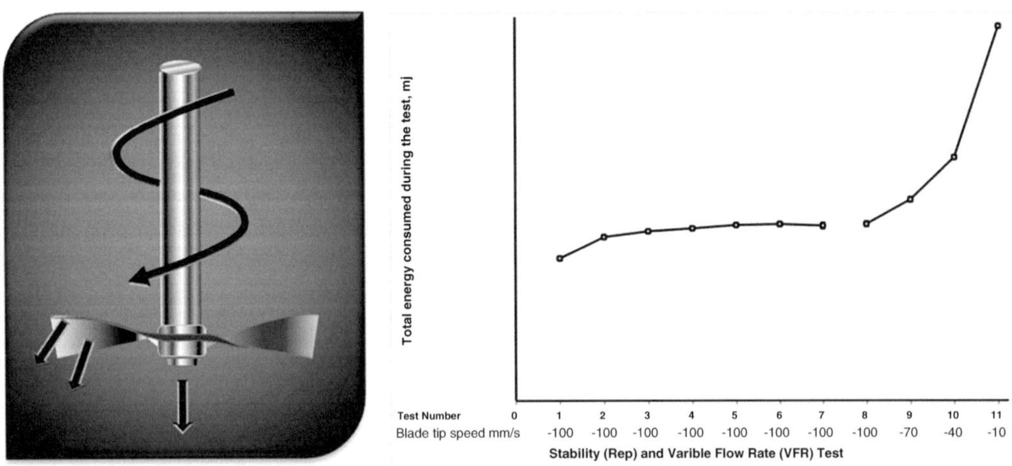

Fig. 8 Courtesy of Freeman Technology, Inc (reprinted with permission). Schematic of helical blade pattern and sample data obtained from the FT4 dynamic test

downward traverse blade pass, also known as the Basic Flow Energy (BFE). The BFE does not account for the changes in mass between materials, and therefore, the Normalized Basic Flow Energy (NBFE) is commonly used.

$$\mathrm{NBFE} = \frac{\mathrm{BFE}}{\mathrm{mass}}$$

A stable, non-friable material has an SI of about 1; the energy consumed over the first blade pass is the same as the seventh pass. An SI value of greater than 1 indicates a material that requires more energy to flow over time. This increase in energy can be due to de-aeration, agglomeration, segregation, moisture uptake, or electrostatic charges. A decrease in the energy required over time, an SI less than 1, can be caused by attrition, de-agglomeration, over blending of an additive, or the coating of the blade and vessel by an additive.

In addition to the energy consumed on the downward traverse, the BFE, the energy consumed during the upwards traverse is also measured, the Specific Energy (SE). The Specific Energy is normalized by the mass of the powder sample. Due to the geometry of the helical blade (Fig. 7) during the upwards traverse, the powder is unconfined. This condition is representative of powder flow in a low stress environment. Therefore, SE is more dependent on cohesive forces and less on compressibility.

Powders that are cohesive and compressible may be sensitive to shear and processing. For example, due to being made to flow or exposure to stress the material could become caked or agglomerated. The FT4 dynamic test also measures the sensitivity of a material's flow energy to a range of flow rates (via variable blade tip speed). The final four repetitions of the conditioning/testing cycle are performed with a tip speed ranging from 10 to 100 mm/s. The Flow Rate Index (FRI) is the ratio between the required flow energy at 100–10 mm/s.

$$\mathrm{FRI} = \frac{E_{11}}{E_8}$$

where E_{11} and E_8 are the BFEs at a blade tip speed of 100 mm/s and 10 mm/s, respectively. Most powders exhibit and FRI value between 1.5 and 3.0, a modest increase in flow energy at higher flow rates. The displacement of the particles due to the blade movement occurs on a timescale dependent on the blade tip speed. At higher flow rates, this occurs much faster than the repacking of the particles resulting in the entrainment of air in the powder bed. This entrained air acts as a lubricant reducing the interparticle contacts and the required flow energy. Conversely, at low flow rates, the blade movement is slower than the repacking of the particles. As a result, the particles become more tightly packed increasing the number of particle contacts and the required flow energy. High FRI values, those larger 3.0, are typically exhibited by very cohesive powders. Cohesive powders tend to be more sensitive to changes in flow rate due to higher air content. Flow rate insensitive powders, those with an FRI value of about 1, usually have larger particle size or surface treatments. Powders that exhibit pseudoplastic or Newtonian behavior, the required flow energy decreasing with increasing flow rate (i.e., an FRI less than 1), usually contain flow additives.

2.5 Hydrophobicity

Hydrophobicity is a quantification of a material's affinity for water. One of the primary quality characteristics in the pharmaceutical industry is the dissolution of the final dosage form (tablet, capsule, etc.). The hydrophobicity is of the blends comprising the dosage forms is therefore critical. The blend properties are a function of the raw material properties as well the processing conditions. Therefore, the hydrophobicity of the raw materials and the effect of processing should be characterized [45, 46].

The wettability of a powder can be measured using a modified Washburn technique [47]. The uptake of water into a powder bed due to capillary action was described by Washburn in 1921 [48]. The volume of water that permeates the powder bed increases linearly with the square root of time. A hydrophobic powder will resist the capillary action resulting in a slower rate of water uptake. The relationship of water uptake to time can also be expressed as a linear relationship between mass of water in the powder bed squared and time [49].

$$t = \frac{\eta}{C\rho^2\gamma\cos\theta}m^2$$

where t is the time, η is the liquid viscosity, C is a geometric factor comprising the effect of powder packing density and particle size, ρ is the liquid density, γ is the liquid surface tension, θ is the contact angle between the liquid and particles, and m is the mass of the liquid in the powder bed column. The slope of this line is considered the hydrophobicity of the material.

The hydrophobicity experimental setup is shown in Fig. 9. The powder sample is loaded into a chromatographic column with a sintered glass bottom. To ensure a constant geometric factor, C, the powder bed plus column assembly is tapped using a tapped density tester. The column is dipped into a large water source. To prevent the dissolution of the power by the large water source, the water is saturated with the material (or materials in the case of a blend). The column is held by a support beam on a load cell. The weight of the water taken into the column is measured as a function of time. To calculate the hydrophobicity, the data is represented as time as a function of mass squared as shown in Fig. 10 [50].

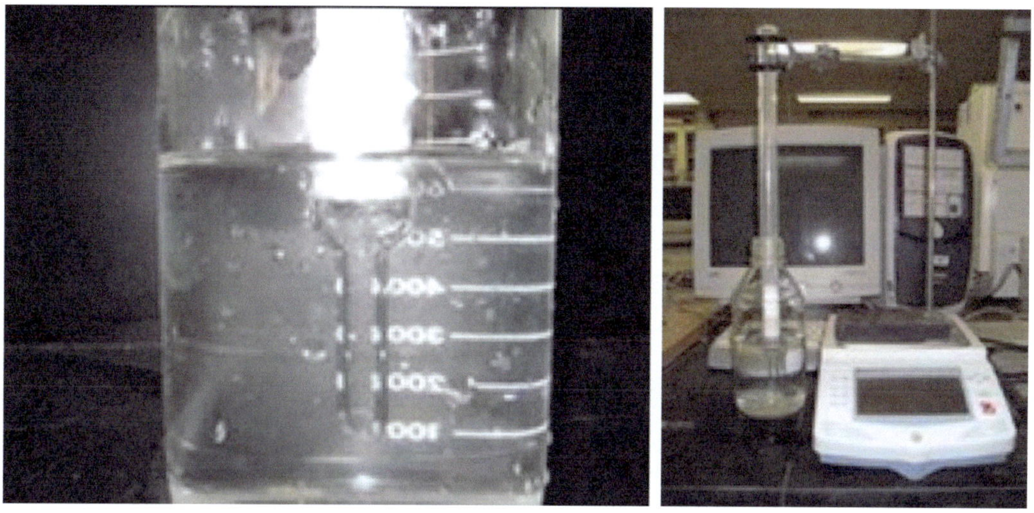

Fig. 9 Experimental setup of the hydrophobicity procedure

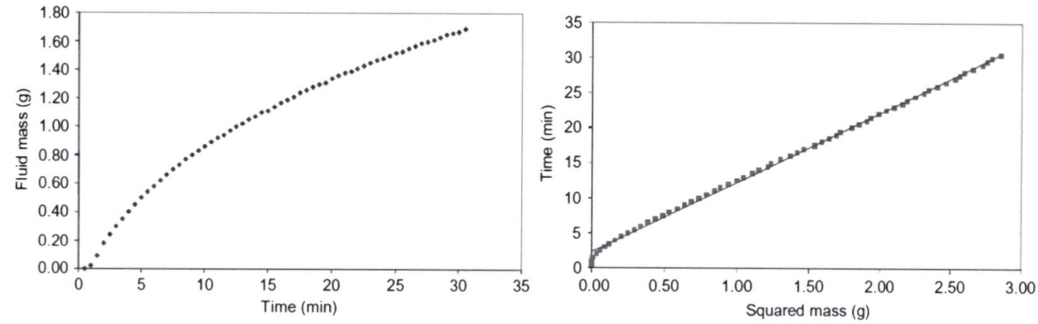

Fig. 10 Data gathered from scale (*left*) and the representation of the data used to calculate the hydrophobicity (*right*)

2.6 Electrostatics

Pharmaceutical processes are ripe with problems such as powder agglomeration, powder sticking to the walls of containers, feeder jamming, poor mixing, and intermittent flow [51]. Figure 11 below shows an example: Avicel 101 sticking to a gravimetric feeder during flow. The root cause of this phenomenon is electrostatics (i.e., charge transfer and charge buildup). Particle charging and electrostatics are complex phenomena. Material properties such as composition, density, and microstructure can affect the electrostatic properties of the material. While several key fundamentals are understood, a complete theory does not exist [52–54]. Numerous techniques can characterize various aspects of electrostatic behavior [55, 56]. Here, the characterization of electrostatic properties by measuring the impedance, resistivity, and dielectric permittivity is detailed. The impedance is the modular ratio of applied

Fig. 11 Avicel 101 sticking to a gravimetric feeder during flow

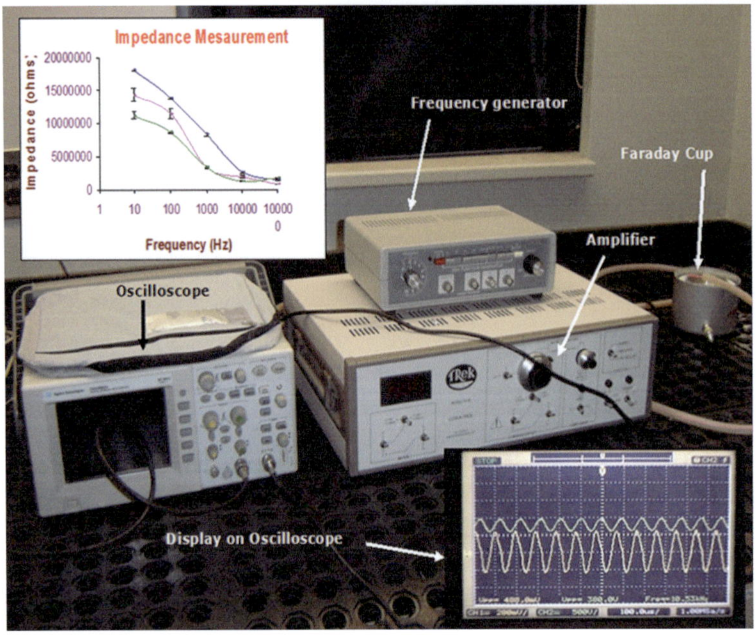

Fig. 12 Experimental setup of the impedance measurement including sample oscilloscope data

voltage to the current flowing through the bed. Knowing the geometry of the testing vessel, the resistivity and dielectric permittivity can be calculated.

The experimental setup and sample data is shown below in Fig. 12. The powder sample is loaded into a test vessel. The vessel is cylindrical and has a conductive bottom with Teflon walls such that it acts as a Faraday cup. A heavy, conductive lid is placed on the sample. The weight of the lid produces a reproducible packing state within the powder bed, necessary for the reliable measurement of impedance, resistivity, and dielectric permittivity [57]. The conductive lid and bottom act as two electrodes. Sinusoidal voltage is applied to the top electrode and the resulting current flowing through the powder bed is measured using an oscilloscope [58].

2.7 Segregation Tendency

Segregation is a phenomenon that is predominantly affected by material properties, rather than processing conditions, although a poorly designed process definitely worsens segregation. Continuous processing involves the flow of materials through channels of various sizes (mixers, pipes, feeders, etc.). For such processes, testing the segregation tendency of the materials can inform the selection of equipment and formulation parameters minimizing potential segregation issues [59]. In practice, sifting can occur during powder flow. The tendency of a material (of either a raw material or a blend) to segregate can be described using a sifting metric, such as with the Jenike Sifting Segregation Tester (originally developed at Rutgers University) that simulates the sifting that can potentially occur during processing [60, 61].

Fig. 13 Experimental setup of the Jenike Sifting Segregation Tester

The experimental setup is shown below in Fig. 13. Two hoppers are placed in a vertical stack. Powder is discharged from the upper hopper into the lower hopper; once or multiple times. Samples are then collected and analyzed. Segregation due to several factors can be characterization in this manner including particle size and chemical composition.

3 Conclusions

Material properties can have a significant impact on process performance and, ultimately, final product quality. As such, it is critical to understand how variation in raw materials, and therefore blends, propagates throughout a manufacturing process. Raw material properties should be studied in concert with the process such that the measurements of the most suitable material characteristics are made.

References

1. Muzzio FJ et al (2002) Powder technology in the pharmaceutical industry: the need to catch up fast. Powder Technol 124:1–7
2. Rietema K (1984) Powders, what are they? Powder Technol 37:5–23
3. Pingali KC et al (2009) Practical methods for improving flow properties of active pharmaceutical ingredients. Drug Dev Ind Pharm 35:1460–1469
4. Faqih AMN et al (2007) Effect of moisture and magnesium stearate concentration on flow properties of cohesive granular materials. Int J Pharm 336:338–345
5. Zhou Q et al (2011) Effect of mechanical dry particle coating on the improvement of powder flowability for lactose monohydrate: a model cohesive pharmaceutical powder. Powder Technol 207:414–421
6. Orband JLR, Geldart D (1995) The use of an antistatic agent to improve powder flowability. Part Part Syst Charact 12:204–206
7. Pingali KC et al (2009) Use of a static eliminator to improve powder flow. Int J Pharm 369:2–4
8. Mullarney MP et al (2011) Applying dry powder coatings to pharmaceutical powders using a comil for improving powder flow and bulk density. Powder Technol 212:397–402
9. Jallo LJ et al (2012) Improvement of flow and bulk density of pharmaceutical powders using surface modification. Int J Pharm 423: 213–225
10. Faqih AN et al (2007) A method for predicting hopper flow characteristics of pharmaceutical powders. Chem Eng Sci 62:1536–1542
11. Vanarase AU et al (2013) Effects of powder flow properties and shear environment on the performance of continuous mixing of pharmaceutical powders. Powder Technol 246:63–72
12. Bouffard J et al (2013) Experimental investigation of the effect of particle cohesion on the flow dynamics in a spheronizer. AIChE J 59:1491–1501
13. He X et al (2013) Assessing powder segregation potential by near infrared (NIR) spectroscopy and correlating segregation tendency to tabletting performance. Powder Technol 236: 85–99
14. Podczeck F (1999) Rheological studies of physical properties of powder used in capsule filling I. Pharm Technol Eur 11:16–24
15. Podczeck F (1999) Rheological studies of physical properties of powder used in capsule filling II. Pharm Technol Eur 11:34–42
16. Nase ST et al (2001) Discrete characterization tools for cohesive granular material. Powder Technol 116:214–223
17. Alexander AW et al (2006) Avalanching flow of cohesive powders. Powder Technol 164:13–21
18. Faqih A et al (2006) An experimental/computational approach for examining unconfined cohesive powder flow. Int J Pharm 324: 116–127

19. Schulze D (2008) Powders and bulk solids: behavior, characterization, storage and flow. Springer, Berlin
20. Hancock BC et al (2004) Development of a Robust procedure for assessing powder flow using a commercial avalanche testing instrument. J Pharm Biomed Anal 35:12
21. Lee YSL et al (2000) Development of a dual approach to assess powder flow from avalanching behavior. AAPS PharmSciTech 1:44
22. Krantz M et al (2009) Characterization of powder flow: static and dynamic testing. Powder Technol 194:239–245
23. Vasilenko A et al (2011) Shear and flow behavior of pharmaceutical blends – method comparison study. Powder Technol 208:628–636
24. Davies CE et al (2004) A new approach to monitoring the movement of particulate material in rotating drums. Dev Chem Eng Miner Process 12:263–275
25. Jenike AW (1964) Storage and flow of solids. Utah Eng Exp Stat Bull 123:1–194
26. Carson JW, Wilms H (2006) Development of an international standard for shear testing. Powder Technol 167:1–9
27. Berry RJ, Bradley MSA (2007) Investigation of the effect of test procedure factors on the failure loci and derived failure functions obtained from annular shear cells. Powder Technol 174: 60–63
28. Pillai JR et al (2007) Comparison between the angles of wall friction measured on an on-line wall friction tester and the Jenike wall friction tester. Powder Technol 174:64–70
29. Abdullah EC, Geldart D (1999) The use of bulk density measurements as flowability indicators. Powder Technol 102:151–165
30. Grey RO, Beddow JK (1969) On the Hausner Ratio and its relationship to some properties of metal powders. Powder Technol 2:323–326
31. Rhodes M (2008) Introduction to particle technology, 2nd edn. John Wiley and Sons, West Sussex
32. Carr RL (1965) Evaluating flow properties of solids. Chem Eng 72:163–168
33. Hausner H (1967) Friction conditions in a massive metal powder. Int J Powder Metall 3:7
34. Freeman R (2007) Measuring the flow properties of consolidated, conditioned and aerated powders — a comparative study using a powder rheometer and a rotational shear cell. Powder Technol 174:25–33
35. Faqih A et al (2006) Flow - induced dilation of cohesive granular materials. AIChE J 52: 4124–4132
36. Vasilenko A et al (2013) Role of consolidation state in the measurement of bulk density and cohesion. Powder Technol 239:366–373
37. Trivedi MR, Dave RH (2014) To study physical compatibility between dibasic calcium phosphate and cohesive actives using powder rheometer and thermal methods. Drug Dev Ind Pharm 40:1585
38. Bemrose CR, Bridgwater J (1987) A review of attrition and attrition test methods. Powder Technol 49:97–126
39. Lindberg N-O et al (2004) Flowability measurements of pharmaceutical powder mixtures with poor flow using five different techniques. Drug Dev Ind Pharm 30:785–791
40. Remon JP, Schwartz JB (1987) Effect of raw materials and processing on the quality of granules prepared from microcrystalline cellulose-lactose mixtures. Drug Dev Ind Pharm 13: 1–14
41. Gohel MC, Jogani PD (2003) Exploration of melt granulation technique for the development of coprocessed directly compressible adjuvant containing lactose and microcrystalline cellulose. Pharm Dev Technol 8:175–185
42. Chevalier E et al (2009) Comparison of low-shear and high-shear granulation processes: effect on implantable calcium phosphate granule properties. Drug Dev Ind Pharm 35:1255–1263
43. Freeman RE et al (2009) Measuring shear properties and normal stresses generated within a rotational shear cell for consolidated and non-consolidated powders. Powder Technol 190:65–69
44. Leturia M et al (2014) Characterization of flow properties of cohesive powders: a comparative study of traditional and new testing methods. Powder Technol 253:406–423
45. Mendez R et al (2012) Effect of feed frame design and operating parameters on powder attrition, particle breakage, and powder properties. Powder Technol 229:253–260
46. Prestidge CA, Tsatouhas G (2000) Wettability studies of morphine sulfate powders. Int J Pharm 198:201–212
47. Llusa M et al (2010) Measuring the hydrophobicity of lubricated blends of pharmaceutical excipients. Powder Technol 198:101–107
48. Washburn EW (1921) The dynamics of capillary flow. Phys Rev 17:273–283
49. Siebold A et al (1997) Capillary rise for thermodynamic characterization of solid particle surface. J Colloid Interface Sci 186:60–70
50. Pingali K et al (2011) Evaluation of strain-induced hydrophobicity of pharmaceutical blends and its effect on drug release rate

under multiple compression conditions. Drug Dev Ind Pharm 37:428–435

51. Eilbeck J et al (2000) Effect of contamination of pharmaceutical equipment on powder triboelectrification. Int J Pharm 195:7–11
52. Harper WR (1951) The Volta effect as a cause of static electrification. Proc Roy Soc Lond Ser A Math Phys Sci 205:83–103
53. Lowell J, Rose-Innes AC (1980) Contact electrification. Adv Phys 29:947–1023
54. Jones TB (1995) Electromechanics of particles. Cambridge University Press, Cambridge
55. Matsusaka S, Masuda H (2003) Electrostatics of particles. Adv Powder Technol 14:143–166
56. Rowley G (2001) Quantifying electrostatic interactions in pharmaceutical solid systems. Int J Pharm 227:47–55
57. Jones TB, King JL, Yablonsky JF (1991) Powder handling and electrostatics: understanding and preventing hazards. CRC Press, Boca Raton, FL
58. Pingali KC et al (2009) An observed correlation between flow and electrical properties of pharmaceutical blends. Powder Technol 192:157–165
59. Liss ED, Glasser BJ (2001) The influence of clusters on the stress in a sheared granular material. Powder Technol 116:116–132
60. Alexander A et al (2000) A method to quantitatively describe powder segregation during discharge from vessels. Pharmaceutical Technology Yearbook. Advanstar Communications, Inc., Santa Monica, CA
61. Prescott JK, Barnum RA (2000) On powder flowability. Pharm Technol 24:60–84
62. Schulze D (2008) Flow properties of bulk solids. In: Powders and bulk solids. Springer, New York, NY, pp 35–74

Chapter 4

Discrete Element Modeling of Solid Dosage Manufacturing Processes

Dana Barrasso and Rohit Ramachandran

Abstract

Solid dosage manufacturing primarily involves powder process operations, such as mixing, granulation, and compaction. A model-based approach can be used to develop a better scientific understanding of these processes and implement Quality by Design. These models must predict the critical quality attributes of the product based on process parameters, equipment geometry, and material properties. Discrete element modeling (DEM) is often used to capture mechanistic, particle-scale information, such as velocity and force profiles. In this chapter, recent literature studies that use the DEM to model solid dosage manufacturing processes will be presented and discussed.

Key words Discrete element modeling, Particulate processes, Solid dosage manufacturing, Quality by Design, Multi-scale modeling

1 Introduction

In order to implement Quality by Design (QbD) in solid dosage manufacturing, a model-based approach can be taken, relating process parameters and material properties to the critical quality attributes (CQAs) of the product. These mathematical models establish the process understanding required to facilitate the definition of the design space.

Pharmaceutical process modeling often employs process-scale models, such as population balance modeling (PBM) and residence time distribution (RTD) models. These models are typically empirical and have limited ability to predict particle-scale behavior. Experimental calibration and validation are needed to estimate unknown parameters, resulting in models that are only valid within the experimental design space. While DEM also requires calibration, its basis on first-principles and particle-scale phenomena results in predictive capabilities beyond those of PBM or RTD models.

Marianthi G. Ierapetritou and Rohit Ramachandran (eds.), *Process Simulation and Data Modeling in Solid Oral Drug Development and Manufacture*, Methods in Pharmacology and Toxicology,
DOI 10.1007/978-1-4939-2996-2_4, © Springer Science+Business Media New York 2016

In contrast, discrete element modeling (DEM) tracks individual particles or granules as they move through space and collide. DEM is a particle scale model that is mechanistic in nature, able to capture velocity profiles and the effects of equipment geometry, particle size and shape distributions, and material properties. While the framework is computationally intensive, it produces detailed results that can be used to build a multi-scale process model. Due to current computational limitations, DEM is often applied to large particles ($>$ 1 mm in diameter) or granules but in theory can be used to model particles of any size.

Ketterhagen et al. [1] reviewed the applications of DEM in the pharmaceutical industry. However, since then the usage of DEM in process modeling of solid dosage manufacturing has expanded greatly. This chapter will discuss the applications of DEM in pharmaceutical operations, emphasizing recent advances in process modeling, multi-scale modeling, and experimental validation.

2 DEM Theory and Background

Fundamentally, DEM uses Newton's laws of motion to solve force balances on granular elements and compute any movement. Each particle is assigned a size and density along with position and velocity vectors. The net external force acting on each particle is calculated to account for any gravitational, electrostatic, fluid, and impact forces, among others when applicable. From Newton's second law of motion, the acceleration of each particle is calculated, resulting in a set of ordinary differential equations for each parcel. Explicit integration is typically used to determine position and velocity vectors at each point in time, while time is incremented in discrete steps.

Despite this simple formulation, detecting contacts and evaluating their forces is not trivial, and various models use different approaches to quantify these events. Further, inter-particle bond forces can be represented, and fluid forces can be implemented using drag models or by coupling DEM with computational fluid dynamics (CFD) models.

Contact models fall into two main categories: hard sphere, and soft sphere approaches. Hard sphere models assume that each collision is instantaneous and elastic, while soft sphere models resolve collisions that can last over several time steps and account for particle deformation by allowing particles to overlap. Soft sphere models, introduced by Cundall and Strack [2] in 1979, are more computationally intensive and require more input parameters than hard sphere models but they are able to capture more complex collision dynamics. As such, soft sphere models are preferred in powder process modeling.

Some soft-sphere contact models to resolve collision behavior and evaluate forces include the elastic Hertz-Mindlin model, the linear spring and dashpot model [2], and the hysteretic linear spring [3]. Detailed discussion and comparison of these models can be found in Renzo and Maio [4], Stevens and Hrenya [5], Kruggel-Emden et al. [6], and Zhu et al. [7]. These primary contact models can be modified to include additional inter-particle forces, such as cohesive forces or bonds between particles. Some of these models account for van der Waals, electrostatic, and liquid bridge forces, among others, and they are discussed in detail in Seville et al. [8] and Zhu et al. [7].

Each contact model requires a set of collision parameters and material properties to be defined. These particle properties typically include the material Young's modulus, a measure of material stiffness, and Poisson's ratio of the material, which describes the strain behavior. Contact parameters must be provided in order to describe the behavior when any two entities (particle types and/or geometry surfaces) come in contact. These parameters typically include the coefficient of friction and the coefficient of restitution, which describes the ratio of the rebound velocity to the collision velocity.

Quantifying these parameters to reproduce the flow behavior of a granular material is an area which is receiving significant research interest. Some studies aim to measure these parameters directly by experiment, using drop tests [9–13] and compression analyses [11, 14]. Other studies determine their values based on measurable properties [15].

Freireich et al. [16] found that the collision-scale information, such as collision rates and impact velocities, depend on the contact parameters of coefficients of restitution and stiffness, even though the velocity profiles were insensitive to these parameters. This observation suggests that experimental validation of the bulk behavior cannot be extended to collision-scale behavior.

A key step in a DEM simulation is contact detection, or the identification of contacts between two particles or a particle and the wall. When the number of particles is large or complex geometry is involved, contact detection can be computationally intensive. To facilitate contact detection for non-spherical particles, a glued-sphere approach is often used to represent these shapes. A rigid assembly of spheres is created to resemble a shape closely, such as a tablet, needle, or cylinder. The effects of this approach on the resulting contact forces are discussed in Kodam et al. [17]. Other approaches for modeling non-spherical granules are discussed in Kodam et al. [18] and Song et al. [19].

Various commercial software for DEM simulations is available, such as EDEM™(DEM Solutions) and STAR-CCM+™(CD-adapco). The open-source DEM software LIGGGHTS (www.cfdem.com) can also be used, and some research groups prefer to use in-house codes for more custom applications. Regardless of the software used, DEM is a highly computationally intensive modeling framework, limiting the number of particles and their sizes, as smaller particles require more time steps to simulate the same physical time. Liu and Hrenya [20] and Berger and Hrenya [21] have discussed some of these limitations in detail.

3 Applications of DEM in Solid Dosage Manufacturing Operations

DEM has been employed to further the understanding of various solid dosage unit operations, including feeding, blending, milling, hopper discharge and storage, granulation, die filling and tablet compaction, and tablet coating. In the following sections, applications and recent advances in DEM simulations of these processes are presented.

3.1 Hoppers and Feeders

Hoppers are used as intermediate storage vessels in solid dosage manufacturing operations. Because they typically contain particles of multiple solid components, including one or more active pharmaceutical ingredients (APIs) and excipients, segregation is a particular concern. Further, hopper discharge rates depend on the flowability of the powder, a result of the size, shape, and other properties of the particles. DEM simulations can be used to model hoppers and their discharge, identifying sources of segregation and characterizing powder flowability.

Beverloo et al. [22] established an empirical correlation to relate the mass discharge rate to the bulk density, particle diameter, geometry of the outlet, and dead space. The Beverloo correlation applied to a rectangular hopper is shown in Eq. (1), where $\dot{W}$ is the mass discharge rate, ρ_{flow} is the flowing bulk density, g is the gravitational acceleration, L_0 is the exit length, W_0 is the exit width, k is the Beverloo constant, and d is the particle diameter [23].

$$\dot{W} = 1.03\rho_{\text{flow}} g^{1/2}(L_0 - kd)(W_0 - kd)^{3/2} \quad (1)$$

The exit length and width refer to the cross-sectional dimensions of the hopper outlet, assuming a rectangular outlet. Several DEM studies aim to modify or validate this correlation, examining more complex behavior, such as segregation, particle interactions, and the effects of particle shape [23–25].

Hoppers are operated in one of two regimes: mass-flow, where all of the powder flows through the hopper regardless of radial position, and funnel-flow, where the particles at the center of the hopper exit freely as material near the wall is retained for a longer period. A Jenike regime map can be used to identify the regime based on the wall friction angle and the hopper wall angle [26].

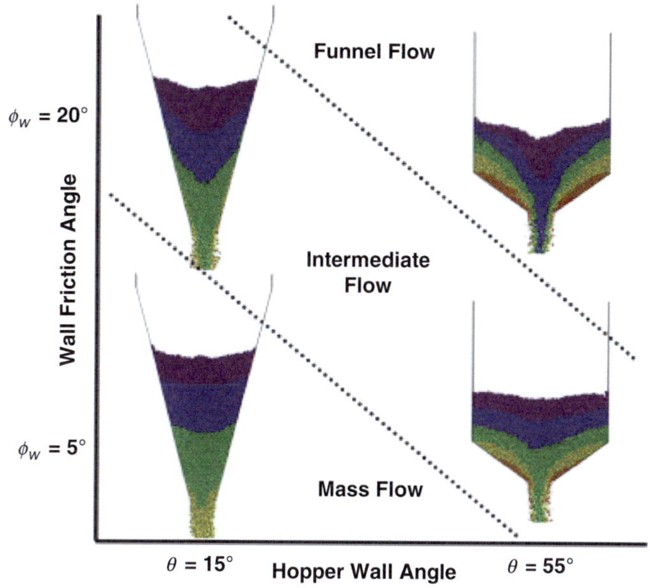

Fig. 1 Wedge-shaped hopper discharge simulation images overlaid onto a standard Jenike design chart, from Ketterhagen et al. [26]

Ketterhagen et al. [26] captured these regimes using DEM simulations with variable friction and hopper angles, as shown in Fig. 1. The mass flow index (MFI), or the ratio of the mean particle velocity at the hopper wall to that at the centerline, was used as a metric for these regimes. Findings were consistent with experimental observations. Yu and Saxen [27] also investigated these flow regimes using DEM simulations.

To characterize segregation, Ketterhagen et al. [28] performed parallel experiments and simulations on a cylindrical hopper with bidisperse spherical glass beads. They found agreement between the experimental and simulated results, noting that the size difference between the two bead types strongly affects the degree of segregation, measured by the variation in mass fraction of the discharge stream over time.

Anand et al. [23] applied these methodologies to a rectangular hopper with wet cohesive particles, using a liquid bridge contact model to account for cohesion. Based on the results, they developed a modified Beverloo correlation, relating the mass discharge rate to the surface tension of the liquid. Anand et al. [29] extended this work to investigate segregation of cohesive materials in hopper discharge, finding that while cohesion reduces the discharge rate, it also reduces the degree of segregation.

Persson et al. [30] experimentally measured the hopper discharge rates of surface modified microcrystalline cellulose granules and performed parallel DEM simulations using non-cohesive and cohesive contact models. They found that lubrication significantly affected the sliding friction coefficient, and the flow behavior was qualitatively different in the presence of a lubricant, discharging clumps of particles rather than a steady stream.

Several studies have investigated the effects of particle shape on hopper discharge. Cleary and Sawley [31] used DEM simulations to predict reductions in flow rates of up to 28 % to non-circular particle shapes. Further DEM studies on particles shape and hopper discharge for pharmaceutical and other applications were performed by Hohner et al. [32] and Tao et al. [33].

Hancock and Ketterhagen [34] used DEM simulations of hoppers for a different purpose, investigating sampling methods within the hopper. They simulated stratified sampling in the hopper and characterized the effects of sample size, number of samples, and sample location. Results showed that size of sample was most important in identifying segregation issues.

Additional studies on hopper discharge examine eccentric hoppers for tablet presses [25], the effect of hopper shape on particle flow patterns [35], and the influence of particle–particle friction [24].

DEM simulations have also been employed to simulate screw feeders and conveyors for powder transport. These devices draw powder from a hopper and mechanically convey it using a rotating screw. Moysey and Thompson [36] proposed using 3D DEM simulations to model transport of solid particles in screw extruders and applied this method to a polymer pellet system with heat transfer.

Owen and Cleary [37] also simulated a screw conveyor using DEM, analyzing the flow rates and power consumption under different operating conditions, such as inclination angle, fill level, and rotational speed.

Hou et al. [38] analyzed the effects of inter-particle cohesion on the flow behavior in a screw feeder using DEM simulations. Cohesion was modeled by applying inter-particle contact forces, finding that cohesion significantly inhibits particle flow, as shown in Fig. 2 [38].

DEM simulations were used by Fernandez et al. [39] to examine the effects of screw design on hopper drawdown in a screw feeder. Results showed that the screw pitch, flight, and diameter significantly influenced the mass flow rate, the evenness of drawdown from the hopper, and the power consumption.

3.2 Blending

Powder blending is critical in solid dosage manufacturing, as inefficient blending can lead to handling and content uniformity issues in later stages of manufacturing and may result in variations in the final tablet composition. Further, the equipment geometry, process parameters, and particle size and shape can have profound effects on the key quality metrics of the blend, such as relative standard deviation (RSD) or Lacey's Index, measures of the degree of mixing.

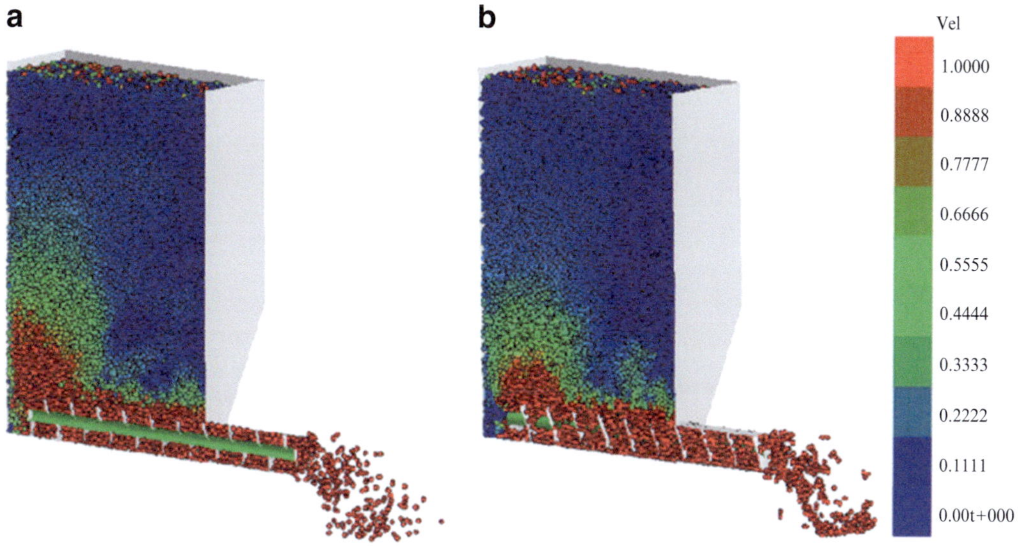

Fig. 2 Solid flow colored by particle velocity at different cohesive forces: (**a**) 0, and (**b**) 15 mg, from Hou et al. [38]

The RSD can be calculated using Eqs. (2) and (3), where σ is the standard deviation of the concentration, n is the number of samples, W_i is the concentration of sample i, and $\bar{W}$ is the mean concentration of the samples [40].

$$\sigma = \sqrt{\frac{1}{n-1}\Sigma_{i=1}^{n}(W_i - \bar{W})^2} \tag{2}$$

$$\mathrm{RSD} = \frac{\sigma}{\bar{W}} \tag{3}$$

A larger RSD represents a wide variation in composition, an indicator of poor mixing. The RSD may be affected by the sample size, and Lacey's Index (M) is a metric that corrects for these effects [40, 41], given in Eq. (4), where the subscript s denotes a completely segregated mixture and the subscript r denotes a perfectly random mixture.

$$M = \frac{RSD_s^2 - RSD^2}{RSD_s^2 - RSD_r^2} \tag{4}$$

$$\mathrm{RSD}_s = \frac{\sigma_s}{\bar{W}} \tag{5}$$

$$\mathrm{RSD}_r = \frac{\sigma_r}{\bar{W}} \tag{6}$$

The standard deviations of the representative mixtures, σ_s and σ_r, are determined from the overall mass fraction of one component in the mixture, p, and the number of particles in the sample, N.

$$\sigma_s = \sqrt{p(1-p)} \tag{7}$$

$$\sigma_r = \sqrt{\frac{p(1-p)}{N}} \tag{8}$$

The actual RSD of the mixture lies between the two extreme values of RSD_r and RSD_s, resulting in a Lacey's index value between zero and one.

Adam et al. [40] presented a systematic approach to QbD using DEM simulations to characterize pharmaceutical mixing processes, with particular focus on risk management, identification of critical process parameters, and quantification of CQAs. DEM simulations have been carried out on various systems to assess mixing and segregation of solid components. A thorough review of DEM studies of various powder mixing devices can be found in Cleary and Sinnott [42], and some recent blending studies are presented here for discussion.

Sarkar and Wassgren [43] performed a parametric study to identify the effects of fill level, impeller speed, and blade spacing of a horizontal continuous blender using DEM simulations, suggesting that DEM simulations be used to improve blade design for optimal mixing of specific types of products.

Two-dimensional DEM simulations were carried out by Xu et al. [44] to qualitatively asses the effects of rotational speed, particle size, density, and friction on the extent of mixing. After validating the simulations by visual comparison with experimental results, they found that large differences in density or size of two solid components can not only inhibit mixing but also induce segregation of initially well-mixed blends. Figure 3 shows profiles of experimental and 2D DEM simulations in the drum mixer [44].

Hassanpour et al. [45] used DEM to simulate a paddle mixer, comparing the internal flow fields and mixing patterns obtained from the simulations to those observed experimentally using positron emission particle tracking (PEPT). PEPT uses a radioactive tracer to detect a single particle as it moves throughout the equipment. Results showed strong qualitative agreement. Further, to model fine powder particles in a computationally efficient manner, larger particles are often used to represent a collection of small particles. Hassanpour et al. [45] quantified the effects of particle scale on mixing and velocity behavior, finding that particle scale has a moderate but systematic effect on average velocity. PEPT was also used successfully to validate DEM simulations of powder blenders by Laurent and Cleary [46], Tahvildarian et al. [47], and Marigo et al. [48].

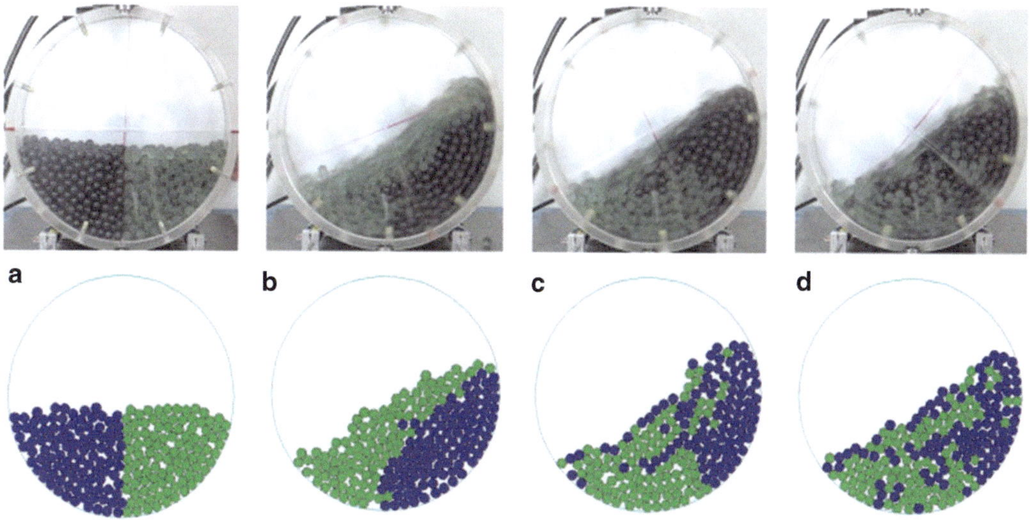

Fig. 3 Profiles of experimental and 2D DEM simulations in a drum mixer obtained by Xu et al. [44]. (**a**) 0 s, 0 rev., (**b**) 1.0 s, 0.33 rev., (**c**) 3.5 s, 1.16 revs., and (**d**) 5.7 s, 1.9 revs.

Hua et al. [49] simulated vertical axis mixers with dense granular materials, predicting the effects of fill level, blade speed, aspect ratio, and material properties on the shaft torque, velocity profiles, solid fraction fields, and particle orientations. The shaft torque was compared to experimental measurements to validate the model. They found that particles with greater aspect ratios exhibited lower velocities, and mixing within the bed improves with an increase in fill level.

DEM simulations have also been used to develop a methodology for the scale-up of a continuous powder mixing process. Gao et al. [50] use periodic sections to isolate the mechanisms of axial and cross-sectional mixing and suggest scaling equations for non-cohesive and cohesive materials.

Sen and Ramachandran [51] also simulated a continuous blending process, using DEM simulations to obtain velocity profiles within a horizontal continuous blender for API and excipient particles. They used these results in a population balance model to develop a one-way coupled model, predicting the RTD, RSD, and component compositions in the PBM.

In some cases, fluid dynamics play an active role in blending. Jovanovic et al. [52] coupled DEM and CFD to model granular flow in Ross and Komax static mixtures, with good quantitative agreement between the experimental and simulated RSDs. They found that the Komax mixer exhibited better blending than the Ross mixer.

Additional recent studies include simulations of a slant cone mixer [53], tetrapodal and V-blenders [54], bladed mixers [55], vibrated beds [56], and tote blenders [57], spouted beds [58], and horizontal continuous blenders [59–61].

3.3 Milling and Breakage

DEM simulations have also been used to understand breakage and attrition processes, either deliberately, as in milling, or as a side effect of blending and handling processes.

Unlike transport and storage processes, milling processes are designed to change the granule properties, particularly to reduce their sizes. This is particularly challenging to simulate using DEM, where particles are given fixed sizes, and additional strategies are required to simulate size changes.

DEM has been used to investigate mechanisms of particle breakage at the micro-scale. Ning et al. [62] used inter-particle forces to simulate an agglomerate of fine particles. The agglomerate was dropped onto a flat surface, and the impact forces shattered the agglomerate, breaking the inter-particle bonds and releasing the fine particles. Breakage was quantified by evaluating the damage ratio, or the ratio of the number of broken bonds after impact to the number of intact bonds prior to impact.

Liu et al. [63] used a similar approach to simulate breakage of agglomerates, analyzing the effects of agglomerate shape. They found that for non-spherical particles, face impacts resulted in more damage than impacts on corners or edges. Hassanpour et al. [64] analyzed agglomerate breakage within a bed of fine particles subjected to shearing. Antonyuk et al. [65] used this method to simulate breakage of needle-shaped particles under compression, presented in Fig. 4.

Additional studies using this methodology were performed by Subero et al. [66], Thornton et al. [67], Mishra and Thornton [68], Moreno et al. [69], Golchert et al. [70], Antonyuk et al. [71], Hassanpour et al. [72], and Tong et al. [73].

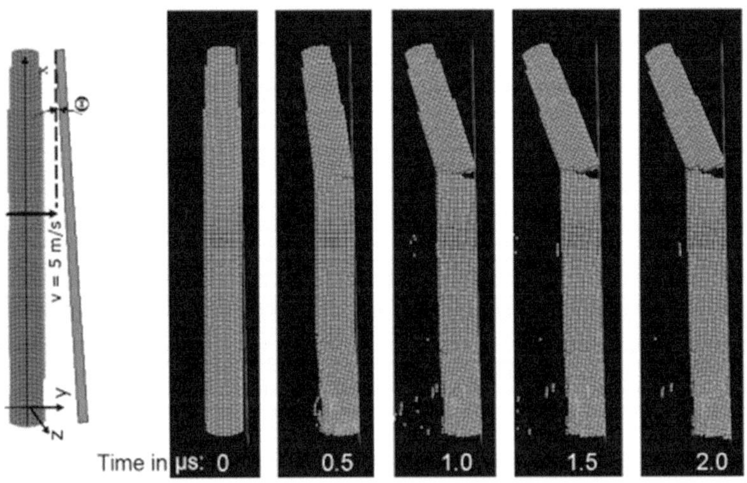

Fig. 4 Stressing conditions of needle shaped particles during the simulation (*left*) and the evolution of cracks during impact of L-threonine crystal on the target (*right*), from Antonyuk et al. [65]

DEM simulations can also be used to collect particle scale information to use in a process-scale breakage model. Several recent studies involve coupling DEM to PBMs of milling equipment, creating a multi-scale breakage model. DEM simulations provide force, velocity, and impact data which are used to evaluate breakage rate expressions in the PBM. An alternative approach divides the particles during the DEM simulations, creating two or more fragments from a larger particle. Other studies do not actually simulate breakage, instead evaluating other process information associated with attrition, such as stress and velocity profiles.

Kwan et al. [74] performed DEM simulations of a ball mill using this approach to determine the milling power, a measure of the collisional energy and frequency. While these simulations furthered the understanding of the particle-scale behavior and breakage rates in the mill, they did not account for particle size changes or predict the evolution of the size distribution throughout the process.

Particle breakage in agitated dryers was investigated by Hare et al. [75] using DEM to evaluate stresses and strains due to shear deformation. Strain rates were validated experimentally using PEPT. They found that stress and strain rates were highest in the lowest region of the bed and at high radial distances from the centerline.

Han et al. [76] performed 2D DEM simulations on a pneumatic conveyor, simulating fragmentation based on impact velocity. Upon impact, if the velocity satisfies criteria, mechanistic correlations for fragment particle size were evaluated. A new spherical particle was created to represent the smaller fragment, and the size of the original particle was reduced.

Ahmadian et al. [77] investigated granule breakage in a rotary drum by analyzing the particle-wall impacts above a velocity threshold, identifying the effects of rotational speed on the power associated with the collisions.

DEM simulations of a vibrational mill were performed by Lee et al. [78] and correlated to breakage parameters in a PBM, which were estimated from experimental data. Further, they proposed a scale-up function for the mill based on the impact energy obtained from DEM simulations and the operating conditions.

Wang et al. [79] used collision, impact, and dissipated energy information within a PBM to predict the evolution of particle size in a tumbling ball mill, comparing results with experimental data. They concluded that the collision energy was primarily linked with particle breakage, and found reasonable agreement between predicted and measured size distributions.

Capece et al. [80] simulated dry milling in a ball mill, using the impact energy distribution obtained via DEM to determine the value of a mechanistic breakage rate constant, finding that the PSD does not affect the breakage rate constant in a first-order

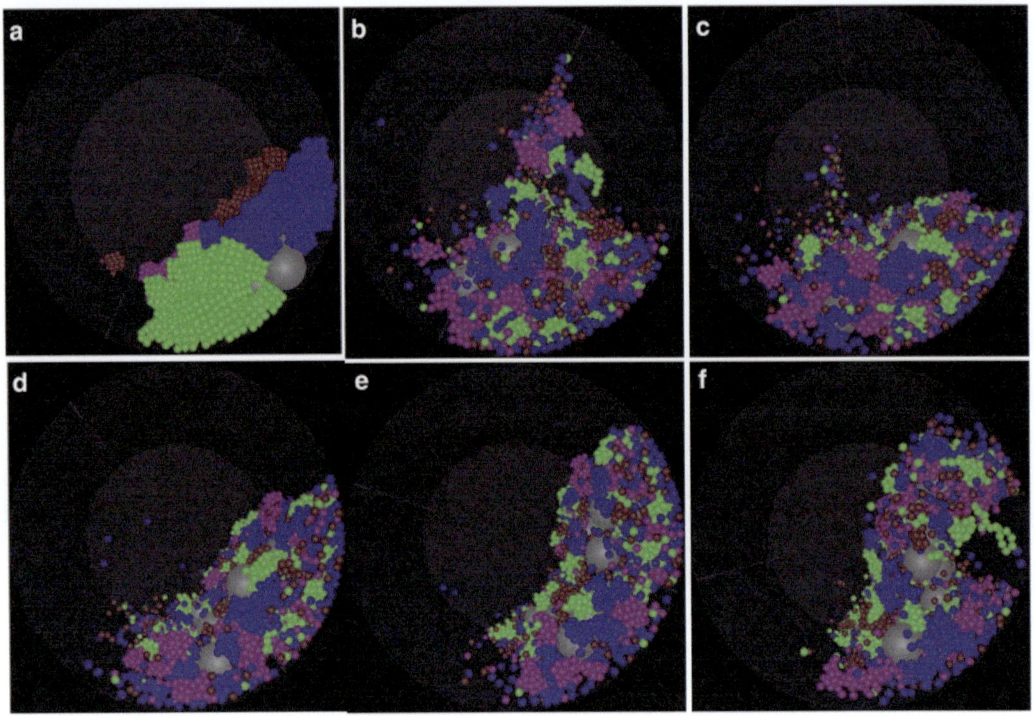

Fig. 5 Snapshots of agglomerate particles breaking within a ball mill over time, from Metzger and Glasser [81]

breakage model. This multi-scale approach takes advantage of the inherent mechanistic nature of DEM in the often empirical PBM.

Finally, Metzger and Glasser [81] combined the micro-scale agglomerate DEM simulations with the equipment-based models, simulating breakage in a rotating batch mill, as shown in Fig. 5. Agglomerates were represented as clusters of 27 smaller particles held together by inter-particle forces. Upon impact, these bonds can be broken, resulting in particle size reduction.

From these recent studies, progress points towards a multi-scale model for breakage that accounts for the process behavior (rotational speeds, fill level, equipment geometry, etc.), but also predicts the CQAs of the product, mainly the evolution of the size distribution. Two approaches towards this goal include simulating breakage within DEM itself or by way of a separate process model, such as PBM, which can make use of DEM predictions to evaluate breakage rates.

3.4 Wet Granulation

Like milling processes, wet granulation processes are particularly challenging to simulate using DEM because the particle sizes and other properties change throughout the process. Various approaches have been demonstrated to model these changes, often creating a multi-scale model through coupling with a process model. Other studies have used DEM simulations to gain insight into spatial and flow behavior without modeling size and property changes.

Talu et al. [82] presented an approach to modeling agglomeration using 2D DEM simulations, where particles are classified as "sticky" or dry. Upon contact between two sticky particles or a sticky and a dry particle, a bond forms, and an inter-particle force connects the particles. Large agglomerates form as additional collisions occur. This computationally intensive approach represents each fine powder particle as an individual entity throughout the simulation.

Hassanpour et al. [72] used the agglomerate particle approach to study the shear elongation and velocity profiles in two high shear granulators. Kafui and Thornton [83] also used interparticle bonds to model aggregates in a study on spray zones in fluidized beds. Further, Hassanpour et al. [84] used this method to simulate seeded granulation.

Alternatively, Goldschmidt et al. [85] developed a novel 2D DEM-based model for wet granulation processes, where particles and droplets collide and coalesce under certain criteria. If wet particles collide and merge, they are immediately removed from the simulation and replaced with a larger particle.

Gantt and Gatzke [86] first used DEM simulations to model changes in particle size distributions using mechanistic criteria for aggregation, using the approach of Goldschmidt et al. [85] to replace coalescing particles with a larger particle during the simulation.

Significant interest lies in coupling DEM with PBM techniques to form a multi-scale model for granulation processes. Challenges and potential frameworks for these efforts are discussed in Ingram and Cameron [87, 88].

Gantt et al. [89] used DEM simulations to evaluate mechanistic aggregation kernels for use in a PBM. The DEM simulations did not represent agglomeration, instead providing collision and velocity data to the PBM. The aggregation rate kernel (β) is represented in terms of a collision rate (C), or frequency with which two particle types (a and b) collide, on a per particle basis, and a collision efficiency (Ψ), or the probability that a collision between two particle types will result in coalescence. As shown in Eq. (9), the product of these two terms is the aggregation rate, which can be used in a PBM to predict the evolution of the particle size distribution.

$$\beta(a, b) = C(a, b)\Psi(a, b) \tag{9}$$

The collision efficiency depends on the velocity associated with the collision and the material properties, such as porosity, surface liquid coverage, and binder viscosity. A similar technique was implemented by Gantt and Gatzke [90].

One drawback of these coupled models is the one-way flow of data from the DEM simulations to the PBM. As the particle size distribution develops and particles pick up liquid and consolidate, the behavior of the DEM simulations is affected. Two-way coupling algorithms have been proposed to address these limitations. Reinhold and Briesen [91] presented a technique that solves a PBM concurrently with a DEM simulation.

Barrasso and Ramachandran [92] presented a two-way coupling algorithm in which a 2D PBM is solved within a DEM simulation. In the DEM simulation, liquid droplets are added as particles that coalesce with powder particles on collision, increasing their liquid content. Collisions are tallied based on the sizes and liquid contents of each of the colliding particle, and the PBM is solved at regular intervals, accounting for aggregation based on this collision frequency. A simple collision efficiency function was assumed such that particles have a constant probability of coalescing upon collision if the liquid fraction of one of the particles exceeds a fixed value. Images of the evolution of the particle liquid content and size are presented in Fig. 6 [92].

This framework was extended by Sen et al. [93], who included a coupled CFD model to account for fluid forces in a fluidized bed granulator. Additionally, Barrasso et al. [94] used DEM results to generate a reduced order model in the form of an artificial neural network (ANN) to predict collision frequencies based on an impeller speed and size distribution. The ANN was solved within a 1D PBM to evaluate the collision frequency term in the aggregation rate, and the results of the PBM-ANN model were compared with the fully coupled PBM-DEM model.

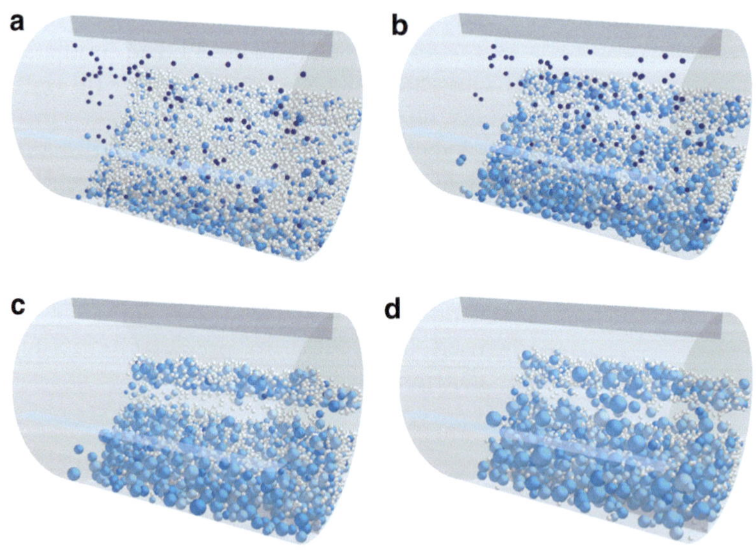

Fig. 6 DEM images of particles at various times, colored by liquid fraction, from Barrasso and Ramachandran [92]. (**a**) 2.5 s, (**b**) 5 s, (**c**) 7.5 s, and (**d**) 10 s

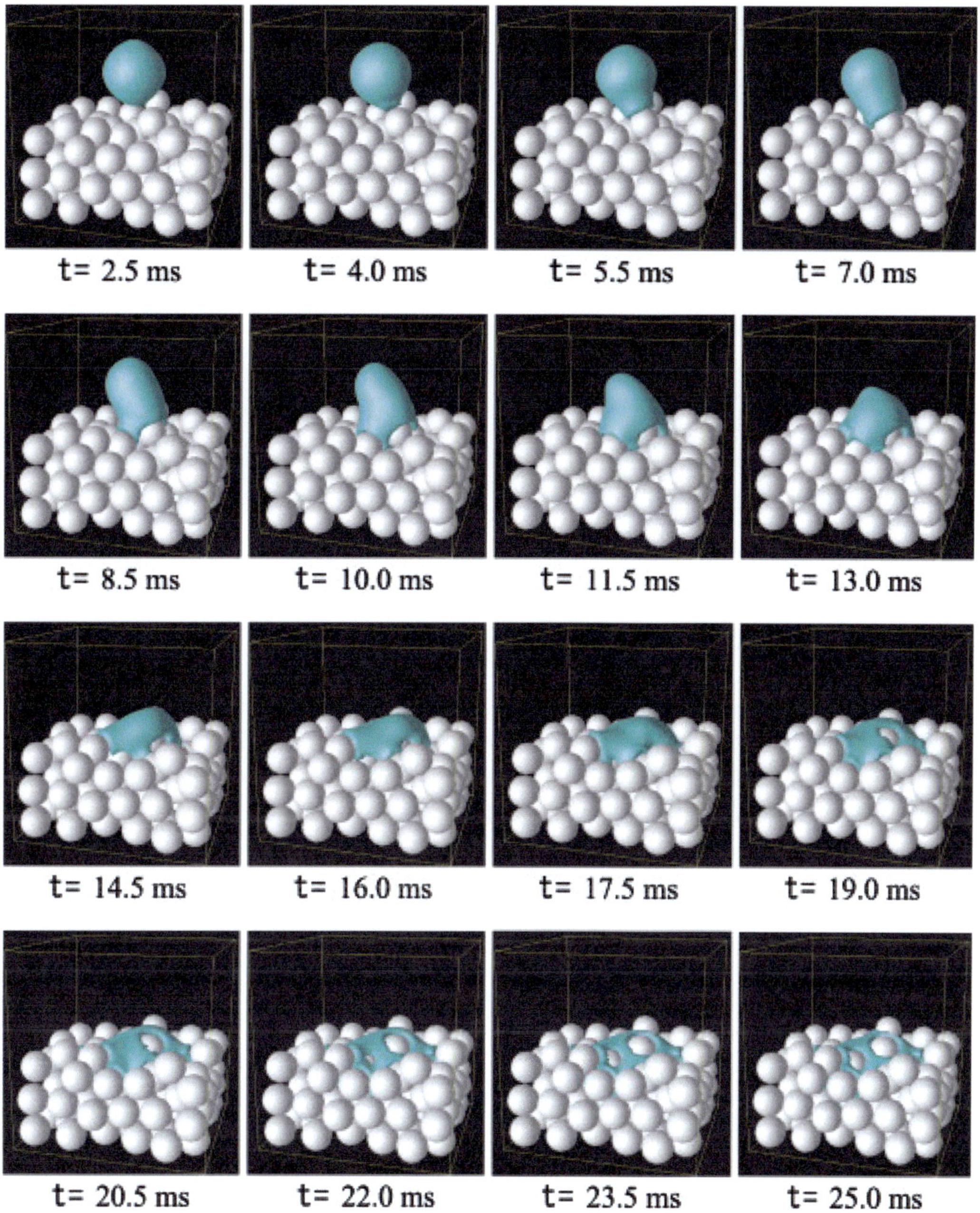

Fig. 7 Snapshots of droplet penetration into a particle bed; liquid and particle phases are shown, from Washino et al. [95]

In a novel study, Washino et al. [95] used a coupled CFD-DEM model to simulate droplet penetration into a bed of fine powder particles, resulting in nucleation, as shown in Fig. 7. Washino et al. [96] also simulated capillary forces and nucleation in a wet

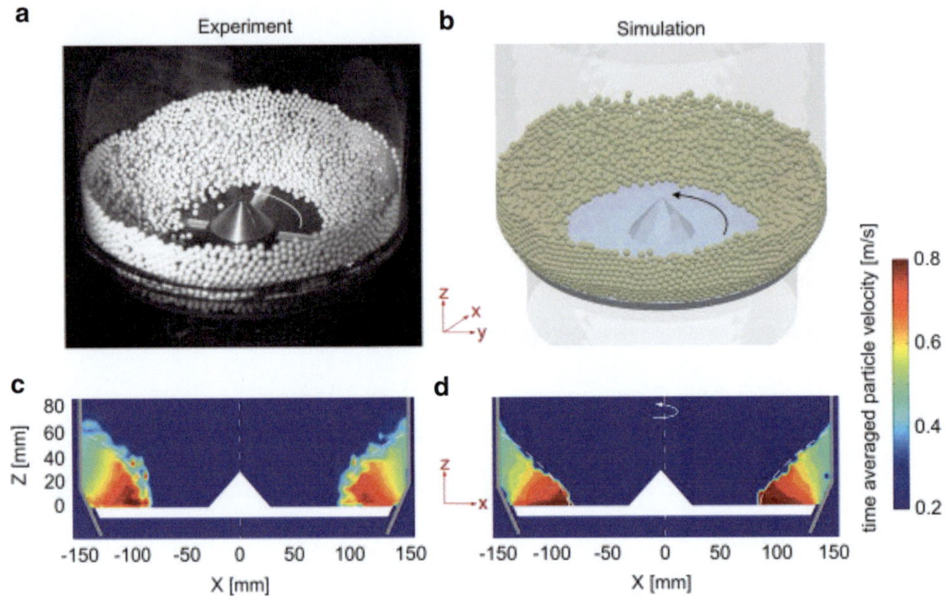

Fig. 8 Snapshots of the granular flow and time averaged particle velocity field (**a**) and (**c**) experiment, (**b**) and (**d**) CFD-DEM simulation, from Neuwirth et al. [97]

granulation process using DEM coupled with a constrained interpolation profile model.

Other studies ignore particle size changes all together, instead focusing on the velocity and force profiles within the equipment, leading to better insight into process behavior. Neuwirth et al. [97] used coupled CFD-DEM simulations to simulate flow in a rotor granulator, validating the model using a magnetic particle tracking technique, as shown in Fig. 8. Fries et al. [98] also coupled DEM with CFD and applied this model to a fluidized bed granulator to analyze particle motion and collision dynamics.

Additional studies on granulation processes utilizing DEM simulations include Link et al. [99], Hassanpour et al. [100], Fries et al. [101], Nakamura et al. [102], Dhenge et al. [103], and Fries et al. [104].

3.5 Die Filling and Tablet Compaction

DEM simulations have been used to investigate two distinct stages of tablet compaction: die filling and compression. In die filling, it is desirable to minimize the degree of segregation. Segregation can result in large variabilities in the final dosage form.

Guo et al. [105, 106] used a coupled DEM-CFD approach to simulate segregation of powder mixtures during die filling. They found that the lighter particles are subjected to drag forces, which reduce the speed with which they fill the die. Therefore, a greater degree of segregation is observed in bidisperse mixtures with large differences in density. Figure 9 demonstrates this phenomenon [106].

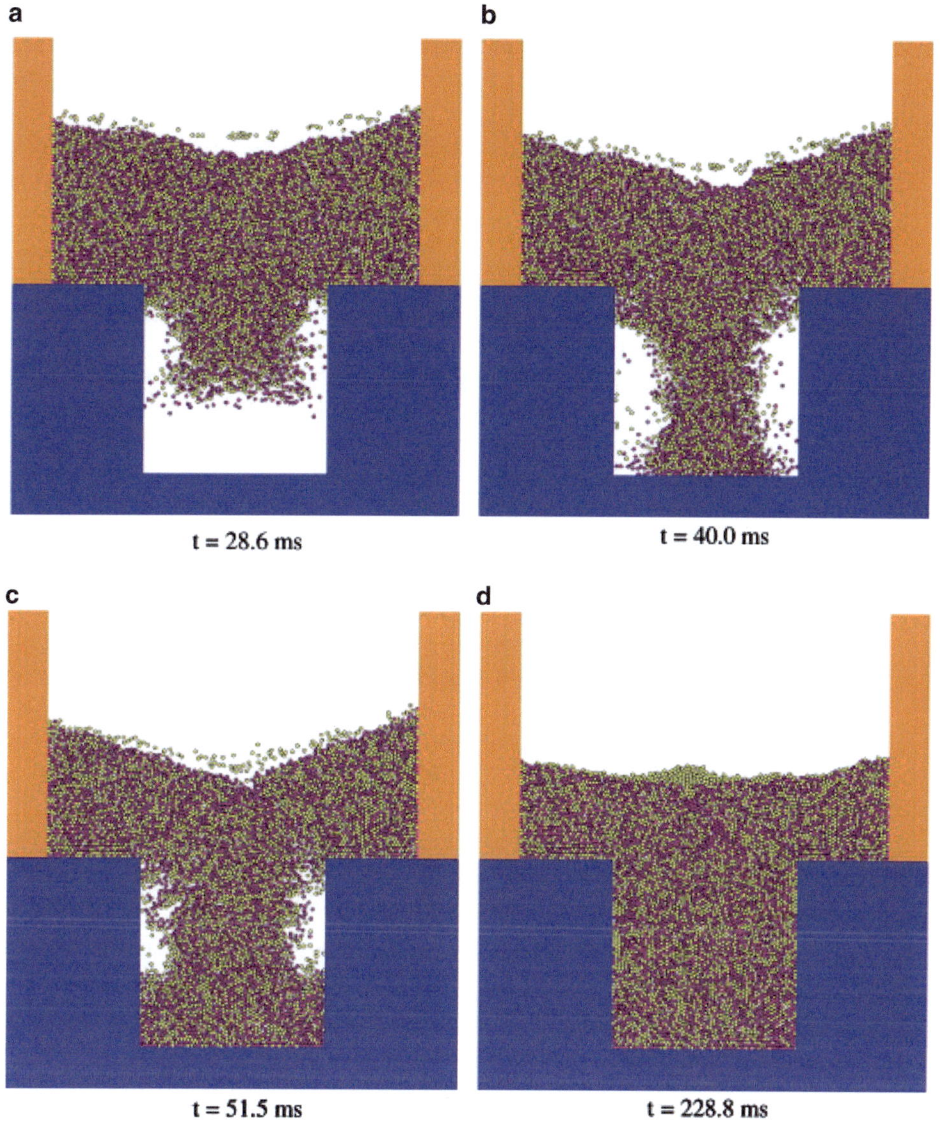

Fig. 9 Die filling with a stationary shoe in air. The *labels* indicate the elapsed time from the start of die filling, from Guo et al. [106]

Mateo-Ortiz et al. [107] studied particle size segregation in a feed frame and during die filling, finding that the paddle wheel speed in the feed frame was important in controlling the degree of segregation. More segregation was observed at lower paddle wheel speeds, and percolation, or the movement of small particles through the void space, was identified as the primary segregation mechanism.

The powder filling rate was analyzed using 2D DEM simulations by Wu [108], who found the critical filling speed at which the die is completely filled when it passes the shoe. At higher speeds, the die is only partially filled. The effects of particle size, shape, number of particles, and height of the bed on the filling ratio and critical filling speed were observed. Results showed that polydisperse blends had higher critical speeds than monodisperse systems, and particle shape has a significant effect on the critical filling speed.

During tablet compaction, empirical models describe relationship between the compression pressure and the bulk density of the tablet. Some recent studies have used DEM simulations to capture this behavior.

Stress–strain relationships and compaction curves relating density to pressure were generated using 3D DEM simulations of compression by Sheng et al. [109]. Results showed three distinct stages of compression. The initial increase in density was the result of particle rearrangement, and subsequent compression was attributed to plastic and elastic deformation of the particles.

Hassanpour and Ghadiri [110] simulated the bulk compression of particles using DEM. A rectangular geometry was loaded with a polydisperse distribution of particles in a random packing pattern. A moving plate was applied to the top of the box, and the relationship between pressure and density was obtained. Results were compared to those of a Heckel analysis, which relates the compaction pressure to the density. Simulations demonstrated that the Heckel model is only valid for certain values of the yield stress and Young's modulus. Samimi et al. [14] performed a similar study on individual and bulk compression of soft granules.

Frenning [111] coupled DEM with finite element analysis to allow particles to deform using a finite element mesh, as shown in Fig. 10. Results showed the relationships between properties of individual particles with the bulk compression behavior, particularly demonstrating the effect of the Young's modulus and yield stress on the compression profiles.

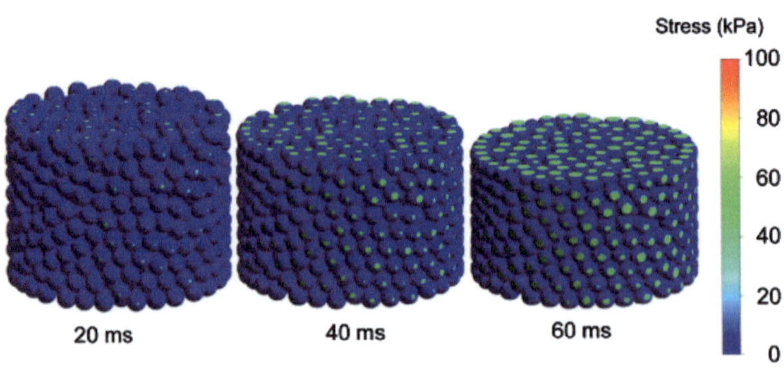

Fig. 10 Examples of configurations during compression of 1000 initially spherical granules at three instants of time (corresponding to axial strains of about 10, 20 and 30 %), from Frenning [111]

Persson and Frenning [112] used experimental compression of single granules to evaluate granule properties, such as Young's modulus and yield stress, to use as inputs in DEM simulations. DEM results on granule beds showed rearrangement and deformation stages in compression that were in agreement with experimental results. A modified contact model was developed to describe the late stages of compaction, introducing a maximum plastic overlap beyond which only elastic deformation can occur.

In additional studies, Martin et al. [113] analyzed particle rearrangement during powder compaction, and Wiacek and Molenda [114] investigated the effects of the particle size distributions on the response under compression.

3.6 Tablet Coating

DEM has also been used extensively to simulate tablet coating processes. They key CQA of coating processes is the coating uniformity or inter-tablet coating variability. This property is often quantified using the coefficient of variation ($\mathrm{CoV}_{\mathrm{inter}}$), shown in Eq. (10), where $\bar{m}$ is the average coating mass and σ is the standard deviation in coating mass.

$$\mathrm{CoV}_{\mathrm{inter}} = \frac{\sigma}{\bar{m}} \tag{10}$$

This parameter is typically related to the time spent in a spray zone and the number of passes into that zone. Based on renewal theory, presented by Mann [115], $\mathrm{CoV}_{\mathrm{inter}}$ can be determined from Eq. (11) [116]:

$$\mathrm{CoV}_{\mathrm{inter}} = \sqrt{\frac{\mu_{ct}}{t}\left[\left(\frac{\sigma_m}{\mu_m}\right)^2 + \left(\frac{\sigma_{ct}}{\mu_{ct}}\right)\right]} \tag{11}$$

The mean and standard deviation of the circulation time are given by μ_{ct} and σ_{ct}, and the mean and standard deviation of the coating mass per tablet per pass are given by μ_m and σ_m. The total coating time is given by t. Using DEM simulations, compartments can be defined and particles can be tracked as they move across compartment boundaries, providing a data set to calculate these statistics.

The intra-tablet coating variability is also of interest as differences in coating thickness may be observed on different faces of the tablet. This CQA is more difficult to evaluate since it pertains to variations on the individual particles, and the circulation time approach is less applicable.

Ketterhagen [116] analyzed the effect of tablet shape on the coating uniformity and compared these results with experimental data. Results showed that the tablet shape had minimal effect on the flow patterns and inter-tablet film variability. However, the intra-tablet coating uniformity was strongly affected by the tablet shape. The intra-tablet coating uniformity was estimated from the orientations of tablets as they passed through the spray zone.

Freireich and Wassgren [117] and Freireich et al. [118] used Monte-Carlo simulations in conjunction with the orientation data obtained from DEM to estimate the intra-particle coating uniformity.

Suzzi et al. [119] used DEM simulations of a continuous tablet coating process to identify effects of tablet shape and fill level on the mixing behavior and RTD.

DEM simulations were also performed by Dubey et al. [120] to assess the effects of rotational speed, fill level, and spray pattern on the inter-particle coating variability of a pan coater. They found that the axial mixing was largely correlated with the coating variability, and increasing the speed and fill level resulted in better mixing. Experimental trends obtained using laser induced breakdown spectroscopy (LIBS) were consistent with the DEM results.

Kumar and Wassgren [121] found that the axial Peclet number can be used to assess inter-particle coating variability in a continuous coater. The Peclet number is the ratio of advection to diffusion, calculated from the axial velocity due to advection, coater length, and diffusivity. They suggested using a Peclet number greater than 2e5 to obtain a coating uniformity of less than 1 %. This study is an example of using DEM simulations to establish design rules, thus implementing QbD.

Several studies have been performed to improve the mechanistic and computational approaches to DEM modeling of coating processes. Song and Turton [122] proposed a model for viscous forces in DEM simulations of a tablet coating process to account for the effects of liquid bridges in tablet coating processes. This contact model was validated using images of tablet–tablet contacts in the presence of silicone oil.

Ketterhagen et al. [123] experimentally determined values for the coefficients of rolling resistance of pharmaceutical tablets, a key input parameter to DEM simulations on tablet coating.

Toschkoff et al. [124] compared three approaches for modeling spray in DEM simulations of coating processes. First, the spray zone approach, defines a spray compartment and tablets at the top of the bed in the spray zone experience an increase in coating thickness. The second method is the discrete drop method, shown in Fig. 11, where droplets are simulated as particles that disappear on contact with tablets, and the coating thickness of each tablet is tracked. Finally, in the ray tracing method, DEM position and orientation data is post-processed, and a spray zone is defined by representing droplet trajectories as rays. The coating thicknesses are calculated by post processing the data. They found that the ray tracing method was fastest and most flexible since it relied on tabulated data instead of being calculated during the simulation.

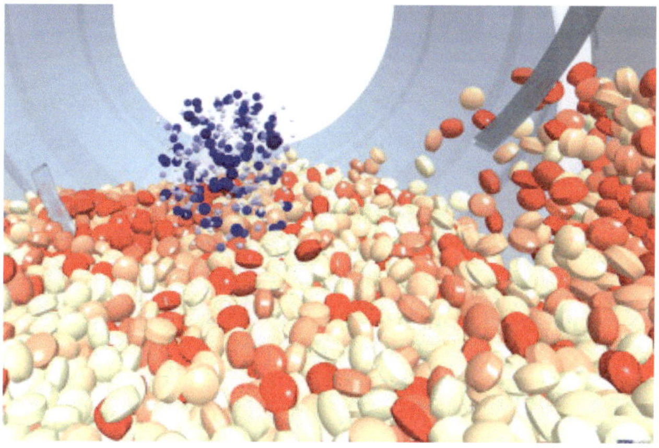

Fig. 11 DEM simulation of a tablet coating process, including the coating spray. The spray droplets are colored *blue*, with *darker shades* indicating bigger droplets. The tablets are colored according to coating mass, from *white* (no coating mass) to *red* (high coating mass), from Toschkoff et al. [124]

DEM studies involving non-spherical shapes typically use a glued-particle approach, where a cluster of bonded spheres represent a single non-spherical body. As an alternative approach, Kodam et al. [125] developed algorithms for detecting contacts between bi-convex shapes. Each tablet can be represented as a single body, reducing the number of particles used in the simulation.

Coupling the DEM simulations to a process-scale model, such as a PBM, has been a subject of recent interest. A DEM-PBM framework was proposed by Li et al. [126] to predict the coating uniformity of a polydisperse system, finding that segregation can result in a greater coating thickness on smaller particles.

Similarly, Freireich et al. [127] coupled DEM results describing velocity distribution to a compartment-based PBM, resulting in an 80 % decrease in computation time compared to DEM alone.

Li et al. [128] developed and experimentally validated a DEM-PBM for a spray coating model, using a 2D PBM to account for distributions in core and coating masses.

Kumar et al. [129] coupled DEM to a compartment based PBM to model particle coating in horizontal rotating drum, creating a multi-scale model that is much faster than DEM alone and can predict the inter-particle coating variability with accuracy.

Additional studies using DEM simulations to analyze tablet coating processes were performed by Pandey et al. [130], Kalbag et al. [131], Kalbag and Wassgren [132], Sahni and Chaudhuri [133], and Sahni et al. [134]. Pandey et al. [135] reviewed applications of DEM and other modeling techniques for pan-coating processes.

4 Summary

DEM simulations have been used extensively to model solid dosage manufacturing processes, producing high-fidelity results describing velocity and force profiles and collision dynamics. Applications of DEM provide qualitative insight into CQAs and simulate complex behavior, such as segregation, powder flow, and effects of particle size and shape. To further utilize these results and quantify process CQAs, recent research focuses on coupling DEM simulations to process-scale models, capturing mechanistic phenomena and the effects of equipment geometry and material properties in a predictive model, facilitating QbD.

References

1. Ketterhagen WR, am Ende MT, Hancock BC (2009) Process modeling in the pharmaceutical industry using the discrete element method. J Pharm Sci 98(2):442–470
2. Cundall P, Strack O (1979) Discrete numerical model for granular assemblies. Geotechnique 29(1):47–65
3. Walton OR, Braun RL (1986) Viscosity, granular-temperature, and stress calculations for shearing assemblies of inelastic, frictional disks. J Rheol 30(5):949–980
4. Renzo AD, Maio FPD (2004) Comparison of contact-force models for the simulation of collisions in DEM-based granular flow codes. Chem Eng Sci 59(3):525–541
5. Stevens A, Hrenya C (2005) Comparison of soft-sphere models to measurements of collision properties during normal impacts. Powder Technol 154(2–3):99–109
6. Kruggel-Emden H, Simsek E, Rickelt S, Wirtz S, Scherer V (2007) Review and extension of normal force models for the Discrete Element Method. Powder Technol 171 (3):157–173
7. Zhu H, Zhou Z, Yang R, Yu A (2007) Discrete particle simulation of particulate systems: theoretical developments. Chem Eng Sci 62(13):3378–3396
8. Seville J Willett C, Knight P (2000) Interparticle forces in fluidisation: a review. Powder Technol 113(3):261–268
9. Iveson S, Litster J (1998) Liquid-bound granule impact deformation and coefficient of restitution. Powder Technol 99(3):234–242
10. Mishra B, Murty C (2001) On the determination of contact parameters for realistic DEM simulations of ball mills. Powder Technol 115(3):290–297
11. Mangwandi C, Cheong Y, Adams M, Hounslow M, Salman A (2007) The coefficient of restitution of different representative types of granules. Chem Eng Sci 62(1–2):437–450
12. Bharadwaj R, Smith C, Hancock BC (2010) The coefficient of restitution of some pharmaceutical tablets/compacts. Int J Pharm 402 (1–2):50–56
13. Hastie D (2013) Experimental measurement of the coefficient of restitution of irregular shaped particles impacting on horizontal surfaces. Chem Eng Sci 101:828–836
14. Samimi A, Hassanpour A, Ghadiri M (2005) Single and bulk compressions of soft granules: experimental study and DEM evaluation. Chem Eng Sci 60(14):3993–4004
15. Weir G, Tallon S (2005) The coefficient of restitution for normal incident, low velocity particle impacts. Chem Eng Sci 60(13): 3637–3647
16. Freireich B, Litster J, Wassgren C (2009) Using the discrete element method to predict collision-scale behavior: a sensitivity analysis. Chem Eng Sci 64(15):3407–3416
17. Kodam M, Bharadwaj R, Curtis J, Hancock B, Wassgren C (2009) Force model considerations for glued-sphere discrete element method simulations. Chem Eng Sci 64(15): 3466–3475
18. Kodam M, Bharadwaj R, Curtis J, Hancock B, Wassgren C (2010) Cylindrical object contact detection for use in discrete element method simulations. Part I – contact detection algorithms. Chem Eng Sci 65(22):5852–5862
19. Song Y, Turton R, Kayihan F (2006) Contact detection algorithms for DEM simulations of tablet-shaped particles. Powder Technol 161 (1):32–40

20. Liu P, Hrenya CM (2014) Challenges of DEM: I. Competing bottlenecks in parallelization of gas-solid flows. Powder Technol 264:620–626
21. Berger KJ, Hrenya CM (2014) Challenges of DEM: II. Wide particle size distributions. Powder Technol 264:627–633
22. Beverloo W, Leniger H, van de Velde J (1961) The flow of granular solids through orifices. Chem Eng Sci 15(3–4):260–269
23. Anand A, Curtis JS, Wassgren CR, Hancock BC, Ketterhagen WR (2009) Predicting discharge dynamics of wet cohesive particles from a rectangular hopper using the discrete element method (DEM). Chem Eng Sci 64 (24):5268–5275
24. Anand A, Curtis JS, Wassgren CR, Hancock BC, Ketterhagen WR (2008) Predicting discharge dynamics from a rectangular hopper using the discrete element method (DEM). Chem Eng Sci 63(24):5821–5830
25. Ketterhagen WR, Hancock BC (2010) Optimizing the design of eccentric feed hoppers for tablet presses using DEM. Comput Chem Eng 34(7):1072–1081
26. Ketterhagen WR, Curtis JS, Wassgren CR, Hancock BC (2009) Predicting the flow mode from hoppers using the discrete element method. Powder Technol 195(1):1–10
27. Yu Y, Saxen H (2011) Discrete element method simulation of properties of a 3D conical hopper with mono-sized spheres. Adv Powder Technol 22(3):324–331
28. Ketterhagen WR, Curtis JS, Wassgren CR, Kong A, Narayan PJ, Hancock BC (2007) Granular segregation in discharging cylindrical hoppers: a discrete element and experimental study. Chem Eng Sci 62 (22):6423–6439
29. Anand A, Curtis JS, Wassgren CR, Hancock BC, Ketterhagen WR (2010) Segregation of cohesive granular materials during discharge from a rectangular hopper. Granul Matter 12 (2):193–200
30. Persson A-S, Alderborn G, Frenning G (2011) Flowability of surface modified pharmaceutical granules: a comparative experimental and numerical study. Eur J Pharm Sci 42(3):199–209
31. Cleary PW, Sawley ML (2002) DEM modelling of industrial granular flows: 3D case studies and the effect of particle shape on hopper discharge. Appl Math Model 26(2):89–111
32. Hohner D, Wirtz S, Scherer V (2013) Experimental and numerical investigation on the influence of particle shape and shape approximation on hopper discharge using the discrete element method. Powder Technol 235: 614–627
33. Tao H, Jin B, Zhong W, Wang X, Ren B, Zhang Y, Xiao R (2010) Discrete element method modeling of non-spherical granular flow in rectangular hopper. Chem Eng Process Process Intensif 49(2):151–158
34. Hancock BC, Ketterhagen WR (2011) Discrete element method (DEM) simulations of stratified sampling during solid dosage form manufacturing. Int J Pharm 418(2):265–272
35. Balevicius R, Kacianauskas R, Mroz Z, Sielamowicz I (2011) Analysis and DEM simulation of granular material flow patterns in hopper models of different shapes. Adv Powder Technol 22(2):226–235
36. Moysey P, Thompson M (2005) Modelling the solids inflow and solids conveying of single-screw extruders using the discrete element method. Powder Technol 153(2): 95–107
37. Owen P, Cleary P (2009) Prediction of screw conveyor performance using the Discrete Element Method (DEM). Powder Technol 193 (3):274–288
38. Hou Q, Dong K, Yu A (2014) DEM study of the flow of cohesive particles in a screw feeder. Powder Technol 256:529–539
39. Fernandez JW, Cleary PW, McBride W (2011) Effect of screw design on hopper drawdown of spherical particles in a horizontal screw feeder. Chem Eng Sci 66(22): 5585–5601
40. Adam S, Suzzi D, Radeke C, Khinast JG (2011) An integrated Quality by Design (QbD) approach towards design space definition of a blending unit operation by Discrete Element Method (DEM) simulation. Eur J Pharm Sci 42(1–2):106–115
41. Lacey PMC (1954) Developments in the theory of particle mixing. J Appl Chem 4(5): 257–268
42. Cleary PW, Sinnott MD (2008) Assessing mixing characteristics of particle-mixing and granulation devices. Particuology 6 (6):419–444. ISSN:1674-2001. Simulation and Modeling of Particulate Systems
43. Sarkar A, Wassgren CR (2009) Simulation of a continuous granular mixer: effect of operating conditions on flow and mixing. Chem Eng Sci 64(11):2672–2682
44. Xu Y, Xu C, Zhou Z, Du J, Hu D (2010) 2D DEM simulation of particle mixing in rotating drum: a parametric study. Particuology 8(2): 141–149. ISSN:1674-2001
45. Hassanpour A, Tan H, Bayly A, Gopalkrishnan P, Ng B, Ghadiri M (2011) Analysis of

particle motion in a paddle mixer using Discrete Element Method (DEM). Powder Technol 206(1–2):189–194

46. Laurent B, Cleary P (2012) Comparative study by PEPT and DEM for flow and mixing in a ploughshare mixer. Powder Technol 228: 171–186
47. Tahvildarian P, Ein-Mozaffari F, Upreti SR (2013) Circulation intensity and axial dispersion of non-cohesive solid particles in a V-blender via DEM simulation. Particuology 11(6):619–626. ISSN:1674-2001
48. Marigo M, Davies M, Leadbeater T, Cairns D, Ingram A, Stitt E (2013) Application of Positron Emission Particle Tracking (PEPT) to validate a Discrete Element Method (DEM) model of granular flow and mixing in the Turbula mixer. Int J Pharm 446(1–2):46–58
49. Hua X, Curtis J, Hancock B, Ketterhagen W, Wassgren C (2013) The kinematics of non-cohesive, sphero-cylindrical particles in a low-speed, vertical axis mixer. Chem Eng Sci 101: 144–164
50. Gao Y, Muzzio FJ, Ierapetritou MG (2013) Scale-up strategy for continuous powder blending process. Powder Technol 235: 55–69
51. Sen M, Ramachandran R (2013) A multi-dimensional population balance model approach to continuous powder mixing processes. Adv Powder Technol 24(1):51–59
52. Jovanovic A, Pezo M, Pezo L, Levic L (2014) DEM/CFD analysis of granular flow in static mixers. Powder Technol 266:240–248
53. Alian M, Ein-Mozaffari F, Upreti SR, Wu J (2014) Using discrete element method to analyze the mixing of the solid particles in a slant cone mixer. Chem Eng Res Des
54. Alizadeh E, Bertrand F, Chaouki J (2014) Discrete element simulation of particle mixing and segregation in a tetrapodal blender. Comput Chem Eng 64:1–12
55. Siraj MS, Radl S, Glasser BJ, Khinast JG (2011) Effect of blade angle and particle size on powder mixing performance in a rectangular box. Powder Technol 211(1):100–113
56. Lu L-S, Hsiau S-S (2008) Mixing in a vibrated granular bed: diffusive and convective effects. Powder Technol 184(1):31–43
57. Arratia P, hang Duong N, Muzzio F, Godbole P, Reynolds S (2006) A study of the mixing and segregation mechanisms in the Bohle Tote blender via DEM simulations. Powder Technol 164(1):50–57
58. Zhu R, Zhu W, Xing L, Sun Q (2011) DEM simulation on particle mixing in dry and wet particles spouted bed. Powder Technol 210 (1):73–81
59. Dubey A, Sarkar A, Ierapetritou M, Wassgren CR, Muzzio FJ (2011) Computational approaches for studying the granular dynamics of continuous blending processes, 1 – DEM based methods. Macromol Mater Eng 296(3–4):290–307
60. Dubey A, Vanarase AU, Muzzio FJ (2012) Impact of process parameters on critical performance attributes of a continuous blender: a DEM-based study. AIChE J 58(12): 3676–3684
61. Sarkar A, Wassgren CR (2012) Comparison of flow microdynamics for a continuous granular mixer with predictions from periodic slice DEM simulations. Powder Technol 221:325–336. Selected papers from 2010 AIChE Annual Meeting
62. Ning Z, Boerefijn R, Ghadiri M, Thornton C (1997) Distinct element simulation of impact breakage of lactose agglomerates. Adv Powder Technol 8(1):15–37
63. Liu L, Kafui K, Thornton C (2010) Impact breakage of spherical, cuboidal and cylindrical agglomerates. Powder Technol 199(2): 189–196
64. Hassanpour A, Antony S, Ghadiri M (2007) Effect of size ratio on the behaviour of agglomerates embedded in a bed of particles subjected to shearing: DEM analysis. Chem Eng Sci 62(4):935–942
65. Antonyuk S, Palis S, Heinrich S (2011) Breakage behaviour of agglomerates and crystals by static loading and impact. Powder Technol 206(1–2):88–98
66. Subero J, Ning Z, Ghadiri M, Thornton C (1999) Effect of interface energy on the impact strength of agglomerates. Powder Technol 105(1–3):66–73
67. Thornton C, Ciomocos M, Adams M (1999) Numerical simulations of agglomerate impact breakage. Powder Technol 105(1–3):74–82
68. Mishra B, Thornton C (2001) Impact breakage of particle agglomerates. Int J Min Process 61(4):225–239
69. Moreno R, Ghadiri M, Antony S (2003) Effect of the impact angle on the breakage of agglomerates: a numerical study using DEM. Powder Technol 130(1–3):132–137
70. Golchert D, Moreno R, Ghadiri M, Litster J (2004) Effect of granule morphology on breakage behaviour during compression. Powder Technol 143–144:84–96
71. Antonyuk S, Khanal M, Tomas J, Heinrich S, Morl L (2006) Impact breakage of spherical

granules: experimental study and DEM simulation. Chem Eng Process Process Intensif 45 (10):838–856
72. Hassanpour A, Antony SJ, Ghadiri M (2007) Modeling of agglomerate behavior under shear deformation: effect of velocity field of a high shear mixer granulator on the structure of agglomerates. Adv Powder Technol 18 (6):803–811
73. Tong Z, Yang R, Yu A, Adi S, Chan H (2009) Numerical modelling of the breakage of loose agglomerates of fine particles. Powder Technol 196(2):213–221
74. Kwan CC, Mio H, Chen YQ, Ding YL, Saito F, Papadopoulos DG, Bentham AC, Ghadiri M (2005) Analysis of the milling rate of pharmaceutical powders using the Distinct Element Method (DEM). Chem Eng Sci 60 (5):1441–1448
75. Hare C, Ghadiri M, Dennehy R (2011) Prediction of attrition in agitated particle beds. Chem Eng Sci 66(20):4757–4770
76. Han T, Levy A, Kalman H (2003) DEM simulation for attrition of salt during dilute-phase pneumatic conveying. Powder Technol 129 (1–3):92–100
77. Ahmadian H, Hassanpour A, Ghadiri M (2011) Analysis of granule breakage in a rotary mixing drum: experimental study and distinct element analysis. Powder Technol 210(2):175–180
78. Lee H, Cho H, Kwon J (2010) Using the discrete element method to analyze the breakage rate in a centrifugal/vibration mill. Powder Technol 198(3):364–372
79. Wang M, Yang R, Yu A (2012) DEM investigation of energy distribution and particle breakage in tumbling ball mills. Powder Technol 223:83–91
80. Capece M, Bilgili E, Dave R (2014) Insight into first-order breakage kinetics using a particle-scale breakage rate constant. Chem Eng Sci 117:318–330
81. Metzger MJ, Glasser BJ (2013) Simulation of the breakage of bonded agglomerates in a ball mill. Powder Technol 237:286–302
82. Talu I, Tardos GI, Khan M (2000) Computer simulation of wet granulation. Powder Technol 110(1–2):59–75
83. Kafui D, Thornton C (2008) Fully-3D DEM simulation of fluidised bed spray granulation using an exploratory surface energy-based spray zone concept. Powder Technol 184 (2):177–188
84. Hassanpour A, Pasha M, Susana L, Rahmanian N, Santomaso AC, Ghadiri M (2013) Analysis of seeded granulation in high shear granulators by discrete element method. Powder Technol 238:50–55
85. Goldschmidt M, Weijers G, Boerefijn R, Kuipers J (2003) Discrete element modelling of fluidised bed spray granulation. Powder Technol 138(1):39–45
86. Gantt JA, Gatzke EP (2005) High-shear granulation modeling using a discrete element simulation approach. Powder Technol 156 (2–3):195–212
87. Ingram GD, Cameron I (2004) Challenges in multiscale modelling and its application to granulation systems. Dev Chem Eng Miner Process 12(3–4):293–308
88. Ingram GD, Cameron IT (2005) Formulation and comparison of alternative multiscale models for drum granulation. In: Puigjaner L, Espuna A (eds) European symposium on computer-aided process engineering-15, 38th European symposium of the working party on computer aided process engineering. Computer aided chemical engineering, vol 20. Elsevier, pp 481–486
89. Gantt JA, Cameron IT, Litster JD, Gatzke EP (2006) Determination of coalescence kernels for high-shear granulation using DEM simulations. Powder Technol 170(2):53–63
90. Gantt JA, Gatzke EP (2006) A stochastic technique for multidimensional granulation modeling. AIChE J 52(9):3067–3077
91. Reinhold A, Briesen H (2012) Numerical behavior of a multiscale aggregation model-coupling population balances and discrete element models. Chem Eng Sci 70:165–175
92. Barrasso D, Ramachandran R (2014) Multi-scale modeling of granulation processes: bi-directional coupling of PBM with DEM via collision frequencies. Chem Eng Res Des. doi:10.1016/j.cherd.2014.04.016
93. Sen M, Barrasso D, Singh R, Ramachandran R (2014) A multi-scale hybrid CFD-DEM-PBM description of a fluid-bed granulation process. Processes 2(1):89–111
94. Barrasso D, Tamrakar A, Ramachandran R (2014) A reduced order PBM-ANN model of a multi-scale PBM-DEM description of a wet granulation process. Chem Eng Sci 119: 319–329
95. Washino K, Tan H, Hounslow M, Salman A (2013) A new capillary force model implemented in micro-scale CFD-DEM coupling for wet granulation. Chem Eng Sci 93: 197–205
96. Washino K, Tan H, Hounslow M, Salman A (2013) Meso-scale coupling model of DEM and CIP for nucleation processes in wet granulation. Chem Eng Sci 86:25–37

97. Neuwirth J, Antonyuk S, Heinrich S, Jacob M (2013) CFD-DEM study and direct measurement of the granular flow in a rotor granulator. Chem Eng Sci 86:151–163
98. Fries L, Antonyuk S, Heinrich S, Dopfer D, Palzer S (2013) Collision dynamics in fluidised bed granulators: a DEM-CFD study. Chem Eng Sci 86:108–123
99. Link J, Godlieb W, Deen N, Kuipers J (2007) Discrete element study of granulation in a spout-fluidized bed. Chem Eng Sci 62(1–2): 195–207
100. Hassanpour A, Kwan C, Ng B, Rahmanian N, Ding Y, Antony S, Jia X, Ghadiri M (2009) Effect of granulation scale-up on the strength of granules. Powder Technol 189(2): 304–312
101. Fries L, Antonyuk S, Heinrich S, Palzer S (2011) DEM-CFD modeling of a fluidized bed spray granulator. Chem Eng Sci 66(11): 2340–2355
102. Nakamura H, Fujii H, Watano S (2013) Scale-up of high shear mixer-granulator based on discrete element analysis. Powder Technol 236:149–156
103. Dhenge RM, Washino K, Cartwright JJ, Hounslow MJ, Salman AD (2013) Twin screw granulation using conveying screws: effects of viscosity of granulation liquids and flow of powders. Powder Technol 238:77–90
104. Fries L, Antonyuk S, Heinrich S, Niederreiter G, Palzer S (2014) Product design based on discrete particle modeling of a fluidized bed granulator. Particuology 12:13–24
105. Guo Y, Wu C-Y, Kafui K, Thornton C (2010) Numerical analysis of density-induced segregation during die filling. Powder Technol 197 (1–2):111–119
106. Guo Y, Wu C-Y, Thornton C (2011) The effects of air and particle density difference on segregation of powder mixtures during die filling. Chem Eng Sci 66(4):661–673
107. Mateo-Ortiz D, Muzzio FJ, Mendez R (2014) Particle size segregation promoted by powder flow in confined space: the die filling process case. Powder Technol 262: 215–222
108. Wu C-Y (2008) DEM simulations of die filling during pharmaceutical tabletting. Particuology 6(6):412–418
109. Sheng Y, Lawrence C, Briscoe B, Thornton C (2004) Numerical studies of uniaxial powder compaction process by 3D DEM. Eng Comput 21(2–4):303–317
110. Hassanpour A, Ghadiri M (2004) Distinct element analysis and experimental evaluation of the Heckel analysis of bulk powder compression. Powder Technol 141(3):251–261
111. Frenning G (2010) Compression mechanics of granule beds: a combined finite/discrete element study. Chem Eng Sci 65(8): 2464–2471
112. Persson A-S, Frenning G (2012) An experimental evaluation of the accuracy to simulate granule bed compression using the discrete element method. Powder Technol 219:249–256
113. Martin C, Bouvard D Shima S (2003) Study of particle rearrangement during powder compaction by the Discrete Element Method. J Mech Phys Solids 51(4):667–693
114. Wiacek J, Molenda M (2014) Effect of particle size distribution on micro- and macromechanical response of granular packings under compression. Int J Solids Struct 51(25–26): 4189–4195
115. Mann U (1983) Analysis of spouted-bed coating and granulation. 1. Batch operation. Ind Eng Chem Process Des Dev 22(2): 288–292
116. Ketterhagen WR (2011) Modeling the motion and orientation of various pharmaceutical tablet shapes in a film coating pan using DEM. Int J Pharm 409(1–2):137–149
117. Freireich B, Wassgren C (2010) Intra-particle coating variability: analysis and Monte-Carlo simulations. Chem Eng Sci 65(3):1117–1124
118. Freireich B, Ketterhagen WR, Wassgren C (2011) Intra-tablet coating variability for several pharmaceutical tablet shapes. Chem Eng Sci 66(12):2535–2544
119. Suzzi D, Toschkoff G, Radl S, Machold D, Fraser SD, Glasser BJ, Khinast JG (2012) DEM simulation of continuous tablet coating: effects of tablet shape and fill level on inter-tablet coating variability. Chem Eng Sci 69(1):107–121
120. Dubey A, Hsia R, Saranteas K, Brone D, Misra T, Muzzio FJ (2011) Effect of speed, loading and spray pattern on coating variability in a pan coater. Chem Eng Sci 66(21): 5107–5115
121. Kumar R, Wassgren C (2014) Inter-particle coating variability in a continuous coater. Chem Eng Sci 117:1–7
122. Song Y, Turton R (2007) Study of the effect of liquid bridges on the dynamic behavior of two colliding tablets using DEM. Powder Technol 178(2):99–108
123. Ketterhagen WR, Bharadwaj R, Hancock BC (2010) The coefficient of rolling resistance (CoRR) of some pharmaceutical tablets. Int J Pharm 392(1–2):107–110

124. Toschkoff G, Just S, Funke A, Djuric D, Knop K, Kleinebudde P, Scharrer G, Khinast JG (2013) Spray models for discrete element simulations of particle coating processes. Chem Eng Sci 101:603–614
125. Kodam M, Curtis J, Hancock B, Wassgren C (2012) Discrete element method modeling of bi-convex pharmaceutical tablets: contact detection algorithms and validation. Chem Eng Sci 69(1):587–601
126. Li J, Wassgren C, Litster JD (2013) Multi-scale modeling of a spray coating process in a paddle mixer/coater: the effect of particle size distribution on particle segregation and coating uniformity. Chem Eng Sci 95:203–210
127. Freireich B, Li J, Litster J, Wassgren C (2011) Incorporating particle flow information from discrete element simulations in population balance models of mixer-coaters. Chem Eng Sci 66(16):3592–3604
128. Li J, Freireich BJ, Wassgren CR, Litster JD (2013) Experimental validation of a 2-D population balance model for spray coating processes. Chem Eng Sci 95:360–365. ISSN:0009-2509
129. Kumar R, Freireich B, Wassgren C (2014) DEM-compartment-population balance model for particle coating in a horizontal rotating drum. Chem Eng Sci. doi:10.1016/j.ces.2014.06.046
130. Pandey P, Song Y, Kayihan F, Turton R (2006) Simulation of particle movement in a pan coating device using discrete element modeling and its comparison with video-imaging experiments. Powder Technol 161(2):79–88
131. Kalbag A, Wassgren C, Penumetcha SS, Perez-Ramos JD (2008) Inter-tablet coating variability: residence times in a horizontal pan coater. Chem Eng Sci 63(11):2881–2894
132. Kalbag A, Wassgren C (2009) Inter-tablet coating variability: tablet residence time variability. Chem Eng Sci 64(11):2705–2717
133. Sahni E, Chaudhuri B (2011) Experiments and numerical modeling to estimate the coating variability in a pan coater. Int J Pharm 418(2):286–296
134. Sahni E, Yau R, Chaudhuri B (2011) Understanding granular mixing to enhance coating performance in a pan coater: experiments and simulations. Powder Technol 205(1–3):231–241
135. Pandey P, Song Y, Turton R (2007) Modelling of pan-coating processes for pharmaceutical dosage forms. In: Salman AD, Hounslow MJ, Seville JPK (eds) Granulation. Handbook of Powder Technology, vol 11, Chap. 8 Elsevier Science B.V., Amsterdam, pp 377–416

Chapter 5

A Procedure for Developing Quantitative Near Infrared (NIR) Methods for Pharmaceutical Products

Rodolfo J. Romañach, Andrés D. Román-Ospino, and Manel Alcalà

Abstract

The pharmaceutical industry uses procedures to describe all the instructions needed to perform a process in a consistent manner. Procedures are based on process knowledge and the desire to consistently meet desired specifications. This chapter proposes a procedure for the development of NIR partial least squares (PLS) calibration models for pharmaceutical applications. The chapter captures the lessons learned for more than a decade in studies where NIR spectroscopy has been used for quantitative determinations of drug concentration, moisture, polymorphs, and other important applications. Two examples of recent initiatives to develop NIR calibration models in a more efficient manner reducing the number of calibration samples are also discussed. The procedure differs from previously published guidelines since it is based on specific examples of method validations for pharmaceutical processes and provides clear instructions for method development and validation. The procedure and tutorial should advance the implementation of near infrared spectroscopic (NIRS) quantitative methods in the pharmaceutical industry.

Key words Calibration, Partial least squares, Validation, Pharmaceutical, Near infrared spectroscopy

1 Introduction

Pharmaceutical engineers are likely to be part of a process development team that includes one or more analytical methods that involve near infrared spectroscopy (NIRS). The identity test of incoming raw materials and active pharmaceutical ingredient in a new product could be performed with NIRS. NIRS is an effective and proven technique to identify incoming raw materials and meet the 21CFR 211.184 cGMP requirements that "At least one test shall be conducted to verify the identity of each component of a drug product." NIRS is considered an established method for identity testing [1] but is also capable of providing valuable information for pharmaceutical processes.

The adoption of NIRS by the pharmaceutical industry has continued to grow to include monitoring the drying of an active pharmaceutical ingredient or a formulation following wet

Marianthi G. Ierapetritou and Rohit Ramachandran (eds.), *Process Simulation and Data Modeling in Solid Oral Drug Development and Manufacture*, Methods in Pharmacology and Toxicology,
DOI 10.1007/978-1-4939-2996-2_5, © Springer Science+Business Media New York 2016

granulation [2–4]. NIRS could be used to monitor and evaluate the adequacy of mixing for a pharmaceutical formulation [5, 6], and drug concentration in the final product. The numerous applications of NIR spectroscopy are not limited to analytical methods required by regulatory agencies, and also include process understanding and troubleshooting [7, 8]. Thus, an understanding of the current state of NIR spectroscopy is useful for the pharmaceutical engineers involved in process development teams and to the regulatory agencies that evaluate NIR methods developed by industrial scientists.

The main motivations for the adoption of NIRS methods is the analysis of samples in their native "as is" condition in the manufacturing area without the need to dissolve the sample in a solvent and the need to transport the sample to the QC laboratory. Traditional Quality Control (QC) analytical methods are usually characterized by sample preparations that involve extraction of an analyte to a solvent, filtration and centrifugation steps to transform a tablet or other pharmaceutical product into a clear solution. Sample preparation accounts for at least 50 % of the analysis time in the traditional QC lab. The traditional analytical methods also require removing the material from the manufacturing site and transport to the QC lab. This grab sampling involves a significant risk that the sample brought to the lab is not representative of the production process, and it is estimated that the sampling error is much greater than the analytical method error [9–11]. NIR spectrometers have been interfaced with manufacturing equipment providing real time measurements that may be used to monitor the production process [2–6].

A NIR spectrum with a high signal to noise ratio may be obtained in as little as 100 ms, but could require 1 min in some applications such as transmission spectra of tablets. NIR measurements may also be performed using fiber optics to bring the NIR light to the process equipment. The use of fiber optics is especially useful to obtain spectra of a product that is hygroscopic, avoiding increases in moisture while spectra are obtained [12]. Fiber optic measurements could also be very useful to protect chemists from a highly toxic or reactive material while the NIR spectra are obtained.

NIRS is the result of the interaction of light from 780 to 2500 nm with matter [13]. The bands observed in the NIR region are overtones and combination bands of fundamental bands observed in the mid-infrared region. Figure 1 shows the spectrum of heptane in the NIR and mid-IR region. The bands in the region of 5800–5600 cm^{-1} are the first overtone of the C-H fundamental bands observed from 2960 to 2850 cm^{-1}. The overtones are seen at frequencies that are almost twice the fundamental frequency. The overtones of the fundamental bands originate from C–H, N–H, O–H, and Si–OH [14] bonds. The bands in the C-H 4400–4200 cm^{-1} region are combination bands that result from fundamental ($\approx$2860–2850 cm^{-1}) and bending modes ($\approx$1460–1370 cm^{-1} [15]). The assignment of NIR

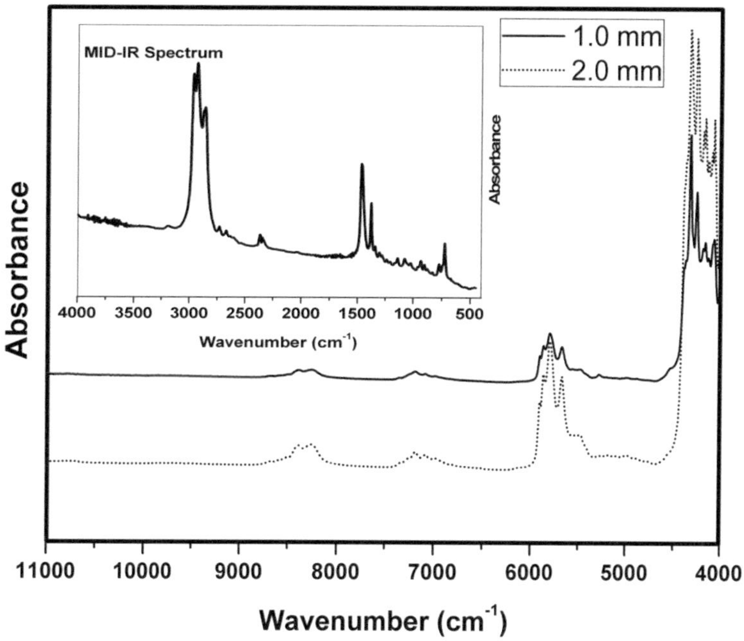

Fig. 1 NIR transflectance spectra of heptane at 1 and 2 mm pathlength, and mid-IR spectrum

bands to the vibrations of functional groups is complicated, because of wide overlapping bands that are commonly observed in this region. These complications are also evident in the spectrum of chloroform, a molecule with only one C-H bond, and six fundamental vibrational bands. The six fundamental modes combine to produce over 30 overtone and combination bands [13]. The wide overlapping NIR bands are 10–100 times weaker than the corresponding mid-IR bands. The intensity of the bands is a function on the change in the dipole moment that occurs during the vibration, and the anharmonicity of the transition. In spite of the wide and overlapping bands, NIR spectra are capable of discriminating spectra with very similar structures [16], and have been extensively used to identify raw materials in different industries [1].

NIR spectra also depend on the physical properties of the materials analyzed. Discerning between the chemical information in a NIR spectrum, and the information related to the material properties is not a simple task and is currently the subject of research efforts in the field [17, 18]. However, the NIR information on the physical properties of materials is extremely valuable to pharmaceutical engineers since the physical properties of materials are the root cause of common problems such as segregation [19]. Figure 2 shows the NIR spectrum of crystalline granulated white sugar, brown sugar, powdered confectioner's sugar, and a sucrose

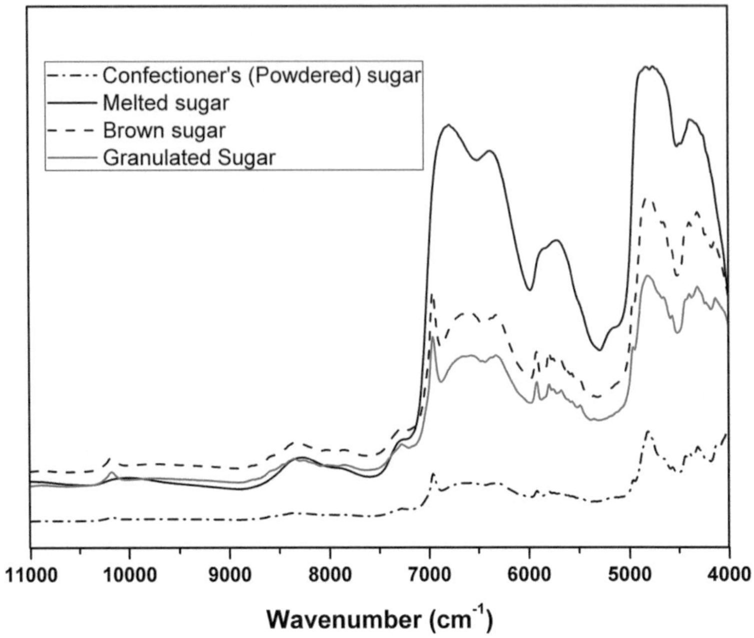

Fig. 2 Spectra of sucrose samples including: confectioner's (powdered), brown, and granulated sugar with the 6960 cm^{-1} associated with the free O–H group. This band is not observed when the crystalline structure is broken and a thick syrup (melted sugar) is produced

solution obtained after melting the sugar with a drop of water. The band at 6960 cm^{-1} associated with the free O–H bond is only observed in the crystalline state, as it disappears in solution where wider bands are observed [20]. Thus, NIRS is a cross-sensitive technique where a spectrum can provide information on both chemical and physical properties of the substance or chemical mixture [21]. The NIR spectrum may be used to predict moisture and drug concentration in a granulation and to identify a formulation [22]. NIR spectra provide valuable qualitative and quantitative information.

The pharmaceutical industry uses procedures to describe everything that is needed to perform a process in a consistent manner. This chapter proposes a procedure for the development of partial least squares (PLS) calibration models for pharmaceutical applications. The chapter seeks to capture the lessons learned for more than a decade in studies where NIR spectroscopy has been used for quantitative determinations of drug concentration, moisture, polymorphs, and other important applications without the need for the labor intensive sample preparations that are characteristic of the currently implemented methods. The chapter also provides two examples of recent initiatives to develop NIR calibration models in a more efficient manner reducing the number of calibration samples needed. The authors recognize that a number of guidelines to the validation of NIR methods have been published in the past [23–25]. This chapter differs from these guidelines in that it examines specific examples of method validations to propose a procedure for the development of calibration models in the pharmaceutical industry. The purpose of this chapter is to advance the implementation of NIR spectroscopy in the pharmaceutical industry by providing a procedure and tutorial for the development of near infrared spectroscopic (NIRS) quantitative methods and a look at two relatively new approaches.

2 Procedure for Development of Calibration Model

Calibration [26] refers to development of a mathematical relationship between the changes in the NIR spectra and drug concentration or property of interest. The calibration may use "empirical data and prior knowledge for determining how to predict unknown quantitative information **Y** from available measurement **X**, via some mathematical transfer function [27]." NIRS calibration models are typically multivariate based on multiple wavelengths or frequencies from the NIR spectrum. These multivariate methods are not common in pharmaceutical manufacturing QC laboratories where single wavelength (univariate) methods are used for HPLC methods. Univariate HPLC calibrations are usually performed every time that the method is run. Multivariate calibration methods include variations in instrument performance and do not require a new calibration every time that the method is used. Multivariate calibration models are developed and used for a certain span of time until it is necessary to update the method. Multivariate calibration methods will require significant changes in the procedures and practices of the QC unit. The principal focus in this chapter is PLS methods since a significant number have been developed for pharmaceutical formulations.

2.1 Planning and Preparation of Calibration Samples

Table 1 provides a list of the steps to develop a PLS calibration model. The first set of steps are related to the planning to prepare the calibration set samples and this plan is the key to the robustness and long term success of the calibration model. As indicated by Kramer calibration samples should: contain all expected components, span the concentration ranges and conditions of interest and be as similar as possible to the unknown samples that it will predict [28]. The recommended first step is to familiarize with the pharmaceutical product. Some products have only one formulation but different doses where the total drug content is changed by increasing or decreasing the tablet or dose unit weight. Other products have a constant unit weight and formulations with different concentrations of the active pharmaceutical ingredient (API). This was

Table 1
Proposed procedure for development of PLS calibration model

Step #	Action
	Part I—Planning and preparation of calibration samples
1	Familiarize with pharmaceutical formulation. Obtain information on drug concentration, and learn whether the concentration is the same for all formulations, or varies.
2	Meet with QC and regulatory affairs personnel. Establish goals for project and validation. Determine method accuracy, precision, and range requirements.
3	Decide how spectra will be obtained: diffuse reflectance, transmittance, transflectance.
4	Request and obtain production samples.
5	Develop a plan to expand the concentration range of the production samples through the use of laboratory/pilot plant/production process samples. Prepare samples that expand the concentration range. This plan should build in the methods' robustness from the start to reduce future needs for updating the calibration model.
6	Prepare calibration samples that expand the concentration range.
	Part II—Obtaining spectra and selection of samples for calibration model
7	Obtain spectra of API, excipients, calibration and production samples.
8	Perform visual evaluation of spectra. Perform initial evaluation of effect of pretreatment methods on spectra. Are any samples identified as possible outliers after spectral pretreatment.
9	Select samples to encompass the variations in drug concentration as well as the chemical and physical variability expected from the process.
	Part III. Calculation of calibration model and evaluation
10	Calculate the calibration model.
11	Obtain NIR spectra of prediction set (first validation set—independent production samples). Predict samples and evaluate results.
	Part IV. Fine tuning of calibration model
12	Evaluate the spectral range for use in the calibration model.
13	Evaluate effect of spectral pretreatment and prepare a table that summarizes all the calibration models developed and their performance, comparing sample sets, spectral range and pretreatment, number of factors, etc.
14	Develop identification method, determine whether sample corresponds to model developed.
15.	Predict samples from a second independent validation set. Evaluate results, and whether samples from these lots need to be included in the model.
	Part V. Validation
16	Does the method meet the preestablished requirements? Perform validation tests and document

the case for a commercially available steroid [29] tablets with nominal contents of 5, 10, 15, 20, and 30 mg of API. Ten tablets were used from each of four different batches for each nominal API content without requiring specially manufactured samples to expand the concentration range. A NIR transmission method was developed with steroid tablets with an API range 2.94 from 17.64 % w/w along with a placebo tablet. Unfortunately, many products are not manufactured with different concentrations making the expansion of the concentration range much more challenging.

The first steps also include deciding whether to obtain NIR spectra in diffuse reflectance or transmission mode for the final product. In-process measurements of powders are carried out in diffuse reflectance. Tablets may be analyzed through both diffuse reflectance and transmission measurements. Transmission measurements are generally considered to provide greater accuracy since a greater sample volume is analyzed. Diffuse reflectance spectra analyze a much lower sample volume and multiple spectra are required to assess the distribution of the analyte of interest [30]. However, it is always easier to obtain diffuse reflectance measurements and these include the entire NIR spectrum whereas transmission spectra include only the higher wavenumber end of the spectrum as shown in Fig. 3. NIR calibration models for transmission are usually in the 11,000–8000 cm^{-1} spectral range. The absorbance below 8000 cm^{-1} is usually 2 or more with very few photons reaching the detector. The selection of transmission or diffuse reflectance measurements for the final product will depend on QC expectations and objectives. However, generally transmission is used when greater accuracy is needed and diffuse reflectance is used for faster in-line measurements.

The scientific literature includes numerous efforts to expand the concentration range as part of method development and

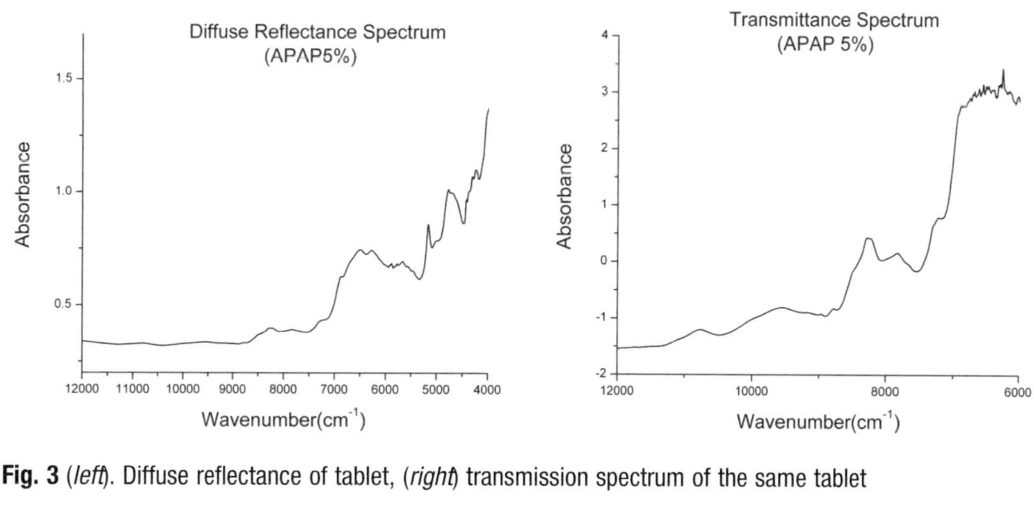

Fig. 3 (*left*). Diffuse reflectance of tablet, (*right*) transmission spectrum of the same tablet

validation efforts. Most of the studies involve calibration models where the analytes of interest varies from 70 to 130 % (w/w) of the target concentration. However, production processes usually provide samples with drug concentrations that vary only from 95 to 105 % (w/w) of the target concentration. Therefore, all method developers have included steps to expand the calibration range. The approaches to obtain calibration samples are classified in this chapter according to the size of the equipment used to prepare the calibration samples. Calibration samples may be prepared in the analytical laboratory, obtained from a pilot plant or from the actual manufacturing process. These approaches are often combined. Table 2 summarizes the approaches to prepare calibration samples.

Table 2
Summary of approaches to increase the concentration range of pharmaceutical samples

	Lab scale	Pilot plant	Obtained from production process
Description	May be as simple as preparing synthetic mixtures of excipient and drug. May also include lab scale granulation, tablets, and coating.	Use smaller volume equipment but with the same operating principle as the production process equipment.	Spectra obtained during the production process.
Advantages	Low amounts of materials	Lower amount of materials than in manufacturing process	Easy to implement with drying and milling processes that constantly change
	No need to use manufacturing area	Does not interrupt manufacturing process.	Variation associated to mechanical effects of the process is included.
	Facilitates performing DOE	Facilitates performing DOE, if product scale up and NIR method development are coordinated.	
	May be combined with the other approaches, for example samples from the process may be "doped" or "diluted."	A good option where suitable laboratory equipment is not available.	
	Samples may be equilibrated to various humidity levels.		
Disadvantages	May not include process variations.	May not include all of the variation associated with mechanical features of system.	High cost, high volume of material

There are advantages and disadvantages to each of these strategies. The selection of one of these strategies will depend on many factors, such as the availability of materials and equipment, drug toxicity, and the properties of the materials.

The preparation of laboratory samples may simply involve dry mixing of the formulation components, or simulating the production process in the laboratory. One of the principal advantages of laboratory samples is that it facilitates the use of design of experiments (DOE) to obtain calibration samples that avoid strong correlations in the concentrations of excipients and API and build in the method's robustness [31, 32]. The laboratory calibration samples could also include excipients and API from various lots received from the companies suppliers. The calibration model would then include the particle size variation and overall variation found in formulation components. Laboratory samples may also be easily equilibrated to various humidity levels. The preparation of laboratory samples is often necessary due to the high cost of using production and pilot scale facilities. Furthermore, production facilities may be in full use and not available for obtaining the samples needed by the scientists developing the NIR method.

The reference values of laboratory prepared samples may be obtained gravimetrically. Powder blends or wet granulations with a given API concentration may be prepared in the laboratory. The calibration model may be developed with samples prepared gravimetrically and then validation samples can be analyzed by HPLC or UV spectroscopy [33, 34]. The preparation of reference samples gravimetrically entails fewer steps than the analysis of samples by HPLC and is expected to have greater accuracy and precision. The HPLC method ultimately depends on the weighing of reference standards, usually in quantities much lower than those in the calibration samples, dissolving the standards and one or more dilutions that could lead to greater errors than the gravimetric reference method. The authors recognize the challenges of working with blends and the fact that the spectrum obtained may not be representative of the entire blend. This challenge is discussed in the next part of the procedure—obtaining spectra and selection of samples for the calibration model. The preparation of laboratory samples with gravimetric reference values will also lead to the development of a first calibration model in less time since it will not be necessary to wait for the HPLC analyses and reduce method development costs.

The number of calibration samples may be somewhat lower in those cases where design of experiments is used to obtain the possible variations in composition of the API and excipients. Only 21 calibration samples were required for real time prediction of drug concentration during a blending process [33] of an API and major excipients. The concentrations of the calibration samples were selected after 10,000 simulations to determine the criteria that reduced the maximum sum of the six pair correlations of the formulation components. Thus, it is impossible to specify the use of a given number of samples in the calibration set as the number of samples needed will depend on the experimental design. A commonly accepted guideline is to use approximately the same number of samples at each calibration level to avoid biasing the results towards a specific concentration level [35].

Laboratory prepared samples may also include the use of "doped" [36–38] samples where blend or granulation from the production process is spiked with drug to increase its concentration, or diluted by adding excipients. The advantage of this approach is that it provides a greater opportunity to include samples in the calibration set that are physically similar to the production samples. The "doped" samples could include material from several production lots to include process variation in the calibration model. Laboratory prepared samples may also simulate a production process, as when granulates are prepared in the laboratory [34].

The development of a method with laboratory samples may also take advantage of the properties of the API or formulation. For example, scientists at Lilly developed a NIR method for moisture in a highly hygroscopic API, where the water content can vary from 0.5 to 14 % (w/w) at a rate of up to 0.02 % per minute as ambient humidity changes [12]. In this application, the powder samples were first dried and then different moisture levels obtained by exposing the samples to humidity between ambient and 80 % R. H. using a humidity-controlled glove box. This approach was used to obtain moisture values in the range 0.5–11.4 % (w/w). The vials were capped and NIR spectra collected prior to the Karl Fischer titration.

Pilot plant equipment that follows the same operating principles as the full scale manufacturing system may also be used to expand the concentration range. This approach requires less materials and investment than using the full scale production equipment. The preparation of laboratory samples could also be difficult if it involves specialized equipment, or a freeze drying step that is difficult to replicate at laboratory scale. In these situations, pilot plant equipment, often at a tenth of the manufacturing scale, could be used to generate samples for the calibration set. The pilot plant samples may also be the best approach, if a NIR method is developed at the same time that a new formulation is being developed. During formulation development industrial pharmacists evaluate different compression forces and different excipient and drug concentrations that will also be helpful to the development of robust calibration models.

The scientific literature has several examples where pilot scale equipment or production equipment has been used to generate samples with varying concentrations. Scheiwe used 50 mg tablets from the commercial production process, and 40, 45, 55, and

60 mg tablets manufactured in smaller scale equipment by the company's R&D personnel [39]. Development lots were used to obtain tablets from 80 to 120 %, and capsules from 70 to 130 % of target concentrations [40]. The calibration models also included samples from the established production process. In another effort, samples varying in drug concentration from 0.5 to 1.0 % (w/w) were obtained using a 16-punch compression machine, and used to investigate the application of NIR spectroscopy to low drug content formulations [41].

The third strategy is to obtain the samples directly from the production process. This strategy is especially suitable for processes where materials are changing constantly such as drying and milling. Researchers at Merck also followed this approach by collecting in line NIR spectra, and using a sample cup that holds the granulation in front of fiber optic window where an air purge is used to clean the probe window [2]. This setup was used to obtain nearly 750 in-line NIR spectra and 119 samples pulled for KF reference analysis. The samples contained a moisture level based on KF ranged that extended from 3.65 to 25.24 % (w/w), and all the calibration set samples were obtained directly from the production process.

2.2 Obtaining Spectra and Selection of Samples for the Calibration Model

The next step is to obtain the spectra for the calibration samples. These spectra should be obtained in a way that maximizes the signal to noise to ratio and requires careful evaluation of the spectral acquisition parameters, for example, determining the number of scans to average when FT-NIR spectrometers are used and the instrument resolution. The signal to noise of FT-NIR spectra is approximately proportional to the square root of the number of scans that are averaged [42]. High resolution FT-NIR spectra might be desirable to capture fine spectral features, but lead to a reduced signal to noise ratio and higher scan time. The spectral acquisition time will also have an effect on the sample volume analyzed when the sampling is a flowing powder or liquid. Thus, a sufficient number of scans have to be averaged to improve the signal to noise ratio, but signal averaging increases the time needed for spectral acquisition and could also impair the method's ability to detect changes in the manufacturing process.

Spectra should be obtained for the API, excipients, calibration and production samples. The future calibration model should include spectral bands of the analytes being measured, such as API spectral bands in drug concentration methods. A number of calibration models have been developed using the entire NIR spectrum [43], while other calibration models have been developed using limited spectral regions [44, 45]. Thus, there is no set guideline or recommendation to the spectral range that should be used in the calibration model but it is generally recognized that it should include spectral bands related to the analyte.

Calibration spectra should be obtained under conditions similar to those under which the method will be deployed. The distance between the probe and the sample should be similar for the calibration samples and the final production samples. In diffuse reflectance measurements the NIR radiation may penetrate up to 5 mm. However, the radiation that penetrates 5 mm will not likely return to the detector. The radiation that is remitted back to the detector is mostly from the top 0.5–1 mm [46–48], although at higher frequencies NIR spectra have been obtained with spectral features of materials 2 mm below the surface. Differences in the probe to sample distance may lead to differences in the sample volume analyzed in calibration and production samples.

The analysis of powder mixtures by NIR spectroscopy is quite challenging because of the heterogeneity of these powder mixtures. In spite of efforts to thoroughly mix the powder mixtures, the mixtures are never homogeneous and a sampling error is always present [9]. There is always the possibility, especially with cohesive powders, of having a drug or excipient aggregate that is not well mixed and provides a spectrum that is not truly representative of the powder mixture. The best approach is to obtain multiple spectra of each powder blend and in one study, the authors took 100 spectra for each concentration in the calibration set [49]. These spectra could be averaged; however, it is best to visually inspect the spectra and evaluate whether any spectra obtained can be considered outliers that are not representative of the desired concentration. The use of first or second derivative spectra to remove baseline differences is usually very helpful in signaling possible outliers as shown in Fig. 4. The different (outlier) spectrum could have a

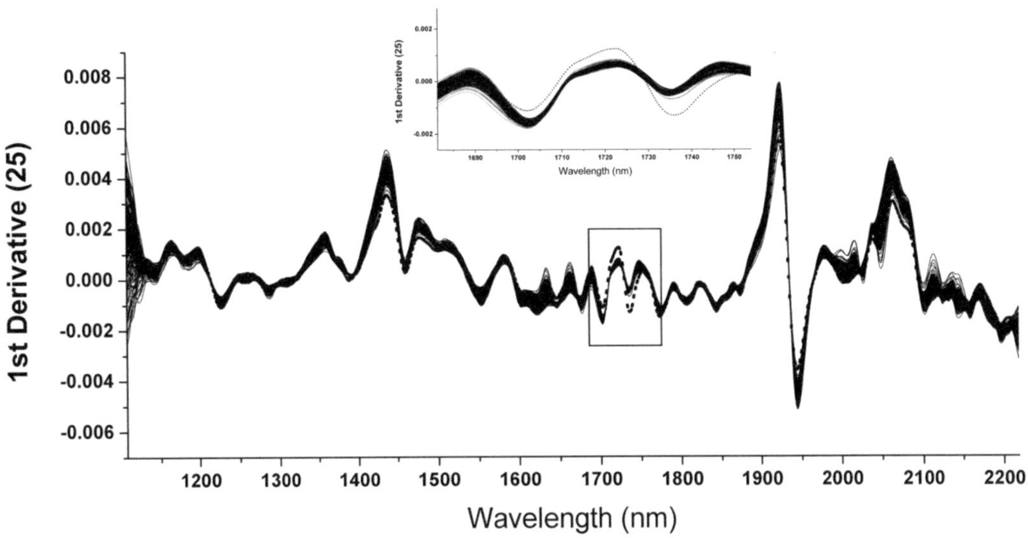

Fig. 4 Calibration spectra with more noise towards the lower wavelength detector cutoff, and a possible outlier (see *inset*)

greater similarity to the spectrum of an excipient or API, than those to other calibration samples indicating that it is the result of an agglomerate. Out of space (OS, orthogonal distance from sample to PCA plane) and in space (IS, distance from PCA data mean (*M*) to projection of the sample) are useful statistics that assess general abnormality of a sample, relative to another set of samples. A useful outlier alarm must use both OS and IS, when Hotelling T2 and Q-residuals exceed the predefined limits [50]. The multiple spectra obtained may also be used to develop a criterion of blend uniformity. The standard deviation of NIR spectra may be evaluated and compared to the standard deviation of samples analyzed by UV or HPLC methods [47, 51]. Visual inspection of the spectra obtained will often reveal higher noise at the higher and lower wavelengths end of the spectrum near the detector cutoff (*see* Fig. 4) that should not be included in the future calibration model.

Visual inspection of spectra should be followed with Principal Component Analysis (PCA) to evaluate whether the calibration set encompasses the variation observed in production samples. A PCA model should be developed with the spectra included in the calibration model. The scientific literature includes examples where the score plot of the first two principal components of the calibration samples encompasses the production samples [52]. However, there are also cases where it was necessary to include production samples in the calibration set. As this chapter progresses, the high value of this PCA model will become more evident. The use of a qualitative method such as PCA for developing a quantitative method might seem a paradox but it is an essential part of the calibration process.

2.3 Calculation of Calibration Model and Evaluation

The next steps are related to the calculation of the PLS regression that will be used to predict future unknown samples. The first PLS model may be performed using a wide spectral area that includes the spectral bands of the analyte of interest. The regression vector obtained should be examined to identify the bands that are highly correlated with the analyte of interest. Some bands will be directly related to the analyte concentration while others will be inversely related. The correlation plot is always very useful, but the selection of only spectral areas with high correlation will not always provide accurate prediction of the unknown samples. The authors' experience is that calibration models developed using several selected spectral areas and spectral pretreatment [29, 41, 52–54] should be summarized in a table to keep track of the calibrations performed.

Methods have been published where the entire spectral range is used [34] in the calibration model while others use a narrower spectral range [44]. The use of a wider spectral range may be considered advantageous in that many more spectral responses are used in the calibration model [55]. The differences in the spectral range used may be related to the scattering of powders and tablets.

In liquids the authors have been able to perform calibration models with very small differences in the errors observed across different spectral regions. However, in pharmaceutical powder mixtures the sample volume analyzed at short wavelengths is greater since the absorption of NIR light is lower and there is greater penetration of the NIR radiation. Higher wavelengths show greater absorption and the NIR radiation is absorbed in the first few micrometers of the particles. Absorbance at higher wavelengths is characteristic of the surface of the particles analyzed while at lower wavelengths a larger sample volume is analyzed.

The analysis of pharmaceutical samples is challenging because these samples are composed of particles with different morphologies and particle size distribution. Pharmaceutical solid oral dosage form formulations do not obey the Lambert–Beer law since this law is defined for non-scattering samples [56]. The radiation remitted (sent back to the detector) increases as the surface area increases. Various studies have shown that spectra of smaller particles show weaker absorbance bands while larger particles with lower surface area remit less radiation and show larger absorbance bands [8]. If a PLS calibration model for analyte concentration was developed with these spectra the spectral changes due to particle size differences would be related to the analyte concentration providing erroneous results. This is the reason why spectral preprocessing such as first and second derivative transforms, multiplicative scattering correction and the standard normal variate (SNV) method are used. The study of the physics of diffuse reflection provides an understanding of the reasons why spectral preprocessing is necessary for method development especially for the highly scattering solid oral dosage forms. However, it is not the only consideration in the development of a PLS method and for this reason a table summarizing the different conditions evaluated to develop the calibration model is always useful [41, 44].

Spectral pretreatment should be performed taking into consideration the objectives of the method. Methods developed to determine the API or moisture concentration need to remove the differences in baseline and the spectral changes related to differences in particle size. In applications that seek to determine the physical properties of materials it is best to leave the effects of surface area and particle size differences in the spectra. This approach was followed in a study on the effect of voiding on particle segregation where baseline correction was used to remove baseline differences [8] but the particle size effects on the spectrum were not removed. Baseline correction was also used to study the effect of compaction on tablet spectra and the subsequent tablet relaxation [57]. A calibration model for prediction of envelope density in roller compacted ribbons was also developed using baseline correction. Method development requires keeping track of the preprocessing methods used by developing a table summarizing the results obtained [41]. This table will be very useful for the selection of the final calibration model.

NIR method development involves the evaluation of a number of calibration models in search of an optimum calibration model. The evaluation of the different calibration models is somewhat like the evaluation of different mobile phases in HPLC method development. Previous knowledge of the concentration of the formulation components and their physical properties is valuable. However, this previous knowledge is not an exemption from developing a number of calibration models and the need to evaluate their predictive ability. The simplest approach is to first develop a calibration model that predicts the same samples used to develop the model. The results of this first evaluation are described by the standard error of calibration where k is the number of PLS factors

$$\mathrm{SEC} = \sqrt{\frac{\sum (\mathrm{Y}_{\mathrm{ref}} - \mathrm{Y}_{\mathrm{pred}})^2}{N - 1 - k}} \qquad (1)$$

This evaluation will give over-optimistic results since it only estimates the model error and not the prediction error. However, if the results are not satisfactory then it is good indication that method parameters should be re-evaluated. A low SEC value will be associated with a high correlation coefficient between the predicted values and the reference values. In this preliminary phase of method development, the authors consider a 5 % relative error as a good accuracy performance for pharmaceutical applications. This first approach is simply doing the simplest evaluation first as an effort to follow an organized path in method development.

Cross-validation is also used frequently in the early stages of method development. Leave one out cross-validation is the simplest form of cross-validation, where all except one of the samples is used to develop the regression equation. The sample left out is predicted by the model calculated with the rest of the calibration samples. The process is performed until all the calibration samples are left out once. Cross-validation is available in all commercial software packages for developing PLS calibration models and also permits an evaluation of the number of PLS factors and the error in the prediction. Thus, in cross-validation the calibration samples are also used to challenge the method providing a preliminary evaluation of its effectiveness in predicting unknown samples. Cross-validation is especially useful when the reference method has a high cost or a limited number of calibration samples are available. Cross-validation should not be the final step in method validation but is valuable in method development and reduces costs by providing a way in which a sample can be used for method development and also to assess the model's predictive ability.

There are many ways to perform cross-validation. Leave one out cross-validation is simple but could also be over optimistic in cases where samples similar to those left out remain in the calibration set. For example, in a calibration model with five spectra at five different concentrations, the removal of one sample for prediction would still leave four similar spectra in the calibration model. This leave one out cross-validation would be very similar to the prediction of the same samples in the calibration set (self leverage). A more realistic and challenging approach is to assign the samples to classes or groups, and perform leave class out cross-validation. Each concentration level could be a different class, and in this way the calibration will set will not include samples similar to those left out and predicted [44]. There are other ways to develop groups or segments that are left out in cross-validations [58, 59] including *v*-fold and approaches to estimate the number of PLS factors based on cross-validation. These approaches provide valuable insight in the early stages of development when the number of samples is limited and are not sufficient to create an independent test set.

Another approach is to develop a calibration model based on laboratory prepared samples and samples from the production process [54]. In this manner the laboratory prepared samples increase the concentration range while the production samples add the variability from the manufacturing process. The calibration model could then be modified to include those samples exhibiting the greatest variability in the first two principal components. A number of samples within the span the PCA score plot could be removed and these would be called a prediction set. The calibration model would then be evaluated based on the results obtained for the prediction set. If the prediction is satisfactory then the calibration model would be used to predict additional samples from the production process. The calibration model could also be improved by adding some of the samples from the prediction set to the calibration model.

2.4 Fine Tuning of Calibration Model

This fourth part includes efforts to fine tune or improve the calibration model. This requires prediction of samples from a completely independent validation set. Sample independency means that samples are not prepared under the same conditions as the calibration set samples. Validation samples should come from the process that will be monitored and be prepared with excipient and API batches that differ from those used in the calibration set. NIR spectra should be obtained under actual conditions of use where the instrument is deployed.

The evaluation of bias in the validation samples is essential. Bias is defined as the average residual between the predicted and reference values [60]. Bias should be close to zero (indicating absence of a systematic error), although there is always some error and it is practically impossible to obtain a value of zero. Bias should be evaluated versus the number of PLS factors used to predict the independent samples. The number of PLS factors that predicts the independent samples with the lowest bias may be different than the number of factors that provided the best cross-validation results. Validation samples provide the definitive challenge of the calibration model.

This stage should also include verification and optimization of a method to identify the samples that will be predicted. NIR methods can be developed and validated to first determine whether unknown samples are suitable for prediction (they are similar to those in the calibration set). This step is important for prediction and is a valuable tool to detect process failures and reduces risk in pharmaceutical manufacturing. Thus, the inclusion of the wrong excipient or API in the formulation can be easily detected by the identification method. Identification can be performed through the correlation coefficient or it could be done through PCA and outlier detection [52, 61]. The identification test is referred to as the Statistical Spectral Quality test in the EMEA guidelines [62].

2.5 Method Implementation

The deliverable from the method development efforts is a regression equation that predicts concentration or other samples properties from NIR spectra. This regression equation will eventually reside within the process control system. There are several commercially available software for real time and laboratory prediction of unknown samples. The software may be used to first identify the samples and make the desired prediction. These software use the regression vector from the calibration models to predict unknown samples, and can also use a PCA model to identify samples. The NIR results may be communicated through OPC (Open Platforms Communication) to process control devices. There are commercially available software that make predictions with the PAT method and also collect data from [63–65] process sensors.

2.5.1 Process Spectrum Approach

As already discussed the success of the PLS method depends on the acquisition of spectra for a suitable calibration that includes the variation of future validation or process samples. The process spectrum is an approach towards including this future variation in a simple efficient manner. The process spectrum approach takes into consideration that during a pharmaceutical manufacturing process the chemical composition is often practically constant and the main changes are in the physical properties of samples [66–68]. The components in the formulation may be mixed or wet granulated increasing the particle size and changing the moisture content. The differences between the NIR spectra obtained at the beginning and end of the process will likely reflect minor concentration changes. This spectral difference may be called the process spectrum as shown in Fig. 5:

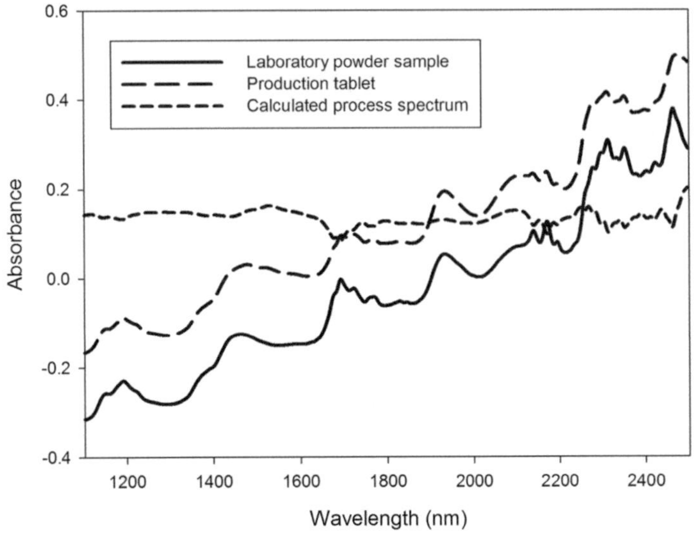

Fig. 5 Process spectrum and spectra required for its calculation

$$\text{Process spectrum} = \text{Abs(end of process)} - \text{Abs(beginning of process)} \quad (2)$$

The "process spectrum" will depend on the changes that the mixture has undergone while being processed. This variability is mainly due to the physical changes that the materials undergo but may also include chemical variation such as the changes of moisture content upon wet granulation. After tablet coating, the chemical composition of the tablets have been slightly modified, as the total weight has been increased by the amount of coating.

The main objective of a NIR calibration model is its use to monitor a pharmaceutical process, with adequate prediction capability. The achievement of this objective is only possible if the model includes all possible sources of variability as already discussed in this chapter. Calibration samples are prepared to include a range of concentrations. However, the main challenge is the inclusion of variability associated with the manufacturing process. The process spectrum addresses this main challenge.

The process spectrum may be developed by preparing laboratory samples that include the variation in the expected concentrations or parameter of interest. The calibration samples may be prepared following the gravimetric method, without the use of HPLC and UV–Vis and the sample preparation errors associated with these techniques. NIR spectra are then obtained at the same concentrations that are expected in the manufacturing process. The process spectrum is then added to the calibration samples. The resulting spectra include the variability from the sample mixed and the variability from the pharmaceutical process.

The process spectrum approach requires that the subtracted spectra correspond to two spectra that correspond to the same API and excipient concentrations. If this requirement is not met, then the process spectrum will include variation that will affect the method's accuracy. The number of process spectra required to perform this process successfully cannot be only one, since the risk of introducing bias into the model would be very high. Thus, a few process spectra are necessary (3–5 spectra are usually sufficient). The process spectra added to the lab samples may be multiplied by a factor *m* (Eq. 3) to obtain the extended spectrum. The factor *m* usually varies from 0.5 to 1.5, according to the predictive capacity of the calibration model and the PCA score plot obtained from the process spectra projected into the calibration samples.

$$\text{Extended spectrum} = \text{Laboratory spectrum} + \text{process spectrum} \times m(m = 0.5 - 1.5) \quad (3)$$

Figure 6 shows the calculation of the process spectra set for a case where the API and excipients are mixed and then compacted to obtain tablets. The laboratory reference sample and the production tablets present the exact chemical composition. It could be a challenging task to assure exactly the same concentration between the laboratory and industrial samples. The basic recommendation is to quantify the API concentration in the industrial samples using the reference method (HPLC, UV) and then preparing the laboratory samples using the analytical balance (max uncertainty 0.1 mg). The number *n* is equal to the number of paired spectra industrial-laboratory samples used for the calculation of process spectrum. The "process variability matrix" is obtained as the result of this first phase describing the process variability between the samples chosen from production and the samples prepared in the lab. Each spectrum in the "process variability matrix" is multiplied by the factor *m*, providing as a result an "extended process variability matrix". The final step is to add the "extended process variability matrix" to the lab prepared calibration spectra. The final matrix is the extended total variability matrix, that includes the calibration and prediction sets for the calculations of the multivariate calibration model.

Figure 7a shows a PCA scores plot of two sets of production tablets (doses of 400 and 600 mg) in the space defined by the lab prepared tablets with an extended variability matrix ($m = 0.5$, 0.75, 1, 1.25). This plot demonstrates that the spectral variability of the laboratory samples after the addition of the process spectra embraces the variability of the production tablets spectra as the production spectra are now included within the score plot of the laboratory spectra. Figure 7b shows the score plots of the lab prepared tablets and the production tablets before the process spectrum approach was used. The calibration model is calculated using the spectra laboratory prepared tablets that include the addition of the extended process variability matrix. The reference values

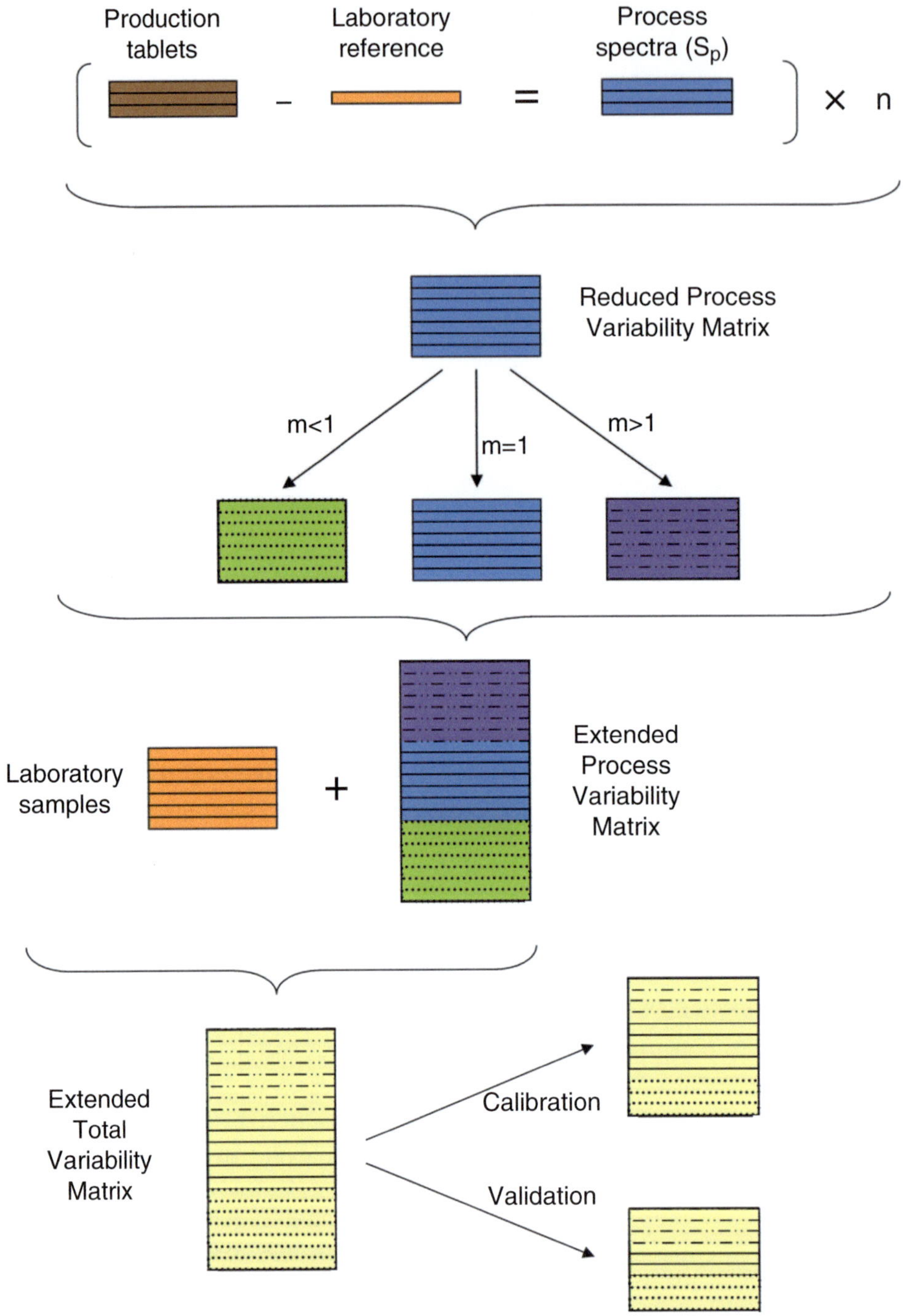

Fig. 6 Proposed process spectrum approach for building calibration set

of the calibration samples are based on gravimetric values. The resulting PLS models are simpler (require a lower number of PLS factors) and provide a greater predictive capability with less error for the tablets from the production process.

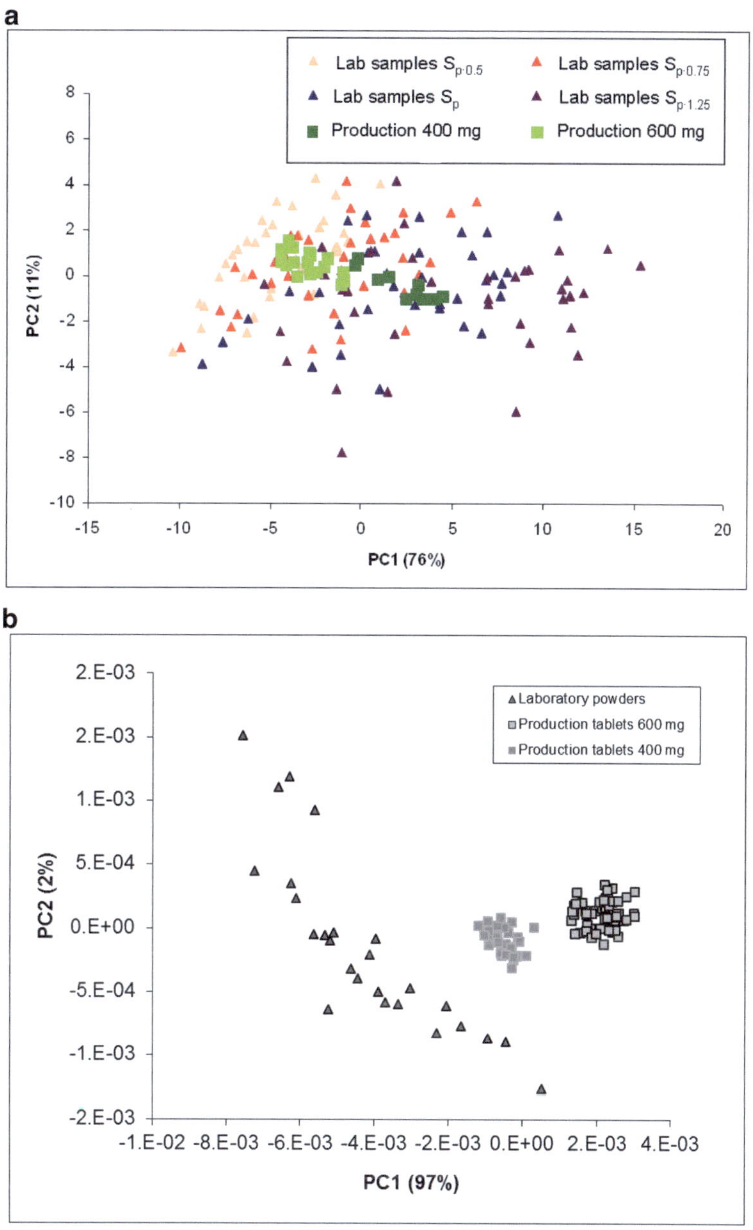

Fig. 7 (**a**) PCA scores plot for laboratory samples modified with the extended process-variability matrix and production samples. Spectral treatment: first derivative + SNV over the wavelength range 1100–2500 nm. (**b**) PCA scores plot for laboratory powder samples and production tablets of 600 and 400 mg. Spectral treatment: first derivative + SNV over the wavelength range 1100–2500 nm

2.5.2 Iterative Optimization Technology (IOT)

The IOT approach is based on the excess of absorbance concept introduced in hydrogen bonding studies of alcohol binary mixtures [69] defined as the difference between the spectra of a mixture and the individual components of the same mixture under identical conditions. The excess of absorbance is different to a simple spectral subtraction because linear features are eliminated and new features emerge. The excess absorbance concept was later associated to the composition of pharmaceutical blends [70]. The minimum in the excess of absorbance was proposed as an estimate of the composition of powder mixtures (*see* Eq. 4).

$$\min r_i = \sum_{n=1}^{N} \left(X_{\text{mix}} - \sum_{i=1}^{i=I} r_i X_{\text{pure},i} \right)^2 \tag{4}$$

X_{mix} = Spectrum of the blend to be predicted
X_{pure} = Spectra of pure components
I = Total number of pure components
r_{i} = Molar fraction of pure components in the blend
N = number of wavelengths

The minimum value of r_{i} that minimizes the function is obtained by an iterative process with constrains to obtain the composition value with sum equal to 1. This linear approach does not consider interaction between the particles. The linear approach has the limitation that pure component spectra are acquired under different conditions than process samples and variation from process strain is not included.

The nonlinear approach proposed [70] consists in a previous transformation by using Box-Cox and perform prediction as described by the linear approach.

$$\min \lambda_n = \left(X_{\text{mixtrain}} - f_n \left(\sum_{i=1}^{i=I} r_{\text{train}} \, X_{\text{pure},i} \right) \right)^2 \tag{5}$$

X_{mixtrain} corresponds to a limited training set and the number of samples required depends on the complexity of the nonlinear characteristics. After the Box-Cox parameter λ_n is identified, the calibration set is not necessary and the mixture composition is calculated iteratively as the linear approach. The advantage of the nonlinear approach is that it facilitates the inclusion in the model of information on molecular interactions such as hydrogen bonding.

3 Conclusion

A significant number of calibration methods have been developed based on PLS for pharmaceutical samples. These methods have overcome the narrow concentration range found in pharmaceutical samples. Several approaches have been developed to increase the concentration range of pharmaceutical samples and develop suitable calibrations. These previous efforts are the basis of the procedure proposed in this chapter. However, there are many more opportunities for improving the development of calibration models for pharmaceutical applications, such as the process spectrum and the Iterative Optimization Technology.

The procedure also shows significant progress in overcoming the challenges of working with samples that differ in particle size and where spectra are often obtained for samples that are flowing or suspended in air as in a fluid bed granulator. Powder mixtures provide significant challenges where in spite of efforts to thoroughly mix the calibration powder mixtures, the mixtures are never homogeneous and a sampling error is always present. NIR methods must also overcome the fact that spectra are the result of the chemical composition of samples and also their physical properties. The study of the physics of diffuse reflection has helped in understanding why spectral pretreatment is necessary for method development especially for the highly scattering solid oral forms.

The knowledge summarized in this procedure and the studies cited in this chapter may be used for a modern quality assurance based on NIR measurements. This modern quality assurance would be based on real time measurements of processes and will not require taking samples to an analytical laboratory removed from the production environment. NIR measurements also take considerably less time than conventional laboratory methods. Thus, drug concentration in tablets could be estimated on the basis of 300 or more units in less time that it typically takes to analyze 30 tablets by HPLC. The advances discussed in this chapter show that NIR methods may be valuable elements within modern pharmaceutical quality assurance.

Acknowledgements

This work was performed thanks to the support of the NSF ERC Structured Organic Particulate Systems EEC-0540855 grant. The authors thank two industrial advisors in this grant: Dongsheng Bu from BMS and Pius Tse from Merck for carefully reviewing the manuscript. Graduate student Krizia Karry and Yusuf Suluf (SABIC Innovative Plastics) are also thanked for their reviews.

References

1. Blanco M, Romero MA (2001) Near-infrared libraries in the pharmaceutical industry: a solution for identity confirmation. Analyst 126 (12):2212–2217
2. Green RL et al (2005) In-line monitoring of moisture content in fluid bed dryers using near-ir spectroscopy with consideration of sampling effects on method accuracy. Anal Chem 77(14):4515–4522
3. Alcala M et al (2010) On-line monitoring of a granulation process by NIR spectroscopy. J Pharm Sci 99(1):336–345
4. Burgbacher J, Wiss J (2008) Industrial applications of online monitoring of drying processes

of drug substances using NIR. Org Process Res Dev 12(2):235–242
5. Sekulic SS et al (1996) On-line monitoring of powder blend homogeneity by near-infrared spectroscopy. Anal Chem 68(3):509–513
6. Shi Z et al (2008) Process characterization of powder blending by near-infrared spectroscopy: blend end-points and beyond. J Pharm Biomed Anal 47(4–5):738–745
7. Dziki W et al (2000) The use of near-infrared spectroscopy to monitor the mobility of water within the sarafloxacin crystal lattice. J Pharm Biomed Anal 22(5):829–848
8. Barajas MJ et al (2007) Near-infrared spectroscopic method for real-time monitoring of pharmaceutical powders during voiding. Appl Spectrosc 61(5):490–496
9. Esbensen KH, Paasch-Mortensen P (2010) Process sampling: theory of sampling – the Missing Link in Process Analytical Technologies (PAT). In: Process analytical technology. John Wiley & Sons, New York, NY, pp 37–80
10. Petersen L, Minkkinen P, Esbensen KH (2005) Representative sampling for reliable data analysis: theory of sampling. Chemom Intell Lab Syst 77(1–2):261–277
11. Petersen L, Esbensen KH (2005) Representative process sampling for reliable data analysis—a tutorial. J Chemometr 19(11-12): 625–647
12. Zhou X, Hines P, Borer MW (1998) Moisture determination in hygroscopic drug substances by near infrared spectroscopy. J Pharm Biomed Anal 17(2):219–225
13. Williams PC, Norris KH (2001) Near-infrared technology: in the agricultural and food industries. American Association of Cereal Chemists, St. Paul, MN, pp 19–37
14. Christy AA (2010) New insights into the surface functionalities and adsorption evolution of water molecules on silica gel surface: a study by second derivative near infrared spectroscopy. Vib Spectrosc 54(1):42–49
15. Weyer LG, Lo SC (2006) Spectra–structure correlations in the near-infrared. In: Chalmers JM, Griffiths PR (eds) Handbook of vibrational spectroscopy. John Wiley & Sons, Chichester
16. Alcalà M et al (2006) Near-infrared spectroscopy in laboratory and process analysis. In: Encyclopedia of analytical chemistry. John Wiley & Sons, New York, NY
17. Dahm DJ (2013) Review: explaining some light scattering properties of milk using representative layer theory. J Near Infrared Spectrosc 21(5):322–339
18. Dahm KD, Dahm DJ (2013) Separating the effects of scatter and absorption using the representative layer. J Near Infrared Spectrosc 21(5):351–357
19. Muzzio FJ et al (2003) Sampling and characterization of pharmaceutical powders and granular blends. Int J Pharm 250(1):51–64
20. Davies AMC, Miller CE (1988) Tentative assignment of the 1440-nm absorption band in the near-infrared spectrum of crystalline sucrose. Appl Spectrosc 42(4):703–704
21. Pasikatan MS, Steele JL, Spillmand CK, Haquee E (2001) Near Infrared Reflectance Spectroscopy for online particle size analysis of powders and ground material. J Near Infrared Spectrosc 9:153–164
22. Blanco M, Villar A (2003) Development and validation of a method for the polymorphic analysis of pharmaceutical preparations using near infrared spectroscopy. J Pharm Sci 92(4): 823–830
23. EMEA (2012) Guideline on the use of Near Infrared Spectroscopy (NIRS) by the pharmaceutical industry and the data requirements for new submissions and variations. EMEA, London
24. Broad N (2002) Guidelines for the development and validation of near-infrared spectroscopic methods in the pharmaceutical industry. In: Chalmers JM, Griffiths PR (eds) Handbook of vibrational spectroscopy. John Wiley & Sons, Chichester
25. ASTM (2002) Standard practice for maintaining constant relative humidity by means of aqueous solutions. ASTM, West Conshohocken, PA
26. Beebe KR, Kowalski BR (1987) An introduction to multivariate calibration and analysis. Anal Chem 59(17):1007A–1017A
27. Martens H, Naes T (1989) Multivariate calibration. Wiley, New York, NY
28. Kramer R (1998) Chemometric techniques for quantitative analysis. Taylor & Francis, Boca Raton, FL
29. Broad NW et al (2001) Application of transmission near-infrared spectroscopy to uniformity of content testing of intact steroid tablets. Analyst 126(12):2207–2211
30. Mateo-Ortiz D et al (2014) Analysis of powder phenomena inside a Fette 3090 feed frame using in-line NIR spectroscopy. J Pharm Biomed Anal 100:40–49
31. Xiang D et al (2009) Robust calibration design in the pharmaceutical quantitative measurements with near-infrared (NIR) spectroscopy: avoiding the chemometric pitfalls. J Pharm Sci 98(3):1155–1166
32. Bondi RW Jr et al (2012) Effect of experimental design on the prediction performance of

calibration models based on near-infrared spectroscopy for pharmaceutical applications. Appl Spectrosc 66(12):1442–1453

33. Sulub Y et al (2009) Real-time on-line blend uniformity monitoring using near-infrared reflectance spectrometry: a noninvasive off-line calibration approach. J Pharm Biomed Anal 49(1):48–54
34. Blanco M, Bautista M, Alcala M (2008) Preparing calibration sets for use in pharmaceutical analysis by NIR spectroscopy. J Pharm Sci 97 (3):1236–1245
35. Williams PC, Norris KH (2001) Near-infrared technology: in the agricultural and food industries. American Association of Cereal Chemists, Paul, MN, pp 145–169
36. Blanco M et al (2001) Influence of the procedure used to prepare the calibration sample set on the performance of near infrared spectroscopy in quantitative pharmaceutical analyses. Analyst 126(7):1129–1134
37. Blanco M, Romero MA, Alcala M (2004) Strategies for constructing the calibration set for a near infrared spectroscopic quantitation method. Talanta 64(3):597–602
38. Xiang D et al (2009) Evaluation of transmission and reflection modalities for measuring content uniformity of pharmaceutical tablets with near-infrared spectroscopy. Appl Spectrosc 63(1):33–47
39. Darvesh AS, Saraf MN, Menon M (1999) Toxicological evaluation of film forming methacrylic acid copolymers. Ind J Pharm Sci 61(3):179–183
40. Ritchie GE et al (2002) Validation of a near-infrared transmission spectroscopic procedure. Part B: Application to alternate content uniformity and release assay methods for pharmaceutical solid dosage forms. J Pharm Biomed Anal 29(1-2):159–171
41. Meza CP, Santos MA, Romanach RJ (2006) Quantitation of drug content in a low dosage formulation by transmission near infrared spectroscopy. AAPS PharmSciTech 7(1):E29
42. Griffiths PR, De Haseth JA (1986) Fourier transform infrared spectrometry. Wiley-Interscience, New York, NY
43. Blanco M et al (2006) A process analytical technology approach based on near infrared spectroscopy: tablet hardness, content uniformity, and dissolution test measurements of intact tablets. J Pharm Sci 95(10):2137–2144
44. Vanarase AU et al (2010) Real-time monitoring of drug concentration in a continuous powder mixing process using NIR spectroscopy. Chem Eng Sci 65(21):5728–5733
45. Alcala M et al (2008) Analysis of low content drug tablets by transmission near infrared spectroscopy: selection of calibration ranges according to multivariate detection and quantitation limits of PLS models. J Pharm Sci 97 (12):5318–5327
46. Iyer M, Morris H, Drennen J III (2002) Solid dosage form analysis by near infrared spectroscopy: comparison of reflectance and transmittance measurements including the determination of effective sample mass. J Near Infrared Spectrosc 10(4):233–245
47. Popo M et al (2002) Blend uniformity analysis using stream sampling and near infrared spectroscopy. AAPS PharmSciTech 3(3):E24
48. Bellamy LJ, Nordon A, Littlejohn D (2008) Real-time monitoring of powder mixing in a convective blender using non-invasive reflectance NIR spectrometry. Analyst 133(1): 58–64
49. Berntsson O et al (2002) Quantitative in-line monitoring of powder blending by near infrared reflection spectroscopy. Powder Technol 123(2–3):185–193
50. Miller CE (2010) Chemometrics in process analytical technology (PAT). In: Process analytical technology. John Wiley & Sons, New York, NY, pp 353–438
51. Momose W et al (2011) Process analytical technology applied for end-point detection of pharmaceutical blending by combining two calibration-free methods: simultaneously monitoring specific near-infrared peak intensity and moving block standard deviation. Powder Technol 210(2):122–131
52. Blanco M, Alcala M, Bautista M (2008) Pharmaceutical gel analysis by NIR spectroscopy. Determination of the active principle and low concentration of preservatives. Eur J Pharm Sci 33(4-5):409–414
53. Blanco M et al (1997) Effect of data preprocessing methods in near-infrared diffuse reflectance spectroscopy for the determination of the active compound in a pharmaceutical preparation. Appl Spectrosc 51(2):240–246
54. Blanco M et al (1999) Development and validation of a method for the analysis of a pharmaceutical preparation by near-infrared diffuse reflectance spectroscopy. J Pharm Sci 88 (5):551–556
55. Duckworth JH (1998) 4 - Spectroscopic quantitative analysis. In: Workman J, Springsteen AW (eds) Applied spectroscopy. Academic, San Diego, CA, pp 93–163
56. Dahm DJ, Dahm KD (2007) Interpreting diffuse reflectance and transmittance: a theoretical

introduction to absorption spectroscopy of scattering materials. NIR, Chichester

57. Ropero J et al (2011) Near-infrared chemical imaging slope as a new method to study tablet compaction and tablet relaxation. Appl Spectrosc 65(4):459–465
58. Burman P (1989) A comparative study of ordinary cross-validation, v-fold cross-validation and the repeated learning-testing methods. Biometrika 76(3):503–514
59. Wold S, Sjöström M, Eriksson L (2001) PLS-regression: a basic tool of chemometrics. Chemom Intell Lab Syst 58(2):109–130
60. Mark H (1991) Principles and practice of spectroscopic calibration. Wiley, New York, NY
61. Bu D, Wan B, McGeorge G (2013) A discussion on the use of prediction uncertainty estimation of NIR data in partial least squares for quantitative pharmaceutical tablet assay methods. Chemom Intell Lab Syst 120: 84–91
62. U.S. Department of Health and Human Services, Food and Drug Administration (2012) Guideline on the use of Near Infrared Spectroscopy (NIRS) by the pharmaceutical industry and the data requirements for new submissions and variations. E.C.C.Q, London, Rev2
63. Markl D et al (2013) Supervisory control system for monitoring a pharmaceutical hot melt extrusion process. AAPS PharmSciTech 14(3): 1034–1044
64. Karande AD, Heng PW, Liew CV (2010) In-line quantification of micronized drug and excipients in tablets by near infrared (NIR) spectroscopy: real time monitoring of tabletting process. Int J Pharm 396(1-2):63–74
65. Liew CV, Karande AD, Heng PWS (2010) In-line quantification of drug and excipients in cohesive powder blends by near infrared spectroscopy. Int J Pharm 386(1–2):138–148
66. Blanco M, Peguero A (2010) Analysis of pharmaceuticals by NIR spectroscopy without a reference method. TrAC Trends Anal Chem 29(10):1127–1136
67. Blanco M, Peguero A (2010) Influence of physical factors on the accuracy of calibration models for NIR spectroscopy. J Pharm Biomed Anal 52(1):59–65
68. Blanco M, Cueva-Mestanza R, Peguero A (2010) Controlling individual steps in the production process of paracetamol tablets by use of NIR spectroscopy. J Pharm Biomed Anal 51 (4):797–804
69. Li Q et al (2008) Excess infrared absorption spectroscopy and its applications in the studies of hydrogen bonds in alcohol-containing binary mixtures. Appl Spectrosc 62(2): 166–170
70. Muteki K et al (2013) Mixture component prediction using iterative optimization technology (calibration-free/minimum approach). Ind Eng Chem Res 52(35):12258–12268

Chapter 6

Advanced Process Decision Making Using Multivariate Latent Variable Methods

Matteo Ottavian, Emanuele Tomba, and Massimiliano Barolo

Abstract

This chapter is intended to show how latent variable modeling techniques can be used to support several pharmaceutical development and manufacturing activities by exploitation of historical databases deriving from experiments, ongoing manufacturing processes or historical products already developed. Basic theoretical concepts about latent variable modeling and latent variable model inversion are first introduced. Then, some applications are reviewed to show how the pharmaceutical industry can benefit from these modeling techniques to support decision-making activities in process development, formulation design, process scale-up, product transfer, process control, and raw materials acceptability assessment.

Key words Latent variable models, Product design, Process understanding, Quality by design, Design space, Process analytical technologies, Principal component analysis, Partial least-squares regression

Notation

A	Number of latent variables selected to build an LV model
b_j	Inequality constraint assigned to the j-th element of $\hat{\mathbf{y}}^{\mathrm{NEW}}$
d_t	Inequality constraint on the t-th element of $\mathbf{x}^{\mathrm{NEW}}$
$\mathbf{E}$	Residual matrix
$\mathbf{e}$	Residual vector
$g_{1,2,3}$	Weights for model inversion
lb^x	Lower physical bound of the domain of $\mathbf{x}^{\mathrm{NEW}}$
lb^y	Lower physical bound of the domain of $\hat{\mathbf{y}}^{\mathrm{NEW}}$
$\mathbf{P}$	Loading matrix for the $\mathbf{X}$ space
$\mathbf{p}$	Loading vector
$\mathbf{Q}$	Loading matrix for the $\mathbf{Y}$ space
R^2	Explained variance by an LV model
$R_{\mathbf{X}}$, $R_{\mathbf{Y}}$	Rank of matrices $\mathbf{X}$ and $\mathbf{Y}$
s_a^2	Variance of the a-th column of $\mathbf{T}$
SPE_i	Squared prediction error of the i-th sample
$\mathbf{T}$	Score matrix for $\mathbf{X}$

Marianthi G. Ierapetritou and Rohit Ramachandran (eds.), *Process Simulation and Data Modeling in Solid Oral Drug Development and Manufacture*, Methods in Pharmacology and Toxicology,
DOI 10.1007/978-1-4939-2996-2_6, © Springer Science+Business Media New York 2016

T^2	Hotelling's T^2
$\mathbf{t}$	Score vector
$\mathbf{t}^{DES}$	Score vector of the solution of model inversion exercise
$\mathbf{t}^{REAL}$	Score vector of the real input variable projections onto the score space
t_a	a-th element of $\mathbf{t}$
ub^x	Upper physical bound of the domain of $\mathbf{x}^{NEW}$
ub^y	Upper physical bound of the domain of $\hat{\mathbf{y}}^{NEW}$
$\mathbf{X}$	Matrix that includes historical samples
$\hat{\mathbf{X}}$	Reconstruction of $\mathbf{X}$ when A components are retained
$\mathbf{x}^{NEW}$	Solution of the optimization problem
$\hat{\mathbf{x}}^{NEW}$	Solution of the model inversion problem
$\mathbf{Y}$	Matrix collecting the quality variables
$\mathbf{y}^{DES}$	Generic set of desired product properties
$\hat{\mathbf{y}}^{NEW}$	Quality attributes corresponding to the solution $\mathbf{x}^{NEW}$
$\mathbf{W}$	Weight matrix of the model for $\mathbf{X}$
$\mathbf{W}^*$	Transformed weight matrix of the model for $\mathbf{X}$

1 Introduction

Pharmaceutical development activities have traditionally relied on extensive experimental campaigns, mainly as a consequence of the challenges the pharmaceutical industry has to face, such as the variety of production paths, the low-volume and multi-product batch productions, product complexity, and, above all, a peculiar regulatory environment [1,2]. These challenges often hinder the possibility of using mechanistic models to reveal the relations between input variables (e.g., raw materials properties, process parameters) and product quality, which represent the founding paradigm of the Quality by Design and Process Analytical Technology initiatives. However, pharmaceutical environments are usually characterized by the availability of production and research databases deriving from experiments, ongoing manufacturing processes or historical products already developed. Latent variable models (LVMs) are powerful tools to extract the information embedded in existing databases. Coupling this information to engineering knowledge about the process represents a formidable tool to advance the knowledge on the underlying physical and chemical phenomena occurring in the process, thus significantly contributing to improve the overall process understanding.

This chapter is intended to show how LVMs can be effectively used to support the decision-making process related to several pharmaceutical manufacturing and development activities. The chapter is organized as follows. Basic theoretical concepts about latent variable modeling are first introduced with respect to standard modeling approaches (principal component analysis, partial least-squares regression) as well as advanced ones (multi block, joint-Y, and L-shape partial least-squares regression). Then, an in-depth analysis of latent variable model inversion techniques is presented, with emphasis on the concept of null space. Finally, some recent applications of latent variable model inversion are reviewed to show how the pharmaceutical industry can benefit from this advanced technique to support decision-making activities in process design, formulation design, process scale-up, product transfer, process control and raw materials acceptability assessment.

2 Latent Variable Modeling Approaches

Latent variable models (LVMs) are statistical models that are intended to analyze large amounts of (usually correlated) data. The idea behind LVMs is that a system can be described by using few underlying factors (called *latent variables*, LVs), which can be expressed as linear combinations of the measurements taken on the system, and can be interpreted based on the knowledge of the physical phenomena involved. LVMs find the driving forces that act on the system and that are responsible for the variability observed in the data, under the assumption that the essential information does not lie in any individual variable, but in how the variables change with respect to one another, i.e., in how they co-vary. Hence, the objective of an LVM analysis is twofold: (1) data compression, and (2) data interpretation.

The available data can be organized into a matrix $\mathbf{X}$ $[I \times N]$, in which the N variables have been observed across I samples/observations (e.g., N properties for I raw materials or N process parameters for I runs). Alternatively, data can be organized in two matrices: a matrix $\mathbf{X}$ of regressors (e.g., input variables as raw material properties and/or process parameters), and a matrix $\mathbf{Y}$ $[I \times M]$ of response variables (e.g., critical to quality attributes). In the former case, the objective of an LVM analysis is to explain the correlation structure of the N variables, in order to understand the relationships among them. Principal component analysis (PCA; [3]) is one of the most useful techniques to this purpose. In the second case, the objective of an LVM analysis is to explain the cross-correlation structure of the variables in $\mathbf{X}$ and in $\mathbf{Y}$, in order to study and quantify the relationships between regressors and response variables. Projection to latent structures (PLS, also called partial least-squares regression; [4]) can be used to this purpose.

Fundamentals on PCA and PLS are reported in the following subsections. More sophisticated LVM approaches are briefly described at the end of this section.

2.1 Principal Component Analysis

Principal component analysis (PCA; [3]) is a multivariate statistical method that summarizes the information embedded in a dataset $\mathbf{X}$

$[I \times N]$ of N correlated variables, by projecting the data through a linear transformation onto a new coordinate system of latent orthogonal variables, which optimally capture the variability of the data and the correlation among the original variables. Each of these new coordinates identifies a latent direction in the original data and is called principal component (PC).

The search for the directions of the new coordinate system can be formulated as an optimization problem, whose analytical solution is given by the eigenvector decomposition of the covariance (or correlation, according to the preprocessing of the data, *see* Sect. 2.1.1) matrix of the original variables [5,6]:

$$\mathbf{X}^{\mathrm{T}}\mathbf{X}\mathbf{p} = \lambda \mathbf{p}, \tag{1}$$

where superscript^{T} indicates the transpose of a matrix, and $\mathbf{p}$ is the $[N \times 1]$ vector of the coefficients of the linear combination (called *loadings*). The loadings of a given PC represent the director cosines of the PC. Vector $\mathbf{p}$ is the eigenvector of the covariance matrix of $\mathbf{X}$ ($\mathbf{X}^{\mathrm{T}}\mathbf{X}$) corresponding to the eigenvalue λ, and λ is a measure of the variance explained by the given PC.

The eigenvector problem in Eq. (1) can be used to determine the N loadings $\mathbf{p}_n$ of the PCA model,[1] which are orthonormal (i.e., $\mathbf{p}_n^{\mathrm{T}}\mathbf{p}_r = 0$ if $n \neq r$, $\mathbf{p}_n^{\mathrm{T}}\mathbf{p}_r = 1$ if $n = r$, with n, $r = 1,\ldots, N$). The *scores* $\mathbf{t}$ $[I \times 1]$, i.e., the projections of the original data along the PC directions, are then given by:

$$\mathbf{t}_n = \mathbf{X}\mathbf{p}_n \tag{2}$$

and are orthogonal (i.e., $\mathbf{t}_n^{\mathrm{T}}\mathbf{t}_r = 0$ if $n \neq r$, with n, $r = 1,\ldots, N$).

Dataset $\mathbf{X}$ can be represented as the sum of the N scores-loadings vectors outer products:

$$\mathbf{X} = \sum_{n=1}^{N} \mathbf{t}_n \mathbf{p}_n^{\mathrm{T}}. \tag{3}$$

As a result of the eigenvector problem in Eq. (1), the PCs are ordered according to the variance of the original dataset $\mathbf{X}$ that they capture. Usually, $A \ll N$ principal components are sufficient to adequately describe $\boldsymbol{X}$, because if two or more original variables are correlated, they identify a common direction of variability that can be described by a single PC. Hence, assuming that only the first A PCs are retained, and defining the score matrix $\mathbf{T}$

[1] Note that the solution of the eigenvector problem Eq. (1) results in the first PCA loading $\mathbf{p}$. In order to evaluate the remaining components, matrix $\mathbf{X}$ has to be deflated as described later in the text.

$[I \times A] = [\mathbf{t}_1, \mathbf{t}_2, \ldots, \mathbf{t}_A]$ and the loading matrix $\mathbf{P}[N \times A] = [\mathbf{p}_1, \mathbf{p}_2, \ldots, \mathbf{p}_A]$, Eq. (3) can be rewritten as:

$$\mathbf{X} = \sum_{a=1}^{A} \mathbf{t}_a \mathbf{p}_a^{\mathrm{T}} + \sum_{a=A+1}^{N} \mathbf{t}_a \mathbf{p}_a^{\mathrm{T}} = \mathbf{TP}^{\mathrm{T}} + \mathbf{E} = \hat{\mathbf{X}} + \mathbf{E}, \qquad (4)$$

where $\mathbf{E}$ is the $[I \times N]$ matrix of the residual generated by the $(N - A)$ discarded PCs of the PCA model when $\mathbf{X}$ is reconstructed (i.e., approximated) by using only the first A PCs (i.e., $\hat{\mathbf{X}} = \mathbf{TP}^{\mathrm{T}}$). Some strategies for the selection of the number of components to retain have been reviewed by Valle et al. [7].

For the computation of the model scores and loadings, either the singular value decomposition (SVD; [8]) of the covariance matrix $\mathbf{X}^{\mathrm{T}}\mathbf{X}$ of $\mathbf{X}$, or the nonlinear iterative partial least-squares algorithm (NIPALS; [9]) can be used. The latter should be preferred as it tolerates a certain amount of missing data into the dataset $\mathbf{X}$ and does not require the calculation of all the N PCs. Details of the algorithms can be found in the above-cited references.

Figure 1 reports the geometrical interpretation of the PCA model parameters in a simplified case, i.e., a $[7 \times 2]$ dataset $\mathbf{X}$. Each available observation is represented as a dot in the plane of the original measurements x_1 and x_2.

As it can be seen, the data follow a trend in the (bidimensional) space of the original variables. If a PCA model is built, the direction of maximum variability of the data is identified by PC1. The model loadings (p_1, p_2) represent the director cosines of PC1. The scores represent the coordinates of the data samples of matrix $\mathbf{X}$ in the new reference system represented by PC1. In Fig. 1, the second principal component (PC2) is reported as a dashed line. It can be seen that PC2 is orthogonal to PC1, and accounts for a very limited variability in the data compared to PC1 (in fact, it accounts for the orthogonal distance of the projection of each measurement from the PC1 direction). In this case, it can be therefore concluded that PC1 is sufficient to adequately describe $\mathbf{X}$.

A graphical representation of the PCA model scores and loadings is often used to gain understanding on the similarity between samples (through the scores) and on the correlations among measured variables (through the loadings).

Given a PCA model, several diagnostic indices can be used to assess its performance. The most common one is the coefficient of determination R^2, which quantifies the amount of variability of the original data that the model explains:

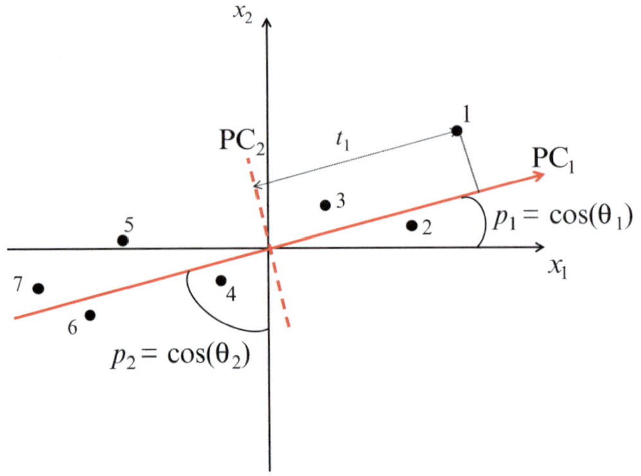

Fig. 1 Geometrical interpretation of the PCA scores and loadings for a dataset with two measured variables (x_1 and x_2) and seven samples

$$R^2 = 1 - \frac{\sum_{i=1}^{I}\sum_{n=1}^{N}\left(x_{i,n} - \hat{x}_{i,n}\right)^2}{\sum_{i=1}^{I}\sum_{n=1}^{N}\left(x_{i,n}\right)^2}, \tag{5}$$

where $\hat{x}_{i,n}$ is the PCA reconstruction of the element $x_{i,n}$ of the original matrix **X**, both properly scaled. If in Eq.(5) the summations are extended only to the I rows of the original matrix, the explained variance per variable is evaluated (R^2_{pv}). Two additional statistics can be evaluated for each sample: the squared prediction error (SPE) and the Hotelling's T^2. SPE measures how well the sample conforms to the PCA model, and is defined for the i-th sample as:

$$SPE_i = \mathbf{e}_i^{\mathrm{T}}\mathbf{e}_i = \mathbf{x}_i^{\mathrm{T}}\left(\mathbf{I} - \mathbf{P}\mathbf{P}^{\mathrm{T}}\right)\mathbf{x}_i, \tag{6}$$

where $\mathbf{e}_i$ is the $[N \times 1]$ residual vector for the reconstruction of the i-th observation $\mathbf{x}_i$ (i.e., the i-th row of the residual matrix **E**), and **I** the identity matrix of appropriate size $[N \times N]$. SPE$_i$ measures the orthogonal distance of the i-th observation from the latent space identified by the model, thus accounting for the model mismatch: samples with a high value of SPE are characterized by a different correlation structure with respect to the one described by the PCA model and, as a consequence, are not represented well by the model [10].

The Hotelling's T^2 statistic [11] measures the overall distance of the projections of a sample of the **X** dataset from the PC latent space origin. Since each PC of the model explains a different percentage of variance of the data, the Mahalanobis distance [10] is used to calculate it:

$$T_i^2 = \sum_{a=1}^{A} \frac{t_{a,i}^2}{\lambda_a}, \tag{7}$$

where $t_{a,i}$ represents the projection of the i-th observation on the a-th PC used to build the model. The T^2 statistic is used to assess the deviation of a sample from the average conditions (the PC space origin) represented in the dataset. A sample with a large Hotelling's T^2 has a large influence on the model and should be handled with care: if it is well-represented by the model (small SPE), the information it provides can be useful to expand the data space and improve model robustness.

For both the SPE and the T^2 statistics, the contribution of each variable of a given sample to their values can be evaluated. These contributions can reveal which variables mainly determine the sample position in the latent space or outside it. A thorough discussion on the use of the variable contributions can be found in Conlin et al. [12].

2.1.1 PCA Data Pretreatment

Before building a PCA model, the $\boldsymbol{X}$ data are usually pretreated. The appropriate pretreatment depends on the characteristics of the data and on the objectives of the analysis, and may include filtering, denoising, nonlinear transformations (e.g., logarithmic), advanced scaling and data compression [13].

In general, when managing simultaneously data with heterogeneous origin and different measurement units (e.g., process measurements such as compositions, temperatures, pressures, torques, formulation recipes, etc.), it is important that all variables are given the same weight. To this purpose, the data are auto-scaled, i.e., the variables are mean-centered and scaled to unit variance [14]. Mean-centering (i.e., subtracting to each column $\mathbf{x}_n$ of $\mathbf{X}$ its mean values) avoids to detect the differences among the mean values of different variables as significant directions of variability; scaling to unit variance (i.e., dividing each column $\mathbf{x}_n$ of $\mathbf{X}$ by its standard deviation) makes the analysis independent of the measurement units, thus enabling the simultaneous analysis of variables with values of very different magnitudes.

2.1.2 Process Monitoring Charts

Once a PCA model has been calibrated on the available dataset $\mathbf{X}$, the model can be used to assess whether a new sample ($\mathbf{x}^{\mathrm{NEW}}$) conforms to the data used to build it or not. The scores ($\hat{\mathbf{t}}^{\mathrm{NEW}}$ $[A \times 1]$) of the new sample can be calculated by projecting $\mathbf{x}^{\mathrm{NEW}}$ onto the reduced latent space of the model:

$$\hat{\mathbf{t}}^{\mathrm{NEW}^{\mathrm{T}}} = \mathbf{x}^{\mathrm{NEW}^{\mathrm{T}}}\mathbf{P}. \tag{8}$$

After evaluation of the scores, the PCA reconstruction of $\mathbf{x}^{\mathrm{NEW}}$ can be computed as:

$$\hat{\mathbf{x}}^{\mathrm{NEW}} = \mathbf{P}\hat{\mathbf{t}}^{\mathrm{NEW}}. \tag{9}$$

Given the values of $\hat{\mathbf{t}}^{\mathrm{NEW}}$ and $\hat{\mathbf{x}}^{\mathrm{NEW}}$, the Hotelling's T^2 and the SPE statistics of the new sample can be evaluated in order to quantify its deviation from the average conditions of the data used to build the model and its deviation from the model space, respectively. Based on the values of the statistics for the data used to build the model (i.e., in the calibration step), confidence limits can be set and monitoring charts built for the Hotelling's T^2 and the SPE [3,15]. When a new sample is available, its statistics can be compared against the relevant confidence limits to detect changes in the mean conditions or in the model representativeness [16].

2.1.3 Multi-way PCA

Although there is currently a strong pressure to move to continuous processing, still nowadays the pharmaceutical industry relies mainly on batch manufacturing. When dealing with time-dependent variables, the dataset to be modeled is represented by a three-way array $\underline{\mathbf{X}}[I \times N \times K]$, where for the i-th sample (the i-th batch out of a total of I batches) each of the N variables is measured at K time instants. Multi-way PCA (MPCA; [17]) is the extension to three-way arrays of the concepts presented in Sect. 2.1. MPCA unfolds the data array $\underline{\mathbf{X}}$ into a two-way matrix $\mathbf{X}$ by grouping two of its dimensions prior to the PCA factorization. Namely, $\underline{\mathbf{X}}$ can be unfolded either batch-wise or variable-wise, respectively into a $[I \times (N{\cdot}K)]$ $\mathbf{X}$ data matrix or into a $[(I{\cdot}K) \times N]$ $\mathbf{X}$ data matrix, as depicted in Fig. 2. The choice of the unfolding strategy is related to the application considered, and more details can be found in Nomikos and MacGregor [17], Wold et al. [18] and Kourti [19].

2.2 Projection to Latent Structures

Projection to latent structures (PLS; [4,20]) is a regression modeling technique that relates a dataset of regressors $\mathbf{X}[I \times N]$ to a dataset of response variables $\mathbf{Y}[I \times M]$ through the projections onto their latent structures. PLS aims at finding a linear transformation of the $\mathbf{X}$ data in order to maximize the covariance of its latent variables with the $\mathbf{Y}$ dataset variables. As for PCA, the optimization problem formalizing the search for the LVs can be converted into an eigenvector problem, namely the eigenvector decomposition of the joint covariance matrix $\mathbf{X}^{\mathrm{T}}\mathbf{Y}\mathbf{Y}^{\mathrm{T}}\mathbf{X}$:

$$\mathbf{X}^{\mathrm{T}}\mathbf{Y}\mathbf{Y}^{\mathrm{T}}\mathbf{X}\mathbf{w} = \lambda\mathbf{w}, \tag{10}$$

being $\mathbf{w}$ the $[N \times 1]$ vector of weights representing the coefficients of the linear combination of $\mathbf{X}$ variables determining the PLS scores $\mathbf{t}$:

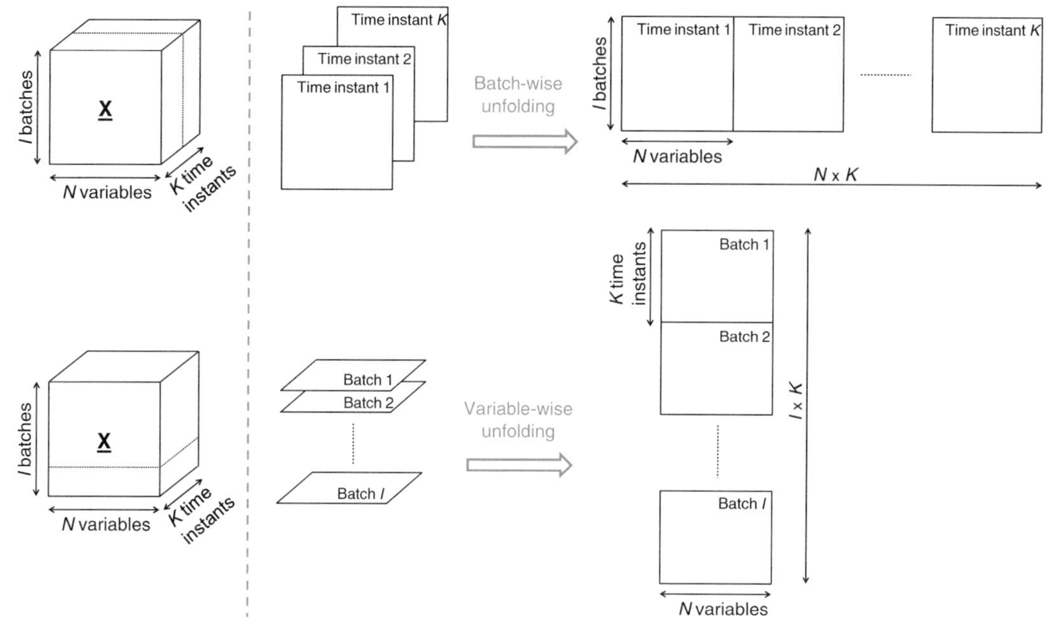

Fig. 2 Batch-wise and variable-wise unfolding of a three-way array $\underline{\mathbf{X}}$

$$\mathbf{t} = \mathbf{X}\mathbf{w}. \tag{11}$$

As shown for the PCA model, assuming that A LVs have been retained, the **X** and **Y** datasets are decomposed and related through their latent structures:

$$\mathbf{X} = \mathbf{T}\mathbf{P}^{\mathrm{T}} + \mathbf{E} \tag{12}$$

$$\mathbf{Y} = \mathbf{T}\mathbf{Q}^{\mathrm{T}} + \mathbf{F} \tag{13}$$

$$\mathbf{T} = \mathbf{X}\mathbf{W}^{*}, \tag{14}$$

where **T** is the $[I \times A]$ score matrix, **P** and **Q** are the $[N \times A]$ and $[M \times A]$ loading matrices, **E** and **F** are the $[I \times N]$ and $[I \times M]$ residual matrices, and $\mathbf{W}^{*}$ is the $[N \times A]$ weight matrix, which is calculated from the weights **W** to allow interpretation with respect to the original **X** matrix:

$$\mathbf{W}^{*} = \mathbf{W}\left(\mathbf{P}^{\mathrm{T}}\mathbf{W}\right)^{-1}. \tag{15}$$

The selection of the number A of LVs to be retained is discussed by Wold [21]. As Eqs. (12)–(14) clarify, PLS modeling returns a model for the correlation structure of **X**, a model for the correlation structure of **Y**, and a model for their mutual relation. Therefore, PLS modeling is suitable to handle reduced-rank datasets including highly correlated and noisy measurements.

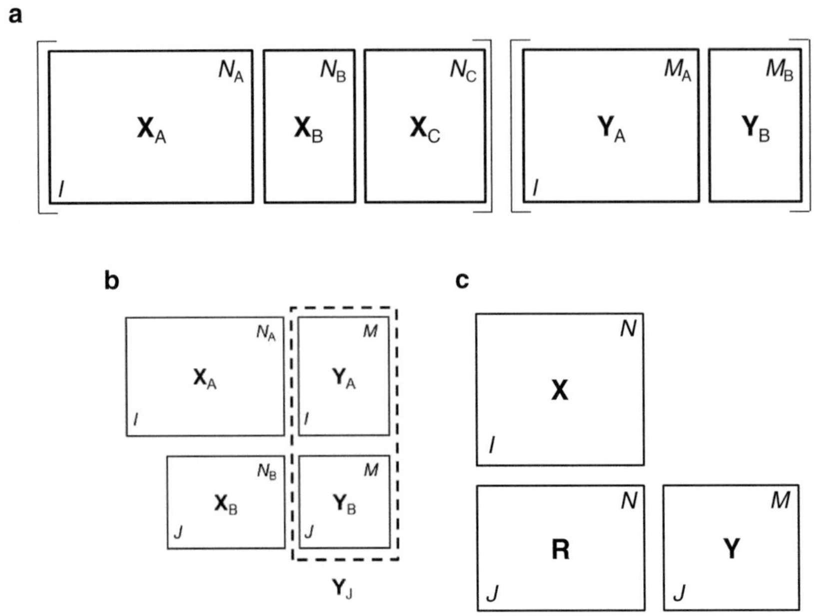

Fig. 3 Schematic of (**a**) multi-block PLS (three **X** blocks and two **Y** blocks), (**b**) joint-Y PLS and (**c**) L-shaped PLS data structures

As for PCA, PLS model scores, loadings and weights can be interpreted to analyze the similarity between samples and the correlation among variables within and between datasets. The considerations in Sects. 2.1.1–2.1.3 regarding data pretreatment, projection of new samples onto an existing model, and multi-way approaches hold true also for PLS models.

A thorough discussion of PLS modeling can be found in Wold et al. [20], Höskuldsson [4] and Burnham et al. [5].

2.3 Other Latent Variable Modeling Approaches

When dealing with more complex datasets (represented by more than two data matrices), more sophisticated LVM approaches may be used. Typical data structures for which these approaches may prove useful are given in Fig. 3. A short review of these advanced modeling approaches is provided in the following subsections.

2.3.1 Multi-block PLS

Multi-block PLS (MB-PLS; Fig. 3a, [22]) is an extension of the PLS method to consider multiple matrices (blocks) of data within a single model. The blocks can be both regressor matrices ($\mathbf{X}_A$, $\mathbf{X}_B$, etc.) and response variable matrices ($\mathbf{Y}_A$, $\mathbf{Y}_B$, etc.). MB-PLS improves the interpretability of the model in all the cases in which it is more convenient to keep variables in separate blocks rather than in a whole dataset, leaving however unchanged (with respect to a traditional PLS model) the prediction capability. "Blocking" the available data can be justified for example by their different origin, by the presence of variables with similar meaning and in different numbers, or by the need of understanding the relationships between variables belonging to different blocks. This makes multi-block models particularly attractive for the analysis of process data in which, for example, data from different plant sections or different unit operations need to be considered separately. For additional details the reader is referred to Kourti et al. [23] and Westerhuis et al. [22].

2.3.2 Joint-Y PLS

Joint-Y PLS (JY-PLS; [24]; Fig. 3b) is a latent variable regression model technique that allows one to relate two (or more) regressor datasets (e.g., an $[I \times N_A]$ dataset $\mathbf{X}_A$ and a $[J \times N_B]$ dataset $\mathbf{X}_B$) through the joint space formed by their corresponding response variables datasets (e.g., an $[I \times M]$ dataset $\mathbf{Y}_A$ and a $[J \times M]$ dataset $\mathbf{Y}_B$). The basic assumption of JY-PLS is that, if the systems A and B are driven by similar mechanisms (physical, chemical, etc.), there will exist a region in the latent space of the joint response matrix $\mathbf{Y}_J$, obtained by joining the response variable datasets ($\mathbf{Y}_J = [\mathbf{Y}_A^T\,\mathbf{Y}_B^T]^T$), in which the LVs of the regressor datasets $\mathbf{X}_A$ and $\mathbf{X}_B$ will be overlapped. This region can be exploited to relate the different datasets and to transfer information between them.

2.3.3 L-Shape PLS and Weighted-Scores PLS

Stemming from the work by Kettaneh-Wold [25], LVMs have been used also to analyze mixture data by regressing a quantitative response of the mixture (**Y**) against the matrix containing the fractions of each component in the mixture [26,27]. Muteki and MacGregor [28] introduced the physical and chemical properties of the raw materials as a third matrix within the model. The resulting L-shape PLS model (Fig. 3c) provides a way to model simultaneously the $[I \times N]$ **X** matrix of the properties of interest for N raw material, the $[J \times N]$ **R** matrix of the blending ratios for J known mixtures, and the $[J \times M]$ **Y** matrix of the M attributes characterizing the mixtures. In order to introduce the processing conditions within the modeling framework, Muteki et al. [29] further proposed an MB-PLS arrangement where the matrix of the processing conditions is used as a regressor together with matrix **RX**, i.e., a weighted average of the physical/chemical properties of the raw materials. García-Muñoz and Polizzi [30] proposed the use of the weighted average of the PCA scores of matrix **X** instead of the weighted average of the physical/chemical properties of the raw materials **RX**. The proposed weighted-scores PLS (WSPLS) model was shown to be more robust than other approaches when dealing with several raw material datasets in which raw materials with different characteristics are collected, and when handling mixture datasets containing missing data.

2.4 Latent Variable Models in the Pharmaceutical Industry

In the pharmaceutical industry, the interest towards LVMs such as PCA and PLS grew tremendously in the last decade. The Quality-by-Design initiative [31] stresses that quality should be built *into* products, rather than tested at the end of the manufacturing process. This can be achieved only if a comprehensive understanding of the characteristics of a drug, of the design and selection of the product components, and of the design of the manufacturing process is obtained. To reach this level of comprehension and develop processes that are able to consistently ensure the desired product quality, appropriate tools need to be employed. To this end, the FDA introduced the Process Analytical Technology (PAT) framework [32]. According to the agency definition, PAT is "a system for designing, analyzing and controlling manufacturing through timely measurements (i.e., during processing) of critical quality and performance attributes of raw and in-process materials and processes, with the goal of ensuring product quality." Within this framework, LVMs (which are considered to be PAT tools) have been used to effectively analyze the large amount of data generated from conventional and advanced measurement systems, and to relate these data to product and process quality. LVMs, in fact, are powerful tools to relate highly correlated input variables (such as spectroscopic, X-ray, chromatography data) to process and product quality [5]. Several reviews on the use of LVMs as tools for product quality estimation in the pharmaceutical industry are available [33–35].

Despite the prominent use of LVMs as PAT tools, García-Muñoz and Oksanen [1] and Tomba [36] stressed on the much more valuable role these models can play in the practical implementation of the QbD paradigm. The effectiveness of LVMs in such activities as process understanding, process monitoring and troubleshooting [16], process control [37–40], process design [41–43], product design [44], and optimization [45] has been proved in several industrial sectors.

3 Latent Variable Model Inversion

The idea of latent variable model inversion, first introduced by Jaeckle and MacGregor [41–43], has been recently generalized by Tomba et al. [46]. The following sections provide an overview of the theoretical aspects beyond this modeling technique.

3.1 Latent Variable Model Inversion Fundamentals

LVMs are commonly used to predict a set of response variables $\hat{\mathbf{y}}^{\mathrm{NEW}}$ starting from an available set of regressor $\mathbf{x}^{\mathrm{NEW}}$. In the model inversion exercise, we assume that an LVM has been built based on a set of historical data, and a set $\mathbf{y}^{\mathrm{DES}}$ $[M \times 1]$ of desired response variables (not included in the historical database) is assigned. The objective is to estimate the set of regressor $\mathbf{x}^{\mathrm{NEW}}$ leading (according

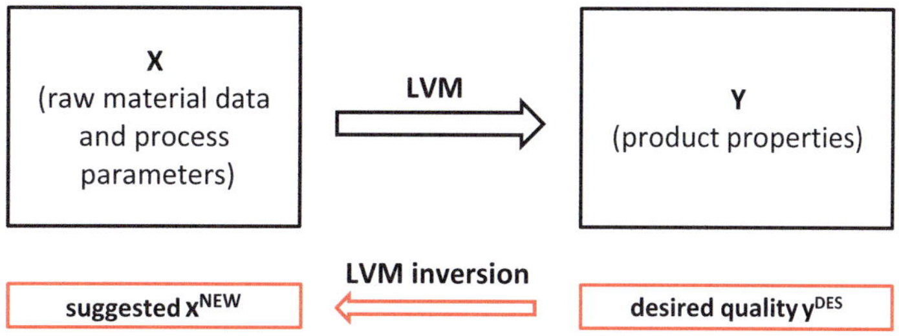

Fig. 4 Schematic of the direct (forward) use of a LVM and of LVM inversion

to the model) to $\mathbf{y}^{DES}$. To achieve this, the model is inverted as sketched in Fig. 4.

According to Jaeckle and MacGregor [41], if $\mathbf{y}^{DES}$ is completely defined (i.e., all product quality variables are assigned as equality constraints, *see* Sect. 4.2), the objective of LVM inversion for a PLS model is that of estimating its projections $\hat{\mathbf{t}}^{DES}$ onto the model latent space:

$$\hat{\mathbf{t}}^{DES} = \left(\mathbf{Q}^{T}\mathbf{Q}\right)^{-1}\mathbf{Q}^{T}\mathbf{y}^{DES}. \tag{16}$$

Given $\hat{\mathbf{t}}^{DES}$ from Eq. (16), the required set of input variables can be obtained using Eq. (9). In this way, $\hat{\mathbf{x}}^{NEW}$ adheres to the covariance structure of the historical data onto which the model was built [41].

The problem with LVM inversion is that, depending on the effective dimension of the latent spaces of $\mathbf{X}$ and $\mathbf{Y}$ (namely the number of significant PCs required to model them, i.e., their statistical rank) and on the number A of retained LVs, the solution to the inversion problem may not be unique. Assuming that $R_{\mathbf{X}}$ is the statistical rank of $\mathbf{X}$ and $R_{\mathbf{Y}}$ is the statistical rank of $\mathbf{Y}$, it usually turns out that:

$$A = \max(R_{\mathbf{X}}, R_{\mathbf{Y}}). \tag{17}$$

Therefore, three cases may arise [41]:

1. $A = R_{\mathbf{X}}$ ($R_{\mathbf{X}} > R_{\mathbf{Y}}$): in this case, some LVs (or their combination) in the latent space of $\mathbf{X}$ are statistically significant to describe the systematic variability in $\mathbf{X}$, but do not contribute in describing the data in $\mathbf{Y}$. Stated differently, part of the variability in $\mathbf{X}$ is not related to $\mathbf{Y}$. In this case (which is the most common one), the inversion exercise moves from the lower dimensional $\mathbf{Y}$ space ($R_{\mathbf{Y}}$) to the higher dimensional $\mathbf{X}$ space ($R_{\mathbf{X}}$).
2. $A = R_{\mathbf{Y}}$ ($R_{\mathbf{Y}} > R_{\mathbf{X}}$): in this case, all the LVs of the $\mathbf{X}$ space explain systematic variability in $\mathbf{Y}$, with part of it remaining

unexplained by **X**. In this case, the inversion exercise moves from the higher dimensional **Y** space ($R_{\mathbf{Y}}$) to the lower dimensional **X** space ($R_{\mathbf{X}}$).

3. $A = R_{\mathbf{X}} = R_{\mathbf{Y}}$ but rank[**XY**] $= R_{\mathbf{XY}} > A$: in this case, despite the statistical ranks of **X** and **Y** are identical, $(R_{\mathbf{XY}} - A)$ latent dimensions do not overlap between the **X** and **Y** spaces. Therefore, this situation is similar to the one where $A = R_{\mathbf{X}}$ $(R_{\mathbf{X}} > R_{\mathbf{Y}})$.

In the first case, the inversion problem is undetermined and the set of solutions is infinite. The direct inversion Eq. (16) returns *one* particular solution to the problem, namely the least-squares one. However, if this solution is moved along the directions of the latent space that do not affect the data in **Y**, *the same* set of desired response variables $\mathbf{y}^{\mathrm{DES}}$ is obtained through the model. The space identified by these directions is called *null space*, and represents the *locus* of the **X** projections not affecting the quality space **Y** [41,43]. In the following subsection, details are given on the evaluation of the null space. It should be observed at this point that the null space concept appears to be strongly linked to the definition of the design space [47] which, according to the FDA, is "the space of the input variable combinations that robustly ensure to obtain a defined product in output."

3.1.1 Null Space Computation

As previously stated, if $R_{\mathbf{Y}} < R_{\mathbf{X}}$, then the direct inversion solution $\hat{\mathbf{x}}^{\mathrm{NEW}}$ falls onto an $R_{\mathbf{Y}}$-dimensional subspace of the $R_{\mathbf{X}}$-dimensional *X* space. Any other solution $\hat{\mathbf{x}}$ defined as

$$\hat{\mathbf{x}} = \hat{\mathbf{x}}^{\mathrm{NEW}} + \hat{\mathbf{x}}^{\mathrm{NULL}}, \tag{18}$$

where $\hat{\mathbf{x}}^{\mathrm{NULL}}$ falls on the remaining $(R_{\mathbf{X}} - R_{\mathbf{Y}})$ directions, yields to the same desired response $\mathbf{y}^{\mathrm{DES}}$ (according to the model). Hence, for Eq. (18) to be a solution of the inversion problem, the following equation should hold true:

$$\mathbf{Q}\mathbf{t}^{\mathrm{NULL}} = \mathbf{0}. \tag{19}$$

Any vector $\mathbf{t}^{\mathrm{NULL}}$ lying in the right kernel of **Q** is a solution of Eq. (19); hence, the null space can be computed from the singular value decomposition of matrix **Q** [42]:

$$\mathbf{Q} = \mathbf{U}_{\mathbf{Q}}\mathbf{S}_{\mathbf{Q}}\mathbf{V}_{\mathbf{Q}}^{\mathrm{T}} = \mathbf{U}_{\mathbf{Q}}\mathbf{S}_{\mathbf{Q}}[\mathbf{G}_1 \vdots \mathbf{G}_2]^{\mathrm{T}}, \tag{20}$$

where $\mathbf{U}_{\mathbf{Q}}$ is the matrix of the left singular vectors of **Q**, $\mathbf{S}_{\mathbf{Q}}$ is the diagonal matrix of the singular values of **Q**, and $\mathbf{V}_{\mathbf{Q}}$ is the matrix of the right singular vectors of **Q**. In particular, the right singular vectors corresponding to the vanishing (zeros) singular values of $\mathbf{Q}$ span its null space. These are included in the columns of matrix $\mathbf{G}_2[A \times (A - R_\mathbf{Y})]$, which therefore defines the null space of the model. Vector $\mathbf{t}^{\mathrm{NULL}}$ can therefore be moved arbitrarily along it, without affecting $\hat{\mathbf{y}}^{\mathrm{NEW}}$, i.e.,:

$$\mathbf{t}^{\mathrm{NULL}^\mathrm{T}} = \boldsymbol{\gamma}^\mathrm{T}\mathbf{G}_2^\mathrm{T}. \quad (21)$$

In Eq. (21), which defines the model null space, $\boldsymbol{\gamma}$ is an $[(A - R_\mathbf{Y}) \times 1]$ vector arbitrary in magnitude and direction.

3.2 General Framework for Latent Variable Model Inversion

Yacoub and MacGregor [45] and García-Muñoz et al. [48,49] formulated the model inversion problem defined in the previous section as an optimization problem. With respect to the direct inversion solution shown in Eq. (16), the optimization approach allows including equality and/or inequality constraints for both the input variables and the response variables in order for the solution $\hat{\mathbf{x}}^{\mathrm{NEW}}$ to be not only coherent with the historical data used to build the underlying model, but also physically reasonable (i.e., with all variables within acceptable practical limits).

A generalized framework for LVM inversion has been proposed recently by Tomba et al. [46]. According to the authors, the inversion problem goes through the following steps:

1. Build the LVM model between (properly preprocessed) $\mathbf{X}$ and $\mathbf{Y}$, after checking their statistical rank to confirm the presence of any null space [48];
2. Determine the desired product specifications ($\mathbf{y}^{\mathrm{DES}}$) in terms of assigned values (i.e., equality constraints), or one- or two-sided constraints (i.e., inequality constraints), and physical bounds[2];
3. Determine the necessary constraints on the solution $\hat{\mathbf{x}}^{\mathrm{NEW}}$ for it to be of practical relevance;
4. If $\mathbf{y}^{\mathrm{DES}}$ is completely specified, verify the model validity by comparing its SPE value with those of the historical samples or with the relevant historical confidence limits (if meaningful, i.e., built considering a sufficiently large number of samples). If $\mathbf{y}^{\mathrm{DES}}$ is not properly described by the model, the inversion is not recommended[3];
5. Solve the appropriate inversion problem.

[2] Physical bounds represent the variable domain in the optimization procedure. Inequality constraints, instead, represent the region inside which the properties (either quality or regressor) are desired to fall, and therefore are subsets of the physical bounds.

[3] The problem of properly selecting the desired product specifications has been addressed in Tomba et al. [50].

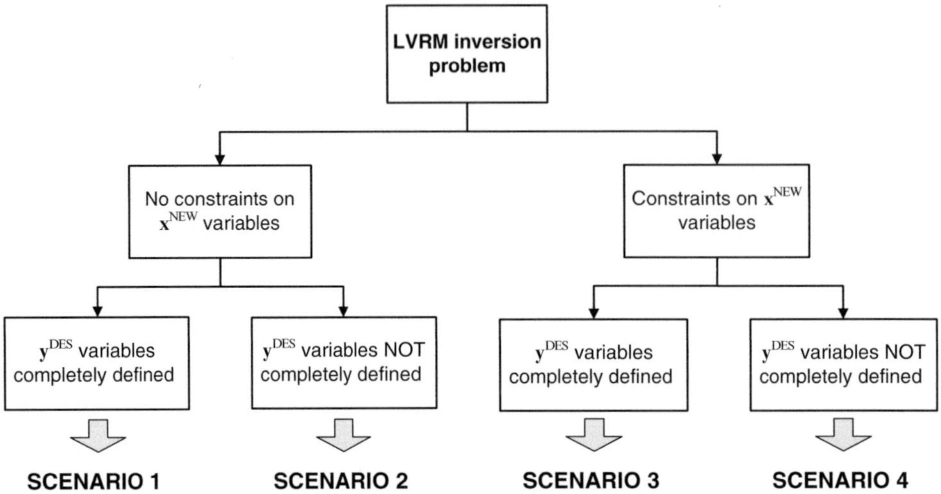

Fig. 5 General framework for LVM inversion (after Tomba et al. [46])

Figure 5 shows the possible scenarios that may be encountered, depending on whether constraints on variables exist or not. The first classification is based on having, or not having, constraints on the regressor vector $\mathbf{x}^{\text{NEW}}$. In the former case, in order for the inversion solution to be close to the historical data, a constraint on the Hotelling's T^2 statistic is sufficient. In the latter case, instead, if too many constraints are assigned, the inversion solution may be forced to move away from the model plane (i.e., SPE > 0). In order to take this occurrence into account, the inversion problem should include a constraint on the SPE statistics.

If no constraints are defined on the regressors and specific values are assigned for all the quality variables (SCENARIO 1), direct model inversion as in Eq. (16) can be applied [41–43]. If the values of some of the elements in $\mathbf{y}^{\text{DES}}$ are not assigned and/or in the presence of a null space, the inversion problem can be defined as follows (SCENARIO 2):

$$\min_{\mathbf{t}} \left[\left(\hat{\mathbf{y}}^{\text{NEW}} - \mathbf{y}^{\text{DES}}\right)^{\text{T}} \boldsymbol{\Gamma} \left(\hat{\mathbf{y}}^{\text{NEW}} - \mathbf{y}^{\text{DES}}\right) + g_1 \cdot \left(\sum_{a=1}^{A} \frac{t_a^2}{s_a^2} \right) \right]$$

$$\begin{aligned} &\text{subject to} \\ &\hat{\mathbf{y}}^{\text{NEW}} = \mathbf{Qt} \\ &\hat{\mathbf{x}}^{\text{NEW}} = \mathbf{Pt} \\ &\hat{y}_j^{\text{NEW}} \le b_j \\ &lb_k^y \le \hat{y}_k^{\text{NEW}} \le ub_k^y \quad lb_l^x \le \hat{x}_l^{\text{NEW}} \le ub_l^x, \end{aligned} \tag{22}$$

where $\mathbf{t}$ is the vector of the decision variables, composed by A scores t_a, s_a^2 is the variance of the a-th column of matrix $\mathbf{T}$, $\hat{\mathbf{y}}^{\text{NEW}}$

is the quality variable vector corresponding to the solution $\hat{\mathbf{x}}^{\mathrm{NEW}}$, b_j is the inequality constraint specified for the j-th element of $\hat{\mathbf{y}}^{\mathrm{NEW}}$ ($\hat{y}_j^{\mathrm{NEW}}$); lb_k^y and ub_k^y are respectively the lower and upper physical bounds for the k-th element of $\hat{\mathbf{y}}^{\mathrm{NEW}}$ ($\hat{y}_k^{\mathrm{NEW}}$), while lb_l^x and ub_l^x are the lower and upper physical bounds for the l-th element of $\hat{\mathbf{x}}^{\mathrm{NEW}}$ ($\hat{x}_l^{\mathrm{NEW}}$). $\boldsymbol{\Gamma}$ is a weight diagonal matrix: variables that are more important for the specific applications should be weighted more heavily. García-Muñoz et al. [48] suggested the use of the fraction of the total variance explained by each variable as a weighting factor, with a value of 0 to be used for those variables for which equality constraints are not assigned. The second term of Eq. (21) is the Hotelling's T^2 of the inversion solution $\hat{\mathbf{x}}^{\mathrm{NEW}}$, and its weight g_1 is intended to properly balance it with respect to the first term. The greater g_1, the more conservative the solution, i.e., $\hat{\mathbf{x}}^{\mathrm{NEW}}$ is closer to the set of historical data.

The model inversion problem when the elements of $\hat{\mathbf{x}}^{\mathrm{NEW}}$ are constrained depends on whether $\mathbf{y}^{\mathrm{DES}}$ is completely defined or not. In the former case (SCENARIO 3), the inversion problem can be formulated as:

$$
\begin{aligned}
&\min_{\mathbf{t}} \left[\left(\mathbf{t} - \hat{\mathbf{t}}^{\mathrm{DES}}\right)^{\mathrm{T}} \boldsymbol{\Sigma} \left(\mathbf{t} - \hat{\mathbf{t}}^{\mathrm{DES}}\right) + g_2 \cdot \mathrm{SPE}_{\mathbf{x}^{\mathrm{NEW}}} \right] \\
&\text{subject to} \\
&\hat{\mathbf{t}}^{\mathrm{DES}} = \left(\mathbf{Q}^{\mathrm{T}}\mathbf{Q}\right)^{-1}\mathbf{Q}^{\mathrm{T}}\mathbf{y}^{\mathrm{DES}} \\
&\hat{\mathbf{y}}^{\mathrm{NEW}} = \mathbf{Q}\mathbf{t} \\
&\hat{\mathbf{x}}^{\mathrm{NEW}} = \mathbf{P}\mathbf{t} \\
&\mathbf{t} = \mathbf{W}^{*\mathrm{T}}\mathbf{x}^{\mathrm{NEW}} \\
&\mathrm{SPE}_{\mathbf{x}^{\mathrm{NEW}}} = \left(\hat{\mathbf{x}}^{\mathrm{NEW}} - \mathbf{x}^{\mathrm{NEW}}\right)^{\mathrm{T}}\left(\hat{\mathbf{x}}^{\mathrm{NEW}} - \mathbf{x}^{\mathrm{NEW}}\right) \le g_3 \cdot \mathrm{SPE}_{\mathbf{X},95\%\,\mathrm{lim}} \\
&x_r^{\mathrm{NEW}} = c_r \\
&x_f^{\mathrm{NEW}} \le d_f \\
&lb_k^y \le \hat{y}_k^{\mathrm{NEW}} \le ub_k^y \qquad lb_l^x \le x_l^{\mathrm{NEW}} \le ub_l^x,
\end{aligned}
\tag{23}
$$

being $\boldsymbol{\Sigma}$ the covariance matrix of the LV scores $\mathbf{T}$ with s_a^2 in the main diagonal [51], c_r the equality constraints for the r-th element of $\mathbf{x}^{\mathrm{NEW}}$, d_f the inequality constraint for the f-th element of $\mathbf{x}^{\mathrm{NEW}}$, g_2 a parameter weighting the importance of the soft constraint for $\mathrm{SPE}_{\mathbf{x}^{\mathrm{NEW}}}$ in the objective function and g_3 a parameter (<1) decreasing the value for the obtained $\mathrm{SPE}_{\mathbf{x}^{\mathrm{NEW}}}$. In the latter case (SCENARIO 4), instead, the inversion problem is formulated as:

$$\min_{\mathbf{x}^{\mathrm{NEW}}}\left[\left(\hat{\mathbf{y}}^{\mathrm{NEW}}-\mathbf{y}^{\mathrm{DES}}\right)^{\mathrm{T}}\boldsymbol{\Gamma}\left(\hat{\mathbf{y}}^{\mathrm{NEW}}-\mathbf{y}^{\mathrm{DES}}\right)+g_1\cdot\left(\sum_{a=1}^{A}\frac{t_a^2}{s_a^2}\right)+g_2\cdot\mathrm{SPE}_{\mathbf{x}^{\mathrm{NEW}}}\right]$$

subject to

$$\hat{\mathbf{y}}^{\mathrm{NEW}}=\mathbf{Qt}$$

$$\hat{\mathbf{x}}^{\mathrm{NEW}}=\mathbf{Pt}$$

$$\mathbf{t}=\mathbf{W}^{*\mathrm{T}}\mathbf{x}^{\mathrm{NEW}}$$

$$\mathrm{SPE}_{\mathbf{x}^{\mathrm{NEW}}}=\left(\hat{\mathbf{x}}^{\mathrm{NEW}}-\mathbf{x}^{\mathrm{NEW}}\right)^{\mathrm{T}}\left(\hat{\mathbf{x}}^{\mathrm{NEW}}-\mathbf{x}^{\mathrm{NEW}}\right)\leq g_3\cdot\mathrm{SPE}_{\mathbf{X},95\%\,\mathrm{lim}}$$

$$\hat{y}_j^{\mathrm{NEW}}\leq b_j$$

$$x_r^{\mathrm{NEW}}=c_r$$

$$x_f^{\mathrm{NEW}}\leq d_f$$

$$lb_k^y\leq\hat{y}_k^{\mathrm{NEW}}\leq ub_k^y\qquad lb_l^x\leq x_l^{\mathrm{NEW}}\leq ub_l^x, \tag{24}$$

with the same meaning for the notation as in the previous scenarios.

For a more thorough discussion on the terms in Eqs. (22–24), the reader is referred to Yacoub and MacGregor [45], García-Muñoz et al. [48,49,51] and Tomba et al. [46,52].

4 Advanced Decision Making by Latent Variable Model Inversion

In the following subsections, some applications of latent variable model inversion for advanced decision-making in the pharmaceutical industry are reviewed. The contributions are grouped according to the main aim of the investigation, though the classification is not rigid:

1. Process design, i.e., the estimation of the optimal processing conditions;
2. Particle engineering and formulation design, i.e., the estimation of the optimal raw materials properties and their ratios in a product formulation;
3. Process scale-up, i.e., the estimation of the optimal operating conditions for a process or an equipment operating at a different scale compared to a given one;
4. Process control, i.e., the estimation of the feedforward or feedback control actions needed to compensate for the variability of the raw materials and/or for process disturbances;
5. Acceptability of raw materials, i.e., how to define whether to accept or to reject a given lot of incoming raw materials.

4.1 Process Design

Liu et al. [53] used an MB–PLS approach to model and optimize an industrial tablet manufacturing line for different API and excipient formulations. The multiblock approach was justified by presence of four different sources of variability for the **X** space, namely the mass fraction of the API, the mass fractions of the excipients (combinations of 11 different excipients were considered), the process settings of the roller compactor and the process settings of the tablet press. The **Y** space was built on a single block, which included three key tablet properties such as average tablet weight, tablet hardness and average disintegration time. The historical database onto which the model was built contained 454 samples. The optimization framework was used considering two different scenarios:

1. For a new formulation (i.e., API and excipient ratios) to be run on the tablet line, determine the process settings for both the roller compactor and the tablet press giving a desired product quality;
2. For a given formulation that was already run through roller compaction process, determine the tablet press settings providing a desired product quality.

While the former scenario represents an example of process design, the latter one is equivalent to the mid-course correction controller discussed by García-Muñoz et al. [51]. Since the **X** space was partially constrained by selecting a specific formulation, the optimization problem solved was equivalent to Eq. (24). The effectiveness of the proposed strategy was demonstrated by replicating the experimental data available for six samples that were not included in the calibration set.

Yacoub et al. [54] used the historical data available for a legacy product (a bilayer tablet) and some additional data obtained from a designed experiment to define a robust design space. As in Liu et al. [53], an MB-PLS approach was used, where the **X** space considered 6 blocks (raw materials, granulation, drying, blending, compression and coating), whereas the **Y** space considered a single block containing the critical to quality attributes (tablet friability, API content uniformity, dissolution and drug-related impurities). For the designed process to be robust, the objective function of the optimization problem was formulated as [45]:

$$\min_{\mathbf{x}^{\mathrm{NEW}}}\left[\left(\hat{\mathbf{y}}^{\mathrm{NEW}}-\mathbf{y}^{\mathrm{DES}}\right)^{\mathrm{T}}\mathbf{Q}_1\left(\hat{\mathbf{y}}^{\mathrm{NEW}}-\mathbf{y}^{\mathrm{DES}}\right)+\frac{\partial\mathbf{y}^{\mathrm{T}}}{\partial\mathbf{d}}\mathbf{Q}_2\frac{\partial\mathbf{y}}{\partial\mathbf{d}}\right], \quad (25)$$

where the first term in the objective function is the same as in Eq. (24) with $\mathbf{Q}_1 = \mathbf{\Gamma}$, while the derivative in the second term represents the sensitivity of the product quality with respect to assigned disturbances **d**. Since this second term penalizes the effect of disturbances on the product quality, it has to be weighted more heavily than the first one to increase the process robustness. The result of the optimization was that a 25 % reduction in the compression speed removed the effect of the variability of the raw materials. In this case, a business decision was made, since reducing the compression speed to tolerate wider specifications reduced also the compression throughput.

Shi et al. [55] proposed and validated the design space for a continuous enantioselective hydrogenation reaction in API synthesis, by formulating an optimization problem as in Eq. (24). A PLS model was built between four process variables (pulse frequency, temperature, gas/liquid ratio, residence time) and two process quality variables (reaction extent and enantiomeric excess). The design space was built on the two-dimensional latent space of the PLS model by iteratively searching the score pairs leading to an extent of reaction greater than 98 % and to an enantiomeric excess greater than 93 %. A series of optimized process parameters with desired quality variables close to 100 and 95 % (respectively for the extent of reaction and the enantiomeric excess) was then used to successfully validate the operating region previously determined.

4.2 Particle Engineering and Formulation Design

The design of a new pharmaceutical solid oral drug product requires making decisions on the physical properties of the incoming material (particle engineering) and their ratios (formulation design). These issues were addressed using LVM inversion by Tomba et al. [46], García-Muñoz and Mercado [56] and Tomba et al. [52].

Tomba et al. [46] used the experimental data reported by Vemavarapu et al. [57], who investigated the role of raw materials properties in the processability and performance of a wet granulated product. Namely, each raw material was characterized and processed at fixed process conditions, providing the information to study the effect of the input properties (H_2O solubility, contact angle, H_2O holding capacity, D[3,2], D90/D10, surface area, and pore volume) on the final product quality (loss on drying, oversize, Δ(Flodex), Δ(compactability), D[3,2], D90/D10, and growth ratio). The objective of the exercise of Tomba et al. [46] was to evaluate the optimal values of the raw materials properties to obtain a product of desired quality through wet granulation. The four scenarios outlined in Sect. 3, exploring different combinations of constraints on both the **X** and **Y** data, were successfully tested by back-calculating the raw materials properties for three different wet-granulated materials after their removal from the calibration dataset.

As an example, some of the results of the first case study discussed by Tomba et al. [46], corresponding to SCENARIO 1 given by the optimization problem (22), are reported in Fig. 6. With reference to Eq. (22), it was assumed that all the product

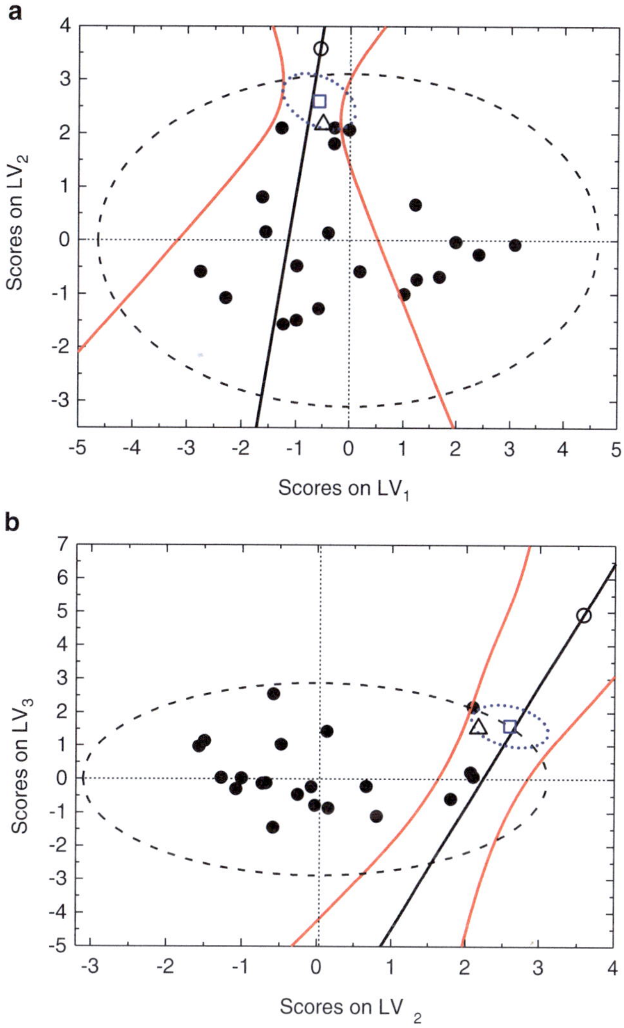

Fig. 6 Projections of the model inversion solutions in the model score plots for maize starch. (**a**) t_1 vs t_2; (b) t_2 vs t_3. In each plot the analytical optimum ($\mathbf{t}^{DES}$, *open circle*), the solution from the optimization in Eq. (22) ($\mathbf{t}$, *open square in blue*) and the real input material variable projections ($\mathbf{t}^{REAL}$, *open triangle*) are reported. The *solid black straight lines* represent the projection of the null space on the considered planes. The *solid red curves* represent the uncertainties in the null space calculation, while the *dotted blue ellipsis* the uncertainty in the optimization solution calculation. After Tomba et al. [46]

properties in $\mathbf{y}^{DES}$ were assigned by the user, while some physical bounds were specified for some of the variables in $\hat{\mathbf{x}}^{NEW}$. Namely:

- The contact angle could vary between 0° and 180°.
- The H_2O holding capacity had to be greater than 0.
- D90/D10 had to be greater than 1.

The model inversion performance was evaluated by comparing the results estimated from the model inversion with the real properties of the material considered after projecting them onto the latent model subspace.

Figure 6 refers to maize starch and reports the projections of the direct inversion solution (○, obtained through Eq. (16)), of the optimized solution (□, obtained by solving the problem in (22)) and of the real input variables (Δ) in the model score spaces of the first and second LVs and of the second and third LVs respectively. The plots also report the scores of the historical samples used to build the model (●), the 95 % confidence ellipse for the sample scores (dashed black line), and the null space projections on the considered score planes (solid black line), together with the 95 % confidence limits of the inversion solution **t** (dotted blue lines) and of the estimated null space (red lines). The latter represent the variation in the solution and null space calculation due to model uncertainty, and were calculated as reported in Tomba et al. [46]. It can be seen that the desired product quality profile ($\mathbf{t}^{DES}$) is out of the range of the historically known product quality. The optimization procedure moves the direct solution of the inversion problem along the null space (or inside its 95 % confidence limits) until it finds a compromise solution between the two terms constituting the objective function in (22). Compared to the direct inversion solution, solution **t** is very similar to the real input material properties projected onto the latent space ($\mathbf{t}^{REAL}$). Other examples are reported in the original reference.

García-Muñoz and Mercado [56] transformed the inversion problem in Eqs. (22)–(24) by adding a set of binary variables to the problem constraints. The nonlinear programming problem was converted into a mixed-integer nonlinear programming problem [58]. The reason for this modification is that, when dealing with formulation design problems, one needs to define variables that explicitly take into account whether a material is chosen (binary variable equal to 1) or not (binary variable equal to 0), and that allows one to limit to a given maximum the number of materials selected. Two successful case studies were presented by the authors. In first one, the aim was that of accelerating the design of a new solid oral drug product with specific targets of tablet hardness and high API concentration. The MB-PLS approach proposed by Muteki et al. [29] was used to model the data from 35 experiments, and LVM inversion was used to simultaneously design the formulation and determine the optimal processing conditions. In the second one, the aim was to minimize the effect on the final product quality of the lot-to-lot variability in the raw materials. The inversion of an L-shape PLS model allowed selecting the amounts of the different available lots to be used (in this case, since a legacy product was considered, the formulation and the process were fixed).

Tomba et al. [52] proposed the inversion of a WSPLS model to successfully design in-silico a new product formulation suitable for direct compression. The available dataset [59] consisted of a matrix $\mathbf{X}_1$ including 45 physical properties and PSD measurements for 64 excipients (including disintegrants), a matrix $\mathbf{X}_2$ including 51 physical properties and PSD measurements for 118 APIs, a matrix $\mathbf{R}$ including the fractions of each considered raw material in 24 historical blended formulations (both active and inactive), and a matrix $\mathbf{Y}$ including 51 physical properties and the PSD measurements for the 24 historical blends. The inversion exercise aimed at selecting the API fractions and the excipient types and fractions to be mixed with the selected APIs in order to obtain a product with the desired values of $B_{MID}80$, compression stress, tensile strength, brittle fracture index and the flow function coefficient. The proposed optimization framework allowed one to handle constraints on the excipient family within which the most suitable excipient should be chosen, constraints on the excipient type itself, on the API dose, or on the final tablet weight. Furthermore, the procedure could deal with different objectives the user may have, such as the maximization of the API dose in the final tablet, or the minimization of the tablet weight.

4.3 Scale-Up and Product Transfer

Scale-up and product transfer activities, though generally referring to different periods during the lifecycle of a product (process development and manufacturing at commercial scale, respectively), both respond to the same question, i.e., how to manufacture a product with given quality attributes in a new equipment that may differ, with respect to a previously used one, in terms of scale, geometry, setup, geographical location, etc., and for which fewer data are usually available.

The first application of LVM and LVM inversion to support pharmaceutical process scale up/product transfer was reported by García-Muñoz [60], who applied a two-step optimization to invert a JY-PLS model to transfer process operating conditions across different scales. This application is discussed deeply in Sect. 4.5.

Liu et al. [61] scaled-up a roller compaction process by inverting a multi-block JY-PLS model to determine how to operate the full-scale unit based on the information available from a similar lab-scale one. Data from a total of 138 (of which 101 at lab scale) runs were available. The three blocks constituting the $\mathbf{X}$ space contained information on the API mass fraction, the excipients mass fraction, and the roller compaction process measurements, whereas the joint $\mathbf{Y}$ space was built considering the ribbon density and the ribbon solid fraction. The effectiveness of the approach was demonstrated by scaling-up seven existing products that had been made at lab scale, and then comparing the results with the closest existing ones already achieved at full scale.

Muteki et al. [62] proposed how to de-risk the scale-up of a high shear wet granulation process using LVM and near-infrared spectroscopy (NIRS), simultaneously accounting for the lot-to-lot variability. The approach goes through the following steps:

1. Build a PLS model on small-scale data from designed experiments, relating API properties ($\mathbf{X}_1$), mechanical process parameters (impeller and chopper speed; $\mathbf{X}_2$) and other process parameters (total amount of water and total batch time; $\mathbf{X}_3$) to the NIRS end point of the operations (i.e., the scores of a PCA model calibrated on the NIR spectra; $\mathbf{Y}$). NIRS was used as it allowed monitoring the process independently from the scale and it provided a better correlation with the downstream properties of interest (particularly the flowability);
2. Estimate the mechanical process parameter of small scale operation that approximate those of the already run large-scale batches using model inversion, keeping the NIRS end-point of large-scale operations constant;
3. Define a suitable target region of NIRS end-points based on the desired downstream properties;
4. For a new API lot to be manufactured at large scale, the NIRS end point are predicted using the PLS model in **step 1** with the API properties, the predicted mechanical process parameters of **step 2** and the other process parameters. If these predictions are outside the target region, the scale-up can be de-risked by adjusting the process parameters.

The effectiveness of the approach was shown using 13 small-scale (10 L volume) and 1 large-scale (70 L volume) batches. The model inversion results showed that an impeller speed of 450 rpm for the 10 L granulator can be effectively used to predict the process end points for the 70 L granulator having an impeller speed of 225 rpm.

Tomba et al. [63] investigated the problem of transferring the production of polymer nanoparticles (to be used as drug carriers for controlled drug delivery) by solvent displacement in passive mixers between two devices of different size. The problem was complicated by the fact that the target device could only be run under a setup that was different from that under which the available historical dataset had been obtained. For the source mixer A, 384 experimental runs were available, whereas for the target mixer B 39 runs were available from previous experimentation; 17 new runs were purposely designed in the target mixer to accomplish the transfer. The selected experiments involved the manipulation of four variables: the polymer concentration in the initial solution (c_{pol}), the inlet water flow rate (*FR*), the anti-solvent-to-solvent flow rate ratio (*W/A*) and the polymer type (*Type*). All experiments were performed using poly-ε-caprolactone (PCL) as a polymer, but considering two lots of polymers of different molecular weight (MW), i.e., PCL_{14} (*Type* = 0) and PCL_{80} (*Type* = 1). The mean particle size (d_p) was the only property considered for the characterization of the nanoparticles obtained from the experiments.

The available data were organized in datasets referring to the three different devices/setups in which experiments were carried out. Namely, $\mathbf{X}_A$, $\mathbf{X}_B$, and $\mathbf{X}_C$ included the operating parameters sets used in the experiments, whereas $\mathbf{Y}_A$, $\mathbf{Y}_B$, and $\mathbf{Y}_C$ included the obtained particle diameter. A JY-PLS model was then fitted on the available data (*see* Sect. 2.3.1) and inverted according to Eq. (24) (SCENARIO 4) to define the operating conditions for mixer B leading to a desired average particle size.

In one of the considered case studies, the objective was to manufacture nanoparticles of mean size y^{DES} = 280 nm in device B under the new experimental setup. The authors evaluated different solutions to the JY-PLS model inversion problem (along the null space), and performed in device B the experiments suggested by these solutions in order to provide the first published experimental validation of the existence of the null space. Four different operating conditions sets $\mathbf{x}_C^{NEW}$ were evaluated. Figure 7a shows the representation of the bidimensional null space in the three-dimensional score space of the JY-PLS model. Figure 7b reports the score space on the first two LVs of the model; this space is represented by the thick line, which corresponds to the intersection between the bidimensional null space and the plane of the first two LV scores. The relevant 95 % confidence limits for the null space are reported as thin lines.

Table 1 reports the four different process operating conditions sets estimated along the null space by inversion of the JY-PLS model. A reasonably wide region of the null space could be explored by assigning different equality constraints to W/A and to the polymer type.

To validate the existence of the null space and to verify that the desired nanoparticle size were actually obtained, the operating conditions estimated by JY-PLS model inversion (Table 1) were actually implemented in a series of experiments on device B under the new experimental setup (i.e., the same used to obtain the data included in datasets $\mathbf{X}_C$ and $\mathbf{Y}_C$). Nanoparticles with a mean size very close to the target one were obtained (last two columns of Table 1), especially for the experiments performed with PCL_{80}; slightly larger errors (yet within the estimated 15 % experimental uncertainty) were observed when using PCL_{14}. Hence, authors concluded that particles of roughly the same size could be obtained by running device B at very different operating conditions.

4.4 Process Control

Compared to other applications, LVM-based process control applications are limited. Outside the pharmaceutical industry, the latent variable model predictive controller proposed by Flores-Cerrillo and MacGregor [39] and recently revised by Yu and Flores-Cerrillo

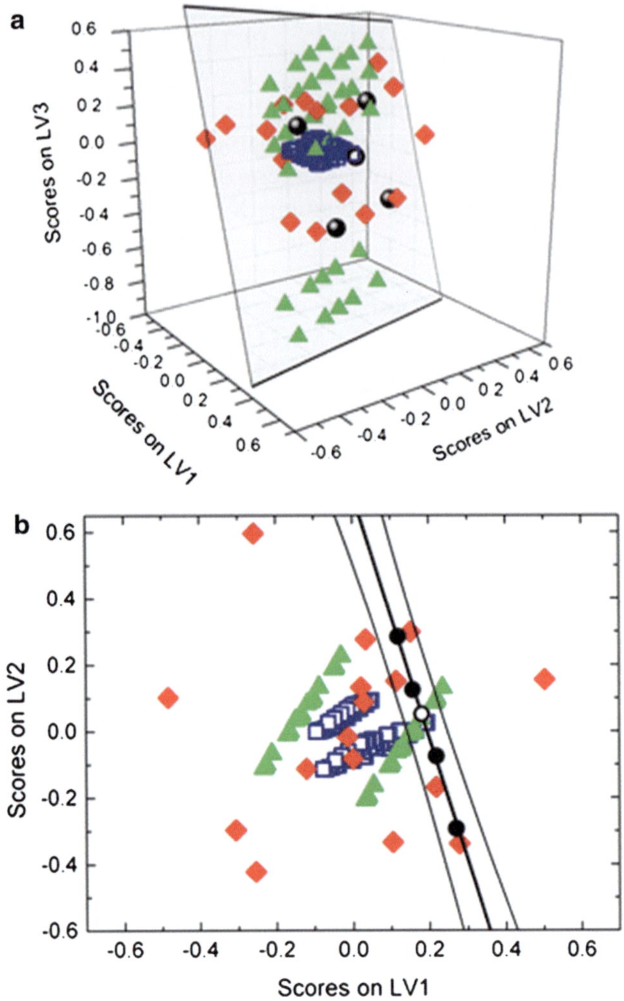

Fig. 7 Representation of the null space for $y^{DES} = 280$ nm, and of the projections (*filled circles*) on the score space of the JY-PLS model of the operating conditions in device B as estimated by model inversion along the null space: (**a**) score space of the three LVs of the model and (**b**) score space of first two LVs of the model. In each plot, the scores of the data in $\mathbf{X}_A$ (*open squares*), $\mathbf{X}_B$ (*filled triangles*), and $\mathbf{X}_C$ (*filled diamond*) are reported together with the scores of the direct inversion solution (*open circle*). The null space is represented in (**a**) by the *gray plane* and in (**b**) by the *thick straight line* with the relevant 95 % confidence limits (*thin curves*). After Tomba et al. [63]

[40] makes use of model inversion to compute the sequence of future control actions to be implemented. The controller, which was tested on batch simulation studies, is based on principal component analysis. In the pharmaceutical industries, the use of LVM-based (feedforward) controllers was pioneered by García-Muñoz et al. [51]. The authors designed four controllers for a high shear

Table 1
Operating conditions in device B determined by inversion of the JY-PLS model to obtain nanoparticles with y^{DES} = 280 nm, and comparison with the mean diameters obtained experimentally. After Tomba et al. [63].

Run no.	c_{pol} [mg/mL]	*FR* [mL/min]	*W/A*	*Type*	d_p^{EXP} [nm]	Error [%]
1	1.5	3	1.00	PCL_{80}	289.5	−3.4
2	2.9	24	2.94	PCL_{80}	287.4	−2.6
3	4.1	53	1.00	PCL_{14}	268.6	+4.1
4	5.0	68	2.84	PCL_{14}	247.9	+11.5

Equality constraints were assigned to variables *W/A and Type*

wet granulation process: a pure feedforward controller based on the raw materials properties (the particle size distribution of the initial dry blend), and three mid-course correction controllers that adjusted the water addition rate and the impeller speed at a given decision point (1/3 of the way through the water addition step). The difference among the three mid-course correction controllers was the set of measurements used as inputs to evaluate the control moves: (1) temperature and power consumption profiles up to the decision point; (2) NIR spectra collected up to the decision points; (3) temperature and power consumption profiles and NIR spectra together. The model was built considering 23 runs (to determine common-cause variability), 10 runs which were parts of a fractional factorial design, and 12 additional runs that were executed as step-tests to identify the model for the mid-course correction controllers. The controllers outputs were evaluated using the inversion scenario given in Eq. (23). The performance of the four controllers were shown to broaden the range of raw materials that could be handled, i.e., the process design space, as observed also by MacGregor and Bruwer [64].

Mid-course correction policies to compensate for the raw materials variability were proposed also in the above-cited work by Liu et al. [53] and in the work by Muteki et al. [65] through the inversion of MB-PLS models. In the latter case, the optimization was proposed to explore the design space simulated using the MB-PLS fitted on data from 11 batches by introducing more than seven millions possible combinations of raw materials attributes obtained from an available database of previously processed lots (microcrystalline cellulose, Fast Flo lactose, and magnesium stearate).

4.5 Raw Materials Acceptability Assessment

The work by García-Muñoz [60] can be of great interest for pharmaceutical industry practitioners. The study was intended to extend the use of LVM to establish multivariate specifications for incoming materials proposed by Duchesne and MacGregor [66], who suggested a set of tests to be carried out to assess whether a given lot of raw materials should be accepted or rejected. As a part of their strategy, they proposed a way to assess the expected variability on final product quality strictly due to changes in process conditions. García-Muñoz [60] extended this work to account for scale differences that characterizes data from pharmaceutical process development. In fact, multiple raw materials are usually tested at the lab or pilot scale providing a rich dataset, whereas fewer experiments are carried out at the larger scale; additionally, different control strategies may be used at different scales. Hence, the problem addressed was how to use data from multiple scales in order to support the specifications of incoming materials at large/commercial scale. The key point was estimating a set of hypothetical process conditions in the full scale plant to obtain the same product quality as in the lab and pilot scales, when using the same set of materials. Given these estimates, application of the Duchesne and MacGregor [66] method was straightforward. Mathematically, the estimates were obtained through a two-step model inversion exercise on a JY-PLS model built considering each scale and operation mode as a different site. The method was illustrated considering data from two scales for the purchase from multiple vendors of a polymeric excipient to be used in the manufacture of oral dosage forms. Twenty-six experiments were carried out at the pilot scale, whereas 77 experiments were carried out at the lab scale, of which 26 operated in closed-loop and 51 in open-loop. The in-vitro dissolution profile of the final product, after being scaled-up with the proposed technique (and hence made scale-independent), was regressed against 13 properties of the polymer, and specifications were established in the multivariate space of the PLS model.

5 Conclusions

Pharmaceutical industries have usually access to large amounts of historical data, resulting from the experimental campaigns that the process and product design activities usually rely on. This chapter showed how latent variable model inversion can be effectively used to extract the information embedded in these historical datasets, and use it for process design and control, particle engineering, formulation design, and scale-up exercises. Several successful applications have been reviewed with the aim of encouraging practitioners to rely on latent variable modeling concepts as a very convenient way to assist the implementation of Quality-by-Design paradigms in pharmaceutical development and manufacturing activities.

Acknowledgment

M.O. and M.B. gratefully acknowledge "Fondazione Ing. Aldo Gini" and "Fondazione CARIPARO" (Project # PARO104725-2010) for the financial support.

References

1. García-Muñoz S, Oksanen CA (2010) Process modeling and control in drug development and manufacturing. Comput Chem Eng 34: 1007–1008
2. Tomba E, Facco P, Bezzo F, Barolo M (2013) Latent variable modeling to assist the implementation of Quality-by-Design paradigms in pharmaceutical development and manufacturing – a review. Int J Pharm 457:283–297
3. Jackson JE (1991) A user's guide to principal components. John Wiley, New York, NY
4. Höskuldsson A (1988) PLS regression methods. J Chemom 2:211–228
5. Burnham AJ, Viveros R, MacGregor JF (1996) Frameworks for latent variable multivariate regression. J Chemom 10:31–45
6. López-Negrete de la Fuente R, García-Muñoz S, Biegler LT (2010) An efficient nonlinear programming strategy for PCA models with incomplete data sets. J Chemom 24:301–311
7. Valle S, Li W, Qin SJ (1999) Selection of the number of principal components: the variance of the reconstruction error criterion with a comparison to other methods. Ind Eng Chem Res 38:4389–4401
8. Meyer CD (2000) Matrix analysis and applied linear algebra. SIAM, Philadelphia, PA
9. Wold H (1966) Estimation of principal components and related models by iterative least squares. In: Multivariate analysis, Academic, New York
10. Mardia KV, Kent JT, Bibby JM (1979) Multivariate analysis. Academic, London
11. Hotelling H (1933) Analysis of a complex of statistical variables into principal components. J Educ Psycol 24:417–441
12. Conlin AK, Martin EB, Morris AJ (2000) Confidence limits for contribution plots. J Chemom 14:725–736
13. Eriksson L, Johansson E, Kettaneh-Wold N, Trygg J, Wikström C, Wold S (2006) Multi- and megavariate data analysis. Part I. Basic principles and applications. Umetrics AB, Umeå, Sweden
14. Wise BM, Gallagher NB, Bro R, Shaver JM, Windig W, Scott Koch R (2006) PLS_Toolbox Version 4.0 for use with MATLAB™. Eigenvector Research, Inc., Wenatchee, WA
15. Qin SJ (2003) Statistical process monitoring: basics and beyond. J Chemom 17:480–502
16. Kourti T, MacGregor JF (1995) Process analysis, monitoring and diagnosis, using multivariate projection methods. Chemom Intell Lab Syst 28:3–21
17. Nomikos P, MacGregor JF (1994) Monitoring batch processes using multi-way principal component analysis. AIChE J 40:1361–1375
18. Wold S, Kettaneh N, Friden H, Holmberg A (1998) Modeling and diagnostics of batch processes and analogous kinetic experiments. Chemom Intell Lab Syst 44:331–340
19. Kourti T (2005) Application of latent variable methods to process control and multivariate statistical process control. Int J Adapt Contr Signal Process 19:213–246
20. Wold S, Martens H, Wold H (1983) The multivariate calibration problem in chemistry solved by the PLS method. Lect Notes Math 973:286–293
21. Wold S (1978) Cross-validatory estimation of the number of components in factor and principal component models. Technometrics 20: 397–405
22. Westerhuis JA, Kourti T, MacGregor JF (1998) Analysis of multiblock and hierarchical PCA and PLS models. J Chemom 12:301–321
23. Kourti T, Nomikos P, MacGregor JF (1995) Analysis, monitoring and fault diagnosis of batch processes using multiblock and multiway PLS. J Process Contr 5:277–284
24. García-Muñoz S, MacGregor JF, Kourti T (2005) Product transfer between sites using Joint-Y PLS. Chemom Intell Lab Syst 79:101–114
25. Kettaneh-Wold N (1992) Analysis of mixture data with partial least squares. Chemom Intell Lab Syst 14:57–69
26. Eriksson L, Johansson E, Wikstrom C (1998) Mixture design – design generation, PLS

analysis, and model usage. Chemom Intell Lab Syst 43:1–24
27. Dingstad GI, Westad F, Naes T (2004) Three case studies illustrating the properties of ordinary and partial least squares regression in different mixture models. Chemom Intell Lab Syst 71:33–45
28. Muteki K, MacGregor JF (2007) Multi-block PLS modeling for L-shape data structures with applications to mixture modeling. Chemom Intell Lab Syst 85:186–194
29. Muteki K, MacGregor JF, Ueda T (2007) Mixture designs and models for the simultaneous selection of ingredients and their ratios. Chemom Intell Lab Syst 86:17–25
30. García-Muñoz S, Polizzi MA (2012) WSPLS – A new approach towards mixture modeling and accelerated product development. Chemom Intell Lab Syst 114:116–121
31. FDA (2004a). Pharmaceutical CGMPs for the 21st century – A risk based approach. Final report. U.S. Department of Health and Human Services. U.S. Food and Drug Administration
32. FDA (2004) Guidance for industry. PAT – A framework for innovative pharmaceutical development, manufacturing and quality assurance. Center for Drug Evaluation and Research, U.S Food and Drug Administration, Rockwille, MD
33. Rajalahti T, Kvalheim OM (2011) Multivariate data analysis in pharmaceutics: a tutorial review. Int J Pharm 417:280–290
34. Pomerantsev AL, Rodionova OY (2012) Process analytical technology: a critical view of the chemometricians. J Chemom 26: $32#299–310
35. Matero S, Van den Berg F, Poutiainen S, Rantanen J, Pajander J (2013) Towards better process understanding: chemometrics and multivariate measurements in manufacturing of solid dosage forms. J Pharm Sci 102: 1385–1400
36. Tomba E (2013) Latent variable modeling approaches to assist the implementation of Quality-by-Design paradigms in pharmaceutical development and manufacturing. PhD Thesis, University of Padova, Italy
37. Flores-Cerillo J, MacGregor JF (2003) Within-batch and batch-to-batch inferential-adaptive control of semibatch reactors: a partial least squares approach. Ind Eng Chem Res 42: 3334–3345
38. Flores-Cerillo J, MacGregor JF (2004) Control of batch product quality by trajectory manipulation using LV models. J Process Contr 14:539–553
39. Flores-Cerillo J, MacGregor JF (2005) Latent variable MPC for trajectory tracking in batch processes. J Process Contr 15:651–663
40. Yu H, Flores-Cerrillo J (2013) Latent variable model predictive control for trajectory tracking in batch processes: internal model control interpretation and design methodology. Ind Eng Chem Res 52:12437–12450
41. Jaeckle CM, MacGregor JF (1998) Product design through multivariate statistical analysis of process data. AIChE J 44:1105–1118
42. Jaeckle CM, MacGregor JF (2000) Industrial application of product design through the inversion of latent variable models. Chemom Intell Lab Syst 50:199–210
43. Jaeckle CM, MacGregor JF (2000) Product transfer between plants using historical process data. AIChE J 46:1989–1997
44. Lakshminarayanan S, Fuji H, Grosman B, Dassau E, Lewin DR (2000) New product design via analysis of historical databases. Comput Chem Eng 24:671–676
45. Yacoub F, MacGregor JF (2004) Robust processes through latent variable modeling and optimization. AIChE J 57:1278–1287
46. Tomba E, Barolo M, García-Muñoz S (2012) General framework for latent variable model inversion for the design and manufacturing of new products. Ind Eng Chem Res 51: $32#12886–12900
47. ICH (2009). ICH harmonised tripartite guide. Pharmaceutical development Q8 (R2)
48. García-Muñoz S, Kourti T, MacGregor JF, Apruzzese F, Champagne M (2006) Optimization of batch operating policies. Part I Handling multiple solutions. Ind Eng Chem Res 45:7856–7866
49. García-Muñoz S, MacGregor JF, Neogi D, Letshaw BE, Mehta S (2008) Optimization of batch operating policies. Part II Incorporating process constraints and industrial applications. Ind Eng Chem Res 47:4202–4208
50. Tomba E, Facco P, Bezzo F, García-Muñoz S (2013) Exploiting historical databases to design the target quality profile for a new product. Ind Eng Chem Res 52:8260–8271
51. García-Muñoz S, Dolph S, Ward HW II (2010) Handling uncertainty in the establishment of a design space for the manufacture of a pharmaceutical product. Comput Chem Eng 34:1098–1107
52. Tomba E, Barolo M, García-Muñoz S (2014) *In silico* product formulation design through latent variable model inversion. Chem Eng Res Des 92:534–544
53. Liu Z, Bruwer MJ, MacGregor JF, Rathore SSS, Reed DE, Champagne MJ (2011)

Modeling and optimization of a tablet manufacturing line. J Pharm Innov 6:170–180
54. Yacoub F, Lautens J, Lucisano L, Banh W (2011) Application of quality by design principles to legacy drug product. J Pharm Innov 6: 61–68
55. Shi Z, Zaborenko N, Reed DE (2013) Latent variables-based process modeling of a continuous hydrogenation reaction in API synthesis of small molecules. J Pharm Innov 8:1–10
56. García-Muñoz S, Mercado J (2013) Optimal selection of raw materials for pharmaceutical drug product design and manufacture using mixed integer nonlinear programming and multivariate latent variable regression models. Ind Eng Chem Res 52:5934–5942
57. Vemavarapu C, Surapanemi M, Hussain M, Badawy S (2009) Role of drug substance material properties in the processability and performance of a wet granulated product. Int J Pharm 374:96–105
58. Grossmann IE (2002) Review of nonlinear mixed integer and disjunctive programming techniques. Optim Eng 3:227–252
59. Polizzi MA, García-Muñoz S (2011) A framework for *in-silico* formulation design using multivariate latent variable regression methods. Int J Pharm 418:235–242
60. García-Muñoz S (2009) Establishing multivariate specifications for incoming materials using data from multiple scales. Chemom Intell Lab Syst 98:51–57
61. Liu Z, Bruwer MJ, MacGregor JF, Rathore SSS, Reed DE, Champagne MJ (2011) Scale-up of a pharmaceutical roller compaction process using a joint-Y partial least squares model. Ind Eng Chem Res 50:10696–10706
62. Muteki K, Yamamoto K, Reid GL, Krishnan M (2011) De-risking scale up of a high-shear wet granulation process using latent variable modeling and near-infrared spectroscopy. J Pharm Innov 6:142–156
63. Tomba E, Meneghetti N, Facco P, Zelenková T, Barresi AA, Marchisio DL, Bezzo F, Barolo M (2014) Transfer of a nanoparticle product between different mixers using latent variable model inversion. AIChE J 60:123–135
64. MacGregor JF, Bruwer MJ (2008) A framework for the development of design and control spaces. J Pharm Innov 3:15–22
65. Muteki K, Swaminathan V, Sekulic S, Reid GL (2011) De-risking pharmaceutical tablet manufacture through process understanding, latent variable modeling, and optimization technologies. AAPS Pharm Sci Tech 12: 1324–1334
66. Duchesne C, MacGregor JF (2004) Establishing multivariate specification regions for incoming materials. J Qual Technol 36:78–94

Chapter 7

Advanced Control of Continuous Pharmaceutical Tablet Manufacturing Processes

Ravendra Singh, Carlos Velazquez, Abhishek Sahay, Krizia M. Karry, Fernando J. Muzzio, Marianthi G. Ierapetritou, and Rohit Ramachandran

Abstract

A novel manufacturing strategy based on continuous processing, integrated with online/inline monitoring tools, coupled with an advanced automatic feedback control system is highly desired for efficient Quality by Design (QbD)-based manufacturing of the next generation of pharmaceutical products with optimal consumption of time, space and resources. In this work, an advanced hybrid MPC-PID control system as well as a simpler PID controller for a direct compaction continuous tablet manufacturing process has been designed and implemented for a pilot-scale pharmaceutical process. An NIR sensor, an online NIR prediction tool, a PAT data management tool, an OPC communication protocol, a standard control platform and control hardware have been used to close the control loop. A systematic methodology to design and implement the control system has been also proposed. A control framework with features such as the option to run the plant in open-loop as well as in a closed-loop scenario has been developed. Furthermore, within the closed-loop scenario, options for a simpler PID, a dead time compensator (Smith predictor) as well as an advanced model predictive controller have been included. The feature to run the control strategy in simulation mode has been added to the control platform to facilitate virtual control system design and performance evaluation. Two case studies involving a direct compaction continuous tablet manufacturing process have been considered to demonstrate the closed-loop operation. Case Study 1 was completed at Rutgers University and constituted the use of a continuous cylindrical blender with a rotating screw. Case Study 2 was based on a continuous tumble mixer and was completed at the University of Puerto Rico—Mayaguez Campus (UPRM).

Key words Control system, Model predictive control, Pharmaceutical, Continuous, Tablet manufacturing

1 Introduction

Currently, pharmaceutical industries are facing enormous challenges to satisfy higher regulatory constraints, flexible market demands, operational complexities, high cost and lengthy time to market (TTM) in new drug product development (approx. $1.2 billion; 10–15 years) [1–4], economical limitations, and relatively

Marianthi G. Ierapetritou and Rohit Ramachandran (eds.), *Process Simulation and Data Modeling in Solid Oral Drug Development and Manufacture*, Methods in Pharmacology and Toxicology,
DOI 10.1007/978-1-4939-2996-2_7,

inefficient Quality by Testing (QBT)-based batch product manufacturing [5]. As a consequence of globalization, the effective patented life of newly discovered drugs has also decreased considerably forcing the manufacturer to minimize the drug development time as well as maximize the throughput and for this, continuous manufacturing has evolved as an efficient alternative [6]. However, better process understanding and controls are vital for minimizing waste due to manufacturing failures and ensure quality product with reduced end product testing [7]. Therefore, a significant opportunity exists for improving product quality and optimizing the production process through the implementation of innovative solutions for online/inline monitoring, analysis and system control [8]. The US Food and Drug Administration (FDA) has supported the application of Process Analytical Technology (PAT) and control systems within manufacturing industries [9]. The application of PAT systems in manufacturing paves the way for continuous process and product improvements through process supervision based on knowledge-based data analysis, QbD concepts and feedback control [9, 10]. The primary goal of PAT is to achieve better understanding of the manufacturing process, and to use that knowledge online to achieve better control of the process and thus consistent product quality. PAT is therefore defined as a system for designing, analyzing, and controlling manufacturing through timely measurements (i.e., during processing) of critical quality and performance attributes of raw and in-process materials and processes with the goal of ensuring final product quality [6, 9, 11].

Batch processing, although traditionally used for drug manufacturing, has a number of disadvantages including larger footprint of equipment, higher equipment and operational costs, and poorer controllability [6]. In recent years, interest has grown rapidly in continuous manufacturing processes that integrate online/inline monitoring tools and advanced model predictive control systems [12]. Continuous manufacturing processes operate at or near steady state, allowing for closed-loop control of the entire operation, which leads to a more robust and reliable process. Moreover, since continuous processes reach the desired steady state in just a few minutes, they enable true QbD-based manufacturing. In addition to its flexibility and time- and cost-saving features, continuous manufacturing is intrinsically steady and therefore easily amenable for model predictive design, optimization, and control methods. These methods have proven to be effective approaches for improving operational efficiency and have been widely used in various process industries. Excitingly, in the pharmaceutical industry, the application of the control system is an open area of study for researchers and technology providers [12].

Continuous processes integrated with online/inline monitoring tools and automatic feedback control systems are desired by the pharmaceutical industry so as to comply with regulatory requirements and meet flexible market demands. Over the last 30 years, process industries such as oil refining and bulk chemical production have made use of model predictive controller (MPC) as an effective control strategy [5]. However, because of different levels of complexities involved (e.g., solid handling, irregular flow) in addition to the unavailability of a suitable process model, the design and implementation of MPC in pharmaceutical manufacturing is still an open area of research. In comparison with the classical PID based control strategy [6], MPC [5] is more complex and requires the development of higher fidelity models and robust optimization strategies. Furthermore, depending on the process dynamics, a MPC scheme that requires a detailed process model and a computationally expensive optimization strategy may not be required, and a more computationally efficient Proportional–Integral–Derivative (PID) control scheme can be used. A hybrid control strategy integrating the advantages of both MPC and PID is therefore highly desirable for the system-wide control of a process comprised of fast and delayed dynamics [5].

There are different levels of complexity involved in the implementation of the control system in pharmaceutical manufacturing involving solid dosage forms. For example: (1) Integration of control hardware, software and sensors with process equipment is a challenging task, (2) Answers to fundamental questions such as, which variables are critical and need to be controlled in a continuous tablet manufacturing process; what the best actuators are for each critical variable; how these critical variables can be monitored online/inline, and how the different process variables interact with each other, are still unclear, (3) Difficulties in real-time online/inline monitoring of the process variables that need to be controlled are another barrier that prevents the implementation of control systems, (4) The most suitable control strategies (PID, PI, MPC, feed forward controller, feedback controller) for tablet manufacturing processes are still unknown, (5) There is no standard control package (commercially available) that can be employed to implement a control system in a pharmaceutical plant. If pharmaceutical companies need to shift from an open-loop process to a closed-loop process, a novel system must be developed to overcome such limitations. Therefore, as proposed in this chapter, a systematic framework through which control systems can be designed, tested and further implemented into the plant with less time and resources is needed.

The four essential steps to move from open-loop to closed-loop operation are: process understanding, control system design, control hardware/software and sensor integration, and control system implementation. For an understanding of the continuous tablet manufacturing process, extensive model-based [13–20] as well as experimental [21–24] studies have been performed. Few attempts have also been made toward the design of a control system for the tablet manufacturing process. Singh et al. [25] suggested a monitoring and control system for a batch tablet manufacturing process. Hsu et al. suggested a control system for a roller compactor, an important unit operation used for a dry granulated continuous tablet manufacturing process [26, 27]. Ramachandran and Chaudhury proposed a control system for a continuous drum granulation process [28]. A detailed review on the control of a fluid bed granulation process has been performed by Burggraeve et al. [29], and discussion has been provided by Bardin et al. [30] on the control aspects for efficient operation of a high shear mixer. Sanders et al. performed extensive control studies using proportional integral derivative (PID) and model predictive control (MPC) methods on an experimentally validated fluidized bed granulation model [31]. An MPC strategy has been proposed for a wet drum granulation process [32–34]. Ramachandran et al. designed a regulatory control system for a continuous direct compaction process with emphasis on blending and tableting processes [35]. Singh et al. developed an advanced MPC system for a direct compaction continuous tablet manufacturing process [5]. Singh et al. designed a control system for the roller compaction route of the continuous tablet manufacturing process [6]. The control system for wet granulation route has been also designed [51]. Singh et al. demonstrated the implementation of the control system into the direct compaction continuous tablet manufacturing pilot plant using NIR as a real time monitoring tool [4, 50].

In this chapter, a systematic framework for design and implementation of a control system for direct compaction tablet manufacturing processes is described. Prior to implementation, the performance of the control system has been evaluated using a process flowsheet model simulated in gPROMS (PSE). The closed-loop operation has been demonstrated using an NIR sensor, PID and model predictive control (MPC) algorithm, an online prediction tool, PAT data management tool, an OPC communication protocol, and a control platform.

2 Process Control Basics

Proportional integral derivative (PID) controllers as well as advanced model predictive controllers have been widely used in the manufacturing industry. These controllers are briefly described in the following sections:

2.1 Proportional Integral Derivative (PID) Controller

PID (also known as a regulatory controller) is the most commonly used controller in the manufacturing industry because of its simplicity. Proportional (P), Integral (I) and Derivative (D) terms can be used individually or in any combination as per requirement of the control loops. The proportional term is based on current deviation of the control variable from the set point. The integral term takes into account past deviations, while the differential term take into account future deviations. In PID controllers, the manipulating variable is calculated as follows [36]:

$$u(t) - u_0(t) = K_C e(t) + \frac{1}{\tau_I}\int_0^t e(t)dt + \tau_D \frac{de(t)}{dt}$$

$$e(t) = Y_{sp}(t) - Y(t)$$

Where $u(t)$ is the manipulating variable, $u_{0(t)}$ is the nominal value of the manipulating variable, $e(t)$ is the deviation of the control variable from the set point, $Y_{sp}(t)$ is the set point, $Y(t)$ is the control variable, K_C is the proportion gain, τ_I is the integral time constant, τ_D is the derivative time constant. K_C, τ_I, and τ_D are tuning parameters that impact the control performance of the system [37].

In feedback control schemes, the process control variable is measured and the signal is sent to the controller that generates the manipulating variable which is sent back to the process. In this case, corrective action occurs as soon as the controlled variable deviates from the set point, regardless of the source and type of disturbances. Feedback control requires minimal knowledge about the process to be controlled; in particular, a mathematical model of the process is not required, although it can be very useful for control system design. This feedback scheme does not provide predictive control action to compensate for the effects of known or measurable disturbances. Also, it is not recommended for processes with large time constant and/or long time delays. In case of dead time dominant process, the dead time compensator needs to be integrated with the PID controller [36].

2.2 Model Predictive Controller (MPC)

MPC refers to a family of control algorithms that employ an explicit model to predict the future behavior of the process over an extended prediction horizon. It is based on the iterative, finite horizon optimization of a plant model. If the linear model is employed then it is called linear MPC, and if the nonlinear model is employed it is called nonlinear MPC. The main concept of MPC dates back to the 1970s, when engineers from Shell demonstrated the real applications of MPC [38, 39]. Over the last 45 years, the MPC has proven to be an efficient control strategy within oil refineries and aerodynamic and chemical industries. MPC has several advantages over the simpler PID controller. For example, MPC can handle multivariable control problems, process constraints (e.g., actuator limitations, constraints on controlled variable, system constraints), process-delayed system disturbances, process variable interactions and equipment (sensor/actuator) failures. The major limitation of MPC includes its design complexity, computationally expensive control algorithm (requires optimization tool),

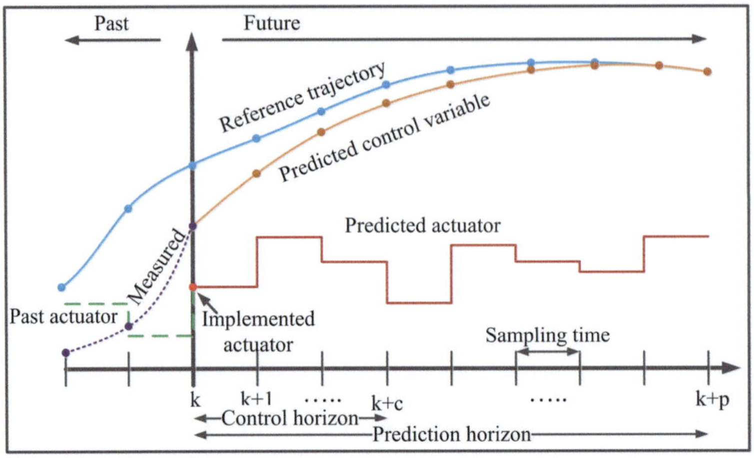

Fig. 1 Model predictive control (moving horizon scheme) (adapted from Singh et al., 2013 [5])

and necessity of additional (linear/nonlinear) process model(s). The performance of MPC depends on the process model, and hence poor process models may lead to poor controller performance [36].

A MPC scheme is shown in Fig. 1. The past value of the manipulated and controlled variables, as well as the predicted actuator and manipulated variable are shown in the figure. The prediction horizon and the sampling time are also shown in the figure. MPC uses the current plant measurements and dynamic state of the process, the MPC models, and the process variable targets and limits to calculate future changes in the independent variables. These changes are calculated to hold the dependent variables close to the target while honoring constraints on both independent and dependent variables. The MPC typically sends out only the first change in each independent variable to be implemented and repeats the calculation when the next change is required. The linear time invariant model (e.g., (output error) OE, Box–Jenkins (BJ), ARX, state space) can be used to compute the predictions in linear MPC. The optimization function that needs to be minimized can be expressed as follows [5]:

$$J = \underbrace{\sum_{i=1}^{P}\sum_{j=1}^{n_y}\left\{w_j^y\left[y_j^{set}(k+i) - y_j(k+i)\right]\right\}^2}_{\text{(I)}}$$

$$+ \underbrace{\sum_{i=1}^{M}\sum_{j=1}^{n_u}\left\{w_j^{\Delta u}\Delta u_j(k+i-1)\right\}^2}_{\text{(II)}}$$

$$+ \underbrace{\sum_{i=1}^{M}\sum_{j=1}^{n_u}\left\{w_j^u\left[u_j(k+i-1) - \overline{u_j}\right]\right\}^2}_{\text{(III)}}$$

(I): the first term represents the weighted sum of squared deviations ($S_y(k)$).

(II): the second term represents the weighted sum of controller adjustments ($S_{\Delta u}(k)$).

(III): the third term represents the weighted sum of manipulated variable deviations ($S_u(k)$).

k: current sampling interval, $k + i$: future sampling interval (within the prediction horizon).

y_j: jth control variable, y_j^{set}: set point of jth variable, u_j: actuator for jth control variable.

P: number of control intervals in the prediction horizon. n_y: number of plant outputs.

$\left[y_j^{set}(k+i) - y_j(k+i)\right]$: predicted deviation for output j at interval $k + i$.

M: number of intervals in the control horizon. n_u: number of manipulated variables.

$\Delta u_j(k+i-1)$: predicted adjustment in manipulated variable j at the future (or current) sampling interval k + i − 1.

w^y_j: weight for output j (the output weights let you dictate the accuracy with which each output must track its set point.

$w_j^{\Delta u}$: rate weight (it penalizes the incremental change rather than the cumulative value and increasing this weight forces the controller to make smaller, more cautious adjustments).

w_j^u: input weight (this weight helps to avoid large deviation from the nominal value of actuators).

$\overline{u_j}$: nominal value of actuator for input j.

The various MPC algorithms may use different cost functions for obtaining the control law [49]. MPC controller design or generation involves many parameters that strongly affect performance and robustness of the controller, making controller tuning a nontrivial task [40]. The optimization based method ITAE (Integral of Time Absolute Error [36]) can be employed for tuning MPC parameters. The output weight (w^y_j) lets you dictate the accuracy with which each output must track its set point. Specifically, the controller predicts deviations for each output over the prediction horizon. It multiplies each deviation by the output's weight value and then computes the weighted sum of squared deviations. The weights must be zero or positive. If a particular weight is large, deviations for that output ($S_y(k)$) dominate. One of the controller's objectives is to minimize $S_y(k)$. Thus, a large weight on a particular output causes the controller to minimize deviations in that output (relative to outputs having smaller weights). Increasing rate weights ($w_j^{\Delta u}$) forces the controller to make smaller, more cautious adjustments. The input weight (w_j^u) helps to avoid large deviation from the nominal value of actuators. However, when a sustained disturbance or set point change occurs, the manipulated variable must deviate permanently from its nominal value. Using a nonzero input weight forces the corresponding input back toward its nominal value. Therefore, in case of variable set point, this weight should be zero. Regarding prediction and control horizon, it is good practice to specify the prediction and control horizons such that $P - M \gg td_{max}/\Delta t$, where P is the prediction horizon, M is the control horizon, td_{max} is the maximum delay, and Δt is the control interval [5].

3 Case Study 1: Direct Compaction Continuous Tablet Manufacturing Process

3.1 Process Description

A direct compaction continuous tablet manufacturing process is shown in Fig. 2. As shown in the figure, there are three gravimetric feeders to provide the necessary lubricant, API and excipient. The feeders contain a hopper that can hold a certain amount of material and a rotating screw to change the flow rate. These feeds are then supplied to a blender to generate a homogeneous mixture. Before the blending process, a milling step can be used for delumping purposes, if required. After the blender, the blended powder is sent directly to the tablet press through a feed frame. The final compacted tablets are obtained from the tablet press, and among them, some tablets are sent for dissolution testing. This process flowsheet has been simulated using the simulation software gPROMS

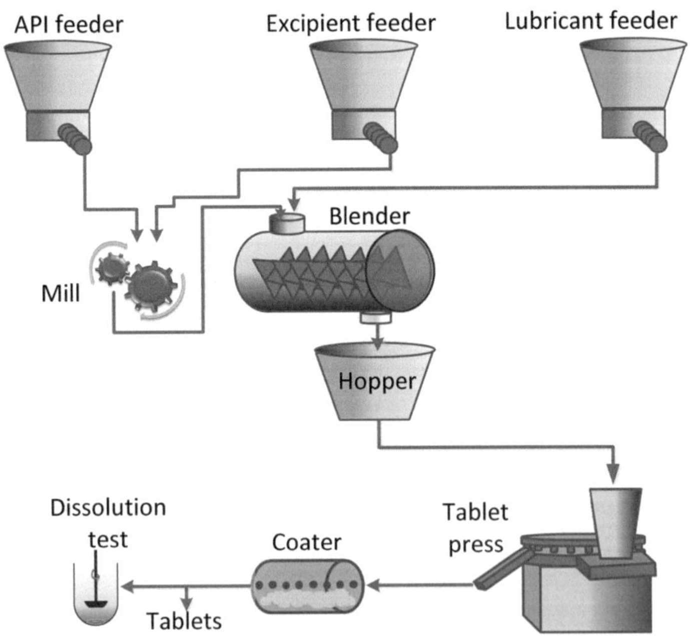

Fig. 2 Continuous direct compaction tablet manufacturing process

(Process Systems Enterprise). The methodology to develop an integrated process flowsheet has been described by Boukouvala et al. [14]. The process conceptually represents a pilot plant situated at the Engineering Research Center for Structured Organic Particulate Systems (ERC-SOPS), Rutgers University in Piscataway, NJ. Some details of the pilot plant have been previously reported [23, 41], and the open-loop [13–24] as well as closed-loop [4–6, 50, 51] operation has been extensively studied. A flexible continuous tablet manufacturing process that incorporates either a direct compaction [5], dry granulation (using roller compactor) [6], or wet granulation (using a wet granulator) route has also been previously developed [41, 42].

3.2 Process Model

A modular flowsheet model of a flexible continuous tablet manufacturing process consisting of the direct compaction (DC), roller compaction (dry granulation or DG), and wet granulation (WG) routes has been developed using gPROMS simulation software (PSE) [5, 6, 13, 14]. The detailed model of each unit operation incorporating Population Balance Models (PBM) [18] and/or the Discrete Element Method (DEM) [16] is developed and saved in a model library which can then be used to draw any combination of the process flowsheet using a modular plug-and-play concept. This flowsheet model provides a powerful virtual experimentation tool to study the different operational scenarios, to perform optimization, to design the plant, to design the control system [4–6, 12, 41, 42], and to provide operator training [13, 14]. The PID controllers can be easily added in the flowsheet model to analyze different controller configurations [6].

The integrated flowsheet model for direct compaction continuous tablet manufacturing process that has been used for design of the control system has been previously reported [5, 6, 13, 14]. The detailed developments of these models are reported elsewhere as summarized here. The mathematical model for powder blending, an important but complex unit operation, has been previously developed [16–18]. The model for the tablet compression process is previously reported in Singh et al. [25]. This model is based on the Kawakita powder compression model [43] and the tablet hardness model described in Kuentz and Leuenberger [44]. The dissolution model was adapted from Kimber et al. [45].

3.3 Pilot Plant

A continuous direct compaction tablet manufacturing pilot plant is situated at ERC-SOPS, Rutgers University, Piscataway, NJ. The snapshot of the pilot plant is shown in Fig. 3 (whole plant is not shown) [41]. The pilot plant is built in three levels at different heights to take advantage of gravitational material flow. The top level is used for feeder placement and powder storage, the middle level is used for delumping and blending, and the bottom level is used for compaction. Each level consists of a 10x10 square feet

Fig. 3 Direct compaction tablet manufacturing pilot plant (adapted from Singh et al., 2012 [41]) (whole plant is not shown): (**1**) Feeders, (**2**) Co-mill, Blender, (**3**) Tablet press

working area. There are three gravimetric feeders—with the capability of adding more—that feed the various formulation components (API, excipient, etc.). A mill (Glatt) is also integrated after the feeder hopper primarily for de-lumping the powders and creating contact between components. The lubricant feeder is added after the mill to prevent over lubrication of the formulation. These feed streams are then connected to a continuous blender (Glatt) within which a homogeneous powder mixture of all the ingredients is generated. Subsequently, the outlet from the blender is fed to the tablet press.

3.4 Design of Control System for a Continuous Tablet Manufacturing Process

The design of a process control system is an iterative procedure that involves the identification of critical controlled variables, coupling of the controlled variables with suitable actuators (manipulated variables), selection of suitable monitoring tools, selection of control strategy followed by tuning of controller parameters, and finally, model-based closed-loop performance assessment [6, 8]. The design control system can then be implemented in the manufacturing plant through the control platform (e.g., Delta V,

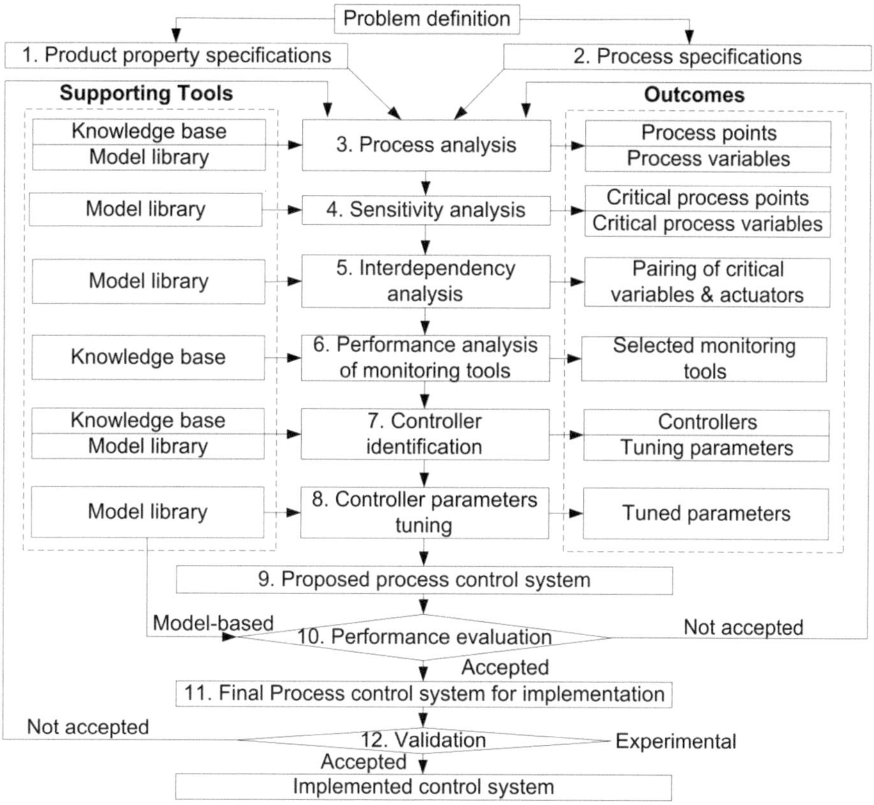

Fig. 4 Systematic methodology for design of control system (adapted from Singh et al., 2012 [6]; 2009 [9])

Emerson) and data communication protocol (e.g., OLE Process Control (OPC)) for experiment-based closed-loop performance evaluation [5]. The systematic methodology for the design of the control system is shown in Fig. 4. A knowledge base and a model library are the important supporting tools for the design of the control system. The knowledge base consists of information related to the critical process variables, corresponding actuator candidates, sensors and controllers [46]. A control expert can also provide such information. The model library consists of the process model of the continuous tablet manufacturing process as described in Sect. 3.2.

Product property specifications and process specifications have been provided as the inputs to the design methodology. The process is then analyzed to make a list of involved unit operations and corresponding process variables. In the next step, the critical control variables have been identified for a direct compaction continuous tablet manufacturing process. These include API composition at blender outlet, powder level of instrumented hopper placed in between blender and tablet press, tablet weight and tablet hardness at tablet press. API composition at blender outlet needs to be controlled in order to assure the drug concentration. Powder level of instrumented hopper needs to be controlled to ensure the presence of powder in front of the NIR sensor. Tablet weight and hardness are the critical quality attributes (CQAs) as these affect tablet dissolution.

For one control variable, there may be many actuator candidates. Therefore, the best actuator for each control variable needs to be selected to make the pair of control variables and actuators. In general, the actuator should have a large, rapid and direct effect on the control variable with minimal process delay time. The relative gain array (RGA) method [47] or dynamic sensitivity analysis method (also called interdependency analysis) [6] can be used to identify suitable actuators. In RGA method, an array of relative gain needs to be prepared. The relative gain is defined as the ratio of open-loop gain and closed-loop gain and can be calculated as follows [47]:

$$\lambda_{\mathrm{ij}} = \frac{(\partial y_i/\partial u_j)_u}{(\partial y_i/\partial u_j)_y} = \frac{\text{open-loop gain}}{\text{closed-loop gain}}$$

Where, y_i is the ith control variable and u_j is the jth actuator candidate. Open-loop gain is obtained by partial differentiation of y_i with respect to u_j while assuming that other actuator candidates are constant. The closed-loop gain is obtained by partial differentiation of y_i with respect to u_j while assuming that other control variables are constant. RGA is then formed as follows [47]:

$$RGA = \begin{array}{c} \begin{array}{cccc} u_1 & u_2 & \dots & u_n \end{array} \\ \begin{bmatrix} \lambda_{11} & \lambda_{12} & \dots & \lambda_{1n} \\ \lambda_{21} & \lambda_{22} & \dots & \lambda_{2n} \\ \dots & \dots & \dots & \dots \\ \lambda_{n1} & \lambda_{n2} & \dots & \lambda_{nn} \end{bmatrix} \end{array} \begin{array}{c} \\ y_1 \\ y_2 \\ . \\ y_n \end{array}$$

In the RGA matrix, the sum of the elements present in each row or in each column must be unity (e.g., $\sum_{j=1}^{n} \lambda_{1j} = 1; \sum_{i=1}^{n} \lambda_{i1} = 1$). The values in the array describe the relationship between the input (actuator) and output variables (control variables). Negative values indicate an unstable relationship. A value of zero indicates no relationship. A value of 1 indicates that specific input variable is the only influence on that output variable. A value between 0 and 1 indicates interactions among the control loops. The RGA method is based on the steady state analysis and does not take into account the dynamic behavior of the process, while the dynamic sensitivity analysis method takes into account the effect of actuator candidates during the duration of process operation. In a dynamic sensitivity analysis method, the actuator candidates are perturbed (e.g., −3 to +3 %) and the absolute percent change in controlled variable $\left(100\left|\frac{Y_0^j(t)-Y_i^j(t)}{Y_0^j(t)}\right|\right)$ is analyzed [6]. The actuator that results in the highest change in control variable is selected as the final actuator. In dynamic sensitivity analysis, the effects of actuator candidates on a controlled variable can be analyzed in the whole operational period. The blending process is considered here as a demonstrative example for actuator selection where one of the control variables is the total flow rate. For total flow rate control, two actuator candidates (rotation speed of API feeder and rotation speed of blender) have been considered for comparison. The actuator candidates have been perturbed (from +3 to −3 % with a step size of 1%) and the absolute percentage changes in controlled variables have been analyzed. The effects of actuator candidates on total flow rate (controlled variable) are shown in Fig. 5a. As shown in the figure, at steady state the rotation speed of the feeder that provides the API is much more sensitive in comparison to blender rotational speed. During startup, the blender rotational speed also has significant sensitivity on controlled variable that makes the blender process control a difficult task. The sensitivity of the actuator candidates on the controlled variable can be compared at steady state through a two dimensional projected view as shown in Fig. 5b. It should be noted that API, excipient and lubricant need to be maintained at a certain ratio through a ratio controller, therefore on changing the API flow rate the excipient and lubricant flow rate will also change. The actuators for other control variables are shown in Fig. 6.

After selecting the final actuator, the next step is to select the sensors for real time measurement of the control variables. Accuracy, precision, response time, operating range, resolution, sensitivity, drift and cost are the important factors that need to be taken into account in selection of the sensors [46]. One alternative for monitoring the controlled variables is shown in Fig. 6. API composition is measured by the NIR sensor, powder flow rate from feeders are measured by load cells (built within feeders), the powder level is measured by a webcam (capacitance sensor is also an option), tablet weight and hardness are measured by a Checkmaster (Fette) and compression force is measured via strain gauges.

After selecting the sensors, the next step is to select the type of controller: PID, MPC or hybrids. In case of PID, subsequently one needs to decide whether to use a single-loop or a cascade one. In cascade mode, the master controller generates a set point for the slave controller. The cascade controller can improve the control system performance over a single-loop controller in many instances. For example, when a large time delay is involved or when disturbances affect a measurable intermediate that directly affects the controlled variable. Also, cascades are preferred when the gain of

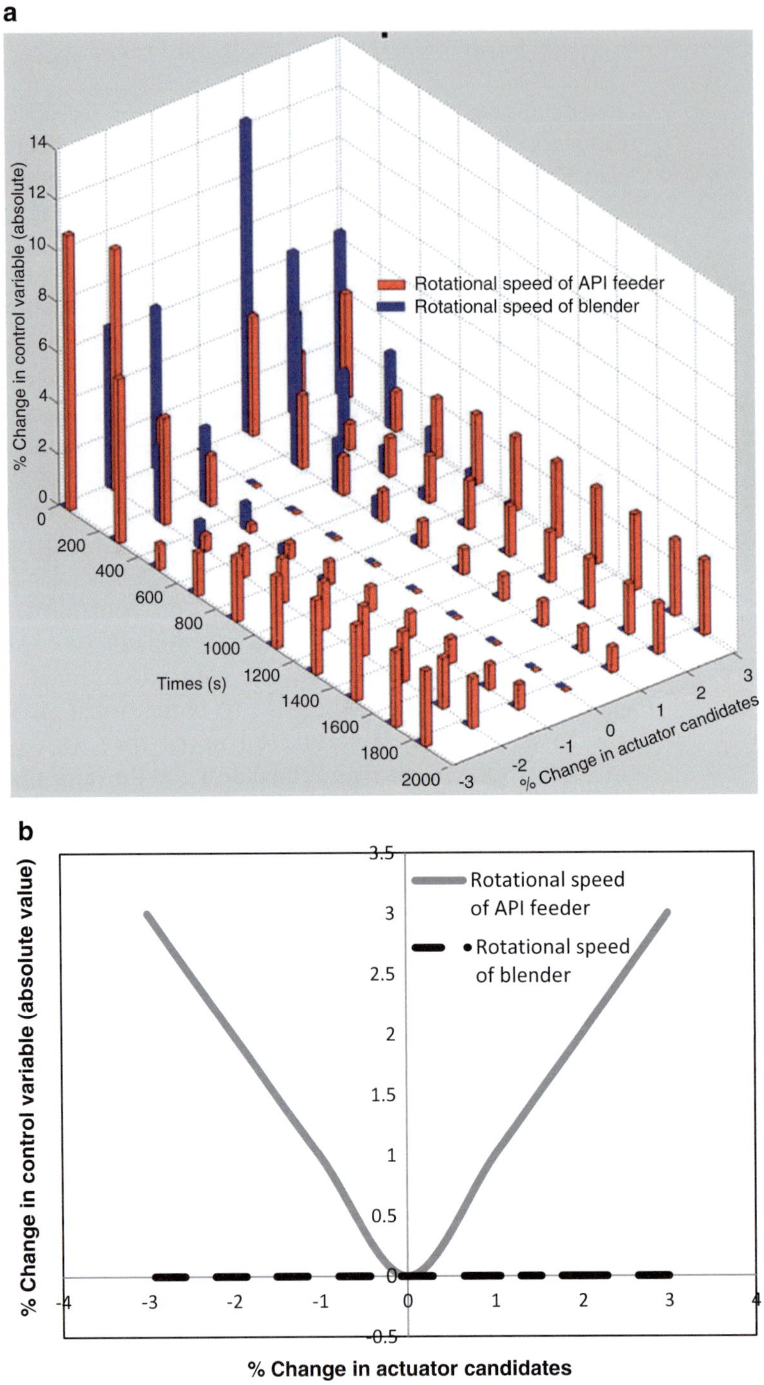

Fig. 5 (**a**) Example of actuator selection (dynamic sensitivity analysis) (**b**) Example of actuator selection (dynamic sensitivity analysis): steady state plot at $t = 1800$ s

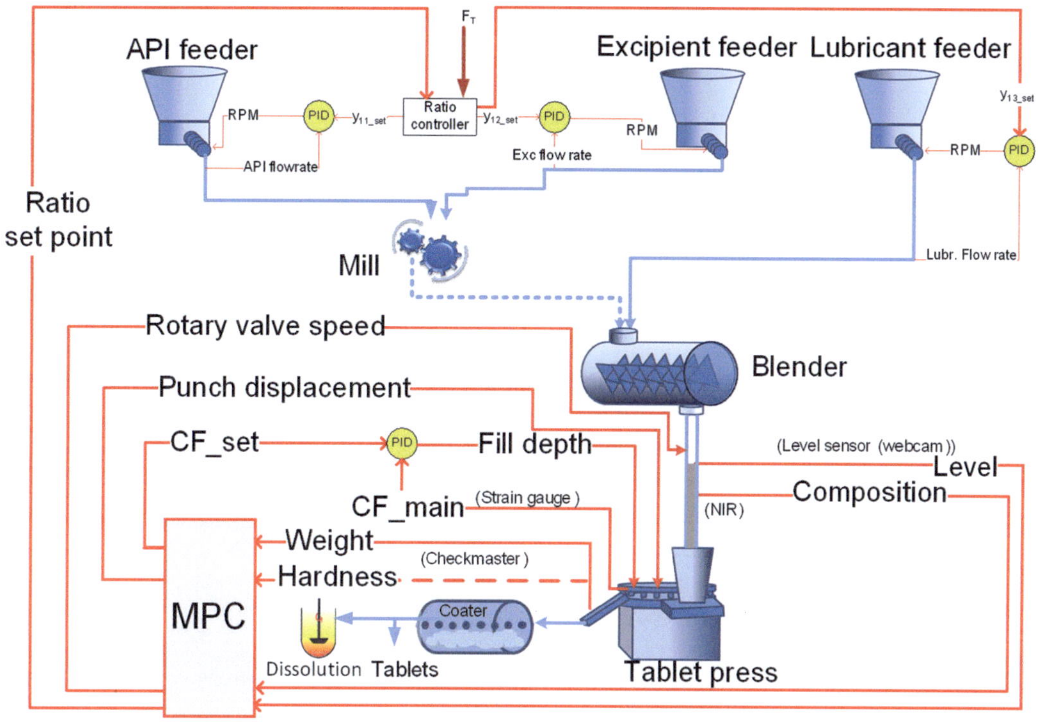

Fig. 6 Closed-loop process flowsheet of direct compaction tablet manufacturing process

the secondary process, including the actuator, is nonlinear and more difficult to tune.

The next step is to tune the controller parameters. Heuristic methods [37] or optimization based methods (e.g., Integral of Time Absolute Error (ITAE)) can be employed to tune the controller parameters [36]. The controller configuration for continuous tablet manufacturing process is shown in Fig. 6. As shown in the figure, a hybrid MPC-PID control scheme in which MPC is placed at supervisory level and PID is placed under MPC as a slave controller has been used. Finally, the controller parameters are tuned. The PID controller parameters are tuned using the Ziegler and Nichols [37] method and ITAE optimization-based method [36]. MPC parameters (prediction horizon, control horizon, output weight, rate weight, input weight) are specifically tuned using the ITAE method.

3.5 Designed Hybrid MPC-PID Control System

A control system for the direct compaction continuous tablet manufacturing process has been designed using the methodology described in earlier sections. The configuration of the designed control scheme is shown in Fig. 6. A combination of model predictive control (MPC) and the more commonly used proportional integral derivative (PID) is used for the control strategy since MPC is better at handling process delay and process variable interactions and can be tuned easily. An NIR sensor is placed at the blender outlet for blend composition measurement. This is the input for the master controller, which generates the feeder ratio set point. Based on this ratio set point and the total powder flow rate, the individual flow rate set points for API, excipients and lubricant feeders are calculated and then controlled by manipulating the respective feeder RPMs using built-in feeder controllers. The powder level in the chute can be controlled by manipulating the rotary valve speed. Alternatively, turret speed can be also manipulated to control the powder level. In the tablet press, the tablet weight is controlled through a cascade control arrangement using a master loop and a slave loop. Master loop provide the set point for the slave controller, which controls the main compression force by manipulating the fill depth. The tablet hardness has been controlled by manipulating the punch displacement. The tablet weight is measured and controlled more frequently. Note that the hardness control loop is activated only when the measured hardness deviates by a certain percentage (e.g., 2 % of set point) from the desired set point.

3.6 Model-Based Performance Evaluation

The performance of the designed control system can be evaluated using the process model. For performance evaluation, either the control system can be implemented in the process model simulated in a simulation tool or the control system can be implemented in a control platform and the control platform can be integrated with the simulation tool (in which the model has been simulated). The latter option is called onsite design of the control system. In onsite design of the control system, the control loops are implemented in a real control platform (e.g., DeltaV, PCS7) and the input and output of each control loop is connected with the process model simulated in a simulation tool [4]. Some of the advantages of onsite design of the control system are that the scenario is more similar to the real plant, control loops do not need to be rebuilt, identified linear models for MPC and controller parameters can be directly used for closed-loop plant operation, and the input and output of the control loops can be easily switched with the input and output of the plant for closed-loop operation. The design and implementation of the control system is an iterative procedure. Therefore, onsite design of the control system can significantly save time and resources. For onsite design, the process model needs to be integrated with the control platform. Commercially available control platforms (e.g., DeltaV (Emerson), PCS7 (Siemens)) can be easily connected with the simulation tools via either MS Excel or any intermediate tool that has the OPC feature. If the control platform and the simulation tool both have the OPC feature (e.g., MATLAB) then they can be directly connected without any intermediate software tool. Most of the simulation tools (e.g., gPROMS (PSE)) can be easily connected with MS Excel and therefore with the control platform. The flowsheet model has been integrated with DeltaV using the gORUN feature of gPROMS, MS Excel and I/GEAR connecting software. The model is simulated in gPROMS simulation software and the control strategy has been implemented in DeltaV control studio. The gORUN feature of gPROMS is used to connect gPROMS with MS Excel, and I/Gear is used to connect MS Excel with DeltaV. The signal of the control variable generated from the model simulation is sent to DeltaV and acts as the input to the controller that generates the actuator. The controller output (actuator) is sent to the process model and acts as the input to calculate the new signal of the control variable. This flowsheet model can also be integrated with the control platform via MATLAB OPC toolbox. The integration of the flowsheet model with MATLAB has been previously demonstrated [5]. The flowsheet model (simulated in gPROMS) is therefore OPC compliant, and can be integrated with any actual operating and control platform (DeltaV, PCS7) to facilitate onsite control system design. This facilitates tuning of controller parameters and identification of the linear model for MPC, as well as performance evaluation of the control system before running the actual plant to reduce cost and improve performance.

The model-based closed-loop performance of the control system is shown in Fig. 7. Figure 7a shows that the API composition at the blender outlet has been controlled at the given set point.

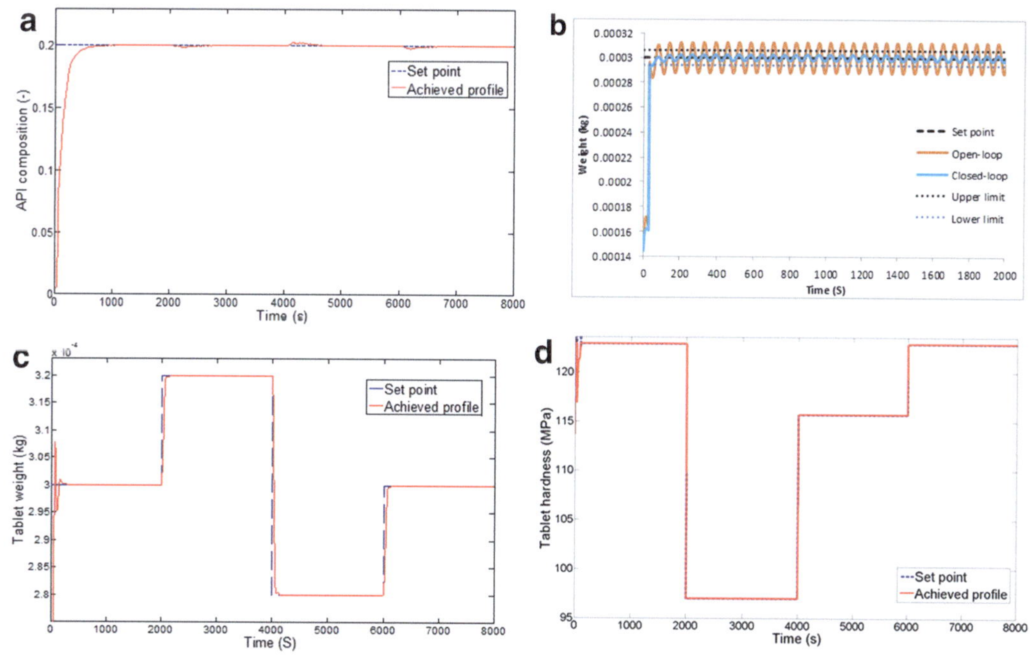

Fig. 7 Closed-loop process performance (simulation-based). (**a**) API composition, (**b**) Tablet weight (disturbances rejection), (**c**) Tablet weight (set point tracking), (**d**) Tablet hardness (set point tracking)

Figure 7b shows the disturbance rejection ability of the tablet weight control loop. A sinusoidal disturbance has been introduced. In the open-loop scenario, the response is oscillatory because of sinusoidal disturbances, while in the closed-loop scenario the response is within the control limits (2 % of the set points) because of the control action. The set point tracking ability of the controllers is demonstrated in Fig. 7c, d. Figure 7c shows the performance of the tablet weight control loop while Fig. 7d shows the performance of the tablet hardness control loop.

3.7 Implementation of Control System

A systematic methodology to implement the control system into the continuous tablet manufacturing process is shown in Fig. 8 through a work-flow diagram (adapted from [4]). The designed control system that includes the control variables, actuators, control strategies and controller parameters are the inputs for the methodology [4, 6, 8].

The monitoring tools (sensors) [46], control hardware and software required for implementation of the control system are selected in **step 1.** In this step, the online monitoring tools for all control variables, the appropriate control platform (e.g., DeltaV, Emerson) in which the control strategy can be implemented, the control hardware needed to communicate data with the plant and the software tools needed to close the loop are selected. Spectroscopic techniques are the preferred option for monitoring the pharmaceutical tablet manufacturing processes because of their non-destructive nature and fast-acquisition times. They can be used to monitor quality attributes such as the API composition of a powder blend, blend uniformity and API tablet potency (total API content of a tablet). The most commonly used spectral

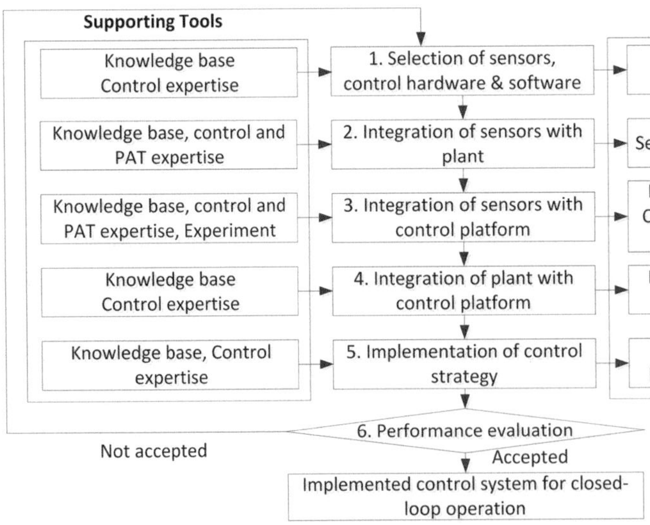

Fig. 8 Systematic methodology for implementation of the control system

techniques in the pharmaceutical industry are near infrared (NIR) and Raman. NIR was used for this study.

In **step 2,** the sensors are integrated with the plant to facilitate real time online/inline monitoring. For continuous manufacturing involving a solid dosage form, the sampling point is normally at the outlet of each unit operation. The sampling point also depends on whether the control mode is feedback or feed forward. An interface (chute) is needed to integrate the inline measuring probe. The design of the chute depends on several factors, including the type of sensor considered, the variables that need to be monitored and the process equipment. For example, if the objective of the NIR sensor is to continuously measure the characteristics of the flowing powder, then a powder layer needs to be present in front of the probe every time spectra is acquired. Therefore, the sampling interface should be specifically designed to provide an area where a new powder layer can continuously form by displacing the old powder layers. After constructing and mounting the interface (chute), the sensor and its respective data acquisition software are integrated into the manufacturing line and the control architecture.

The measured signal is then sent to the control platform in **step 3.** For spectroscopic sensors, this step includes the development of a calibration model [22], the integration of the sensor operating software with the online prediction tools, the integration of the online prediction tools with the PAT data management tool and the integration of the PAT data management tool with the control platform via the OPC communication protocol. Through this step, the measured signal and any other signals (e.g., alarms, warnings, etc.) can be sent to the control platform to be recorded in the historian, and any data from the control platform can be sent back to the PAT data management tool for data storage, inspection and auditing purposes. A multivariate data analysis tool (e.g., Unscrambler X, CAMO) was used to compute principal component analysis (PCA) and partial least square (PLS) models to further develop a calibration model required for prediction of desired variable from the spectra. PLS is a quantitative regression method that looks for correlation between spectral data (X-matrices) and the independent variable of interest (Y-vector). An online prediction tool (e.g., OLUPX, CAMO) is used for real time prediction of the control variable. A PAT data management tool (e.g., synTQ, Optimal) is used to store the data for plotting, analysis, auditing and regulatory inspections and OPC communications. Alternatively, MATLAB OPC toolbox can also be used to communicate the measured signal to the control platform. The PAT data management tool is then integrated with the control platform using the OPC communication protocol. The integration of other type of sensors (e.g., thermocouple) is easier and can be directly connected with the control platforms.

The plant needs to be integrated with a control platform so that it can be operated through a centralized user interface and receive actuator signals. At a minimum, all the actuators need to be integrated with the plant and control platform. The plant is integrated with the control platform in **step 4.** In this step, the plant hardware/unit operations are connected with the control platform so that they can be operated through the control interface. Standard industrial communication protocols such as Fieldbus or EtherNet/IP have been used to make these connections [48]. The control hardware (controller) is connected to the operating computer in which the control software is already installed. The required drivers are added to the control platform, so that the plant equipment can be accessed through the control platform. The control software communicates with the control hardware, which communicates with the plant via fieldbus/serial ports. Feeders are connected with the control hardware through DeviceNet and control hardware is connected with the control platform. The signals of blender RPM can also be directly sent to the blender through a serial port.

To connect the plant input (actuators) with the plant output (control variable), the control strategy is implemented in **step 5.** The input for the controller comes from the sensors and the output from the controller goes to the plant. The control studio (toolbox) of the control platform has been used to add the controller. The controller input has been connected with the measured signal (control variable) and the control output has been connected with the actuator. The linear process model needed for MPC has been identified. Step changes in the actuator (input) need to be introduced and a linear model needs to be generated based on the output (control variable) response. The parameters for PID (K_C, K_I, K_D) and MPC have also been identified.

Finally, the performance of the control system is evaluated in **step 6** through closed-loop operation. If the performance is satisfactory, then the implemented control system can be considered as the final control system for closed-loop manufacturing. Otherwise **steps 1–6** need to be repeated until satisfactory performance is obtained.

3.8 Closed-Loop Operation of a Direct Compaction Continuous Tablet Manufacturing Process

The developed graphical user interface for closed-loop operation of the direct compaction continuous tablet manufacturing process is shown in Fig. 9. In the graphical user interface, three feeders (for the API, excipient and lubricant), a mill, a blender and a tablet press have been included. Each unit can be also operated individually. As shown in the figure (see right hand corner), the user has the option to run the plant in a closed-loop or an open-loop scenario. In the closed-loop scenario, there are three options for the control strategy. These options are PID, PID with Smith predictor, and MPC. There is also an option to run the control strategy in

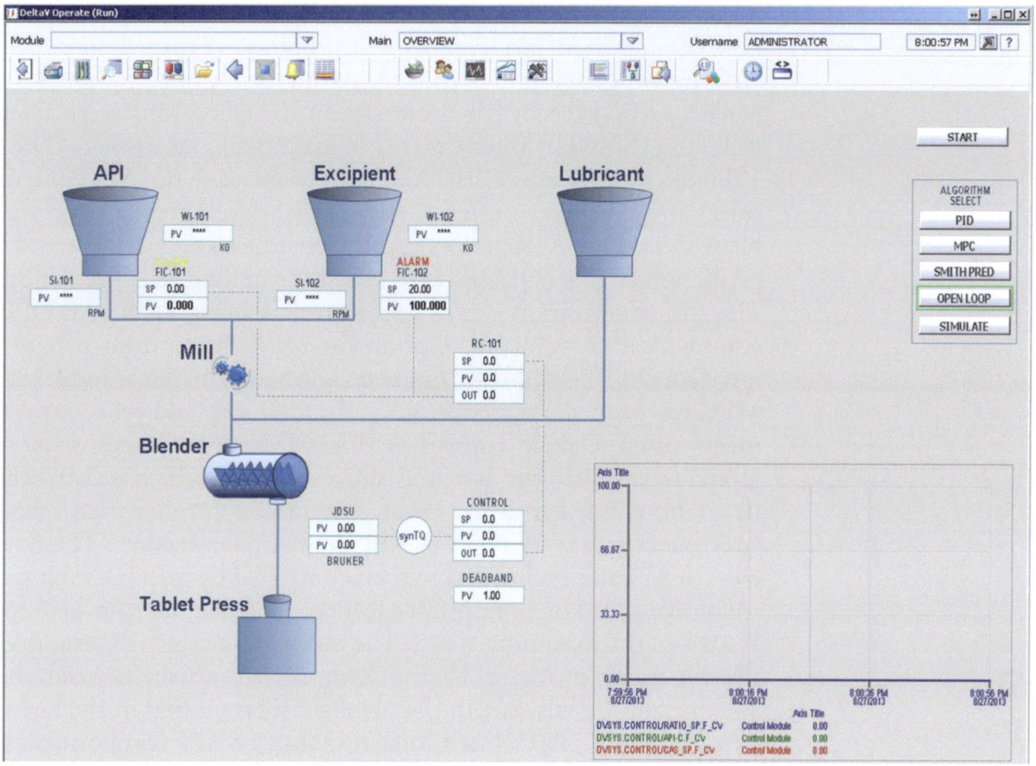

Fig. 9 Graphical user interface for operation of continuous tablet manufacturing process. Flexible control system: PID, PID with Smith predictor, MPC

simulation mode. In simulation mode, the input and output signals of the controller are linked with the process model instead of the plant. This option is useful for onsite control system design, tuning of controller parameters, development of a linear model for MPC and training purposes. The response of the control variable can also be seen directly at the user interface.

The developed framework (*see* Fig. 9) has been used for closed-loop operation of the direct compaction continuous tablet manufacturing process. The control of the API composition through MPC has been considered here as a demonstrative example. There is a cascade control arrangement where the master controller is MPC and the slave controllers are inbuilt PIDs. The input signal for MPC is the API composition (control variable) and the output signal from MPC is the ratio set point. The ratio set point acts as the input for a ratio controller. The ratio controller provides the flow rate set points for the slave PID controllers built into the feeders to control the API, excipient and lubricant flow rates. The final actuators are the rotational speeds of the feeders.

Finally, the plant has been run in closed-loop scenario to validate the different integrations and proposed control framework. One feeder contains the API (APAP (acetyl-para-aminophenol)) while a second feeder has the excipient (Silicified Microcrystalline Cellulose (SMCC)) mixed with 1 % magnesium stearate. A JDSU 1700 micro NIR sensor has been used to measure the API composition at the blender outlet. A PLS model developed in UnscramblerX is used for NIR prediction. Unscrambler Process Pulse and a prediction engine (OLUPX) are used for real time NIR prediction. The API composition is then sent to DeltaV using MATLAB OPC communication protocol. Option for MPC mode from the user interface (*see* Fig. 9, MPC) has been selected. The linear model for MPC has been developed in silico through step and pulse experiments using DeltaV control platform [50]. MPC uses a linear model to calculate the actuator (ratio set point) which is then sent to a ratio controller as the input. The ratio controller then calculates the flow rate set point of API and excipient feeders. The flow rates of API and excipients are then controlled by manipulating the rotational speed of its respective feeders. The blender speed is kept at 30 % of the maximum speed. The different variables for example, the set point provided by the user, signals obtained from the sensors, and signals sent to the plant are then plotted in the historian of DeltaV. The closed-loop response of API composition is shown in Fig. 10 [50]. The figure shows the API composition set point, API composition measured by NIR, filtered API composition signal and the actuator response. Filtered API composition signal is the input for the model predictive controller. As shown in the figure, the API composition is controlled at 0.17 set point then a step change is introduced from 0.17 to 0.12. The figure shows that the controller is able to track the step change in set point. The actuator response is also reasonable.

The performance of the hybrid MPC-PID control system has been compared with the performance of base level PID controller in Fig. 11. In this study, the auto-tuning capability of DeltaV has been used to tune the PID controller parameters (Blevins et al., 2013). The set point has been changed from 0.17 (17 % APAP) to 0.12 (12 % APAP). The acceptable control limits have been specified to be ±0.15 of set point. The figure shows that the MPC response is within the control limits at 17 % API composition while the PID response violates the control limits. When the step change is introduced from 17 to 12 % API composition the MPC brings the signal faster to the new set point in comparison to PID controller, meaning that MPC has less rise time compared to PID. At the new set point, the MPC response is within the control limits except a very minor violation at two points. PID controller has been selected for comparison to analyze the advantages of advanced hybrid MPC-PID control scheme in comparison to basic level controller. Note that the performance of both hybrid

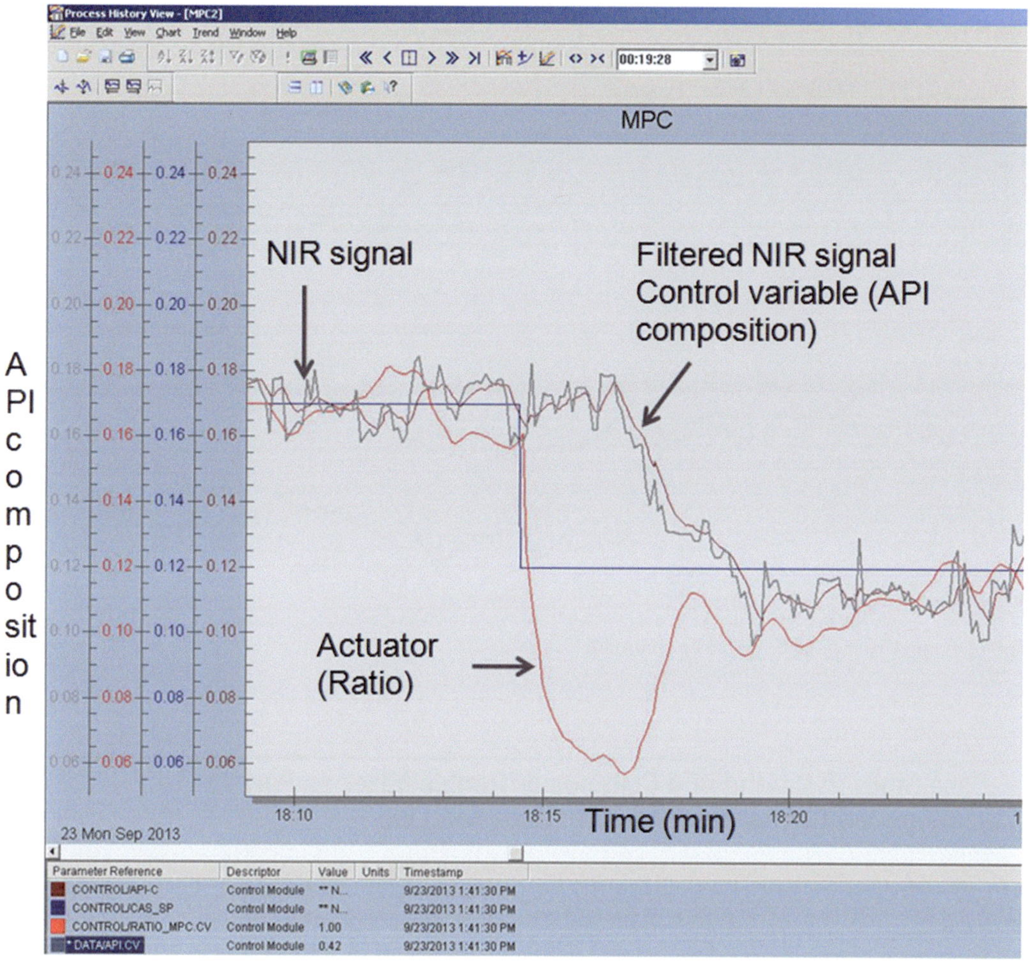

Fig. 10 Closed-loop process operation (control of API composition at blender outlet)

MPC-PID and PID control schemes can be further improved by integrating the other performance improvement techniques and tools and by reducing the measurement noise. The deviation of achieved profile from the set point is shown in Fig. 11. As shown in the figure, the MPC error is within the acceptable limit except for a short duration where step change has been made. The powder at blender outlet need to be diverted for this short duration. In case of PID, the error violates the acceptable limits during most of the operating period [50].

The results presented here validate that the NIR sensor has been integrated with the continuous plant and the sensor output has successfully communicated with the control platform (DeltaV) via OPC communication protocol. Real time online prediction via NIR signal has been accomplished. The controller output has communicated with the plant [4, 50]. The performance of MPC has been found to be better than PID controller.

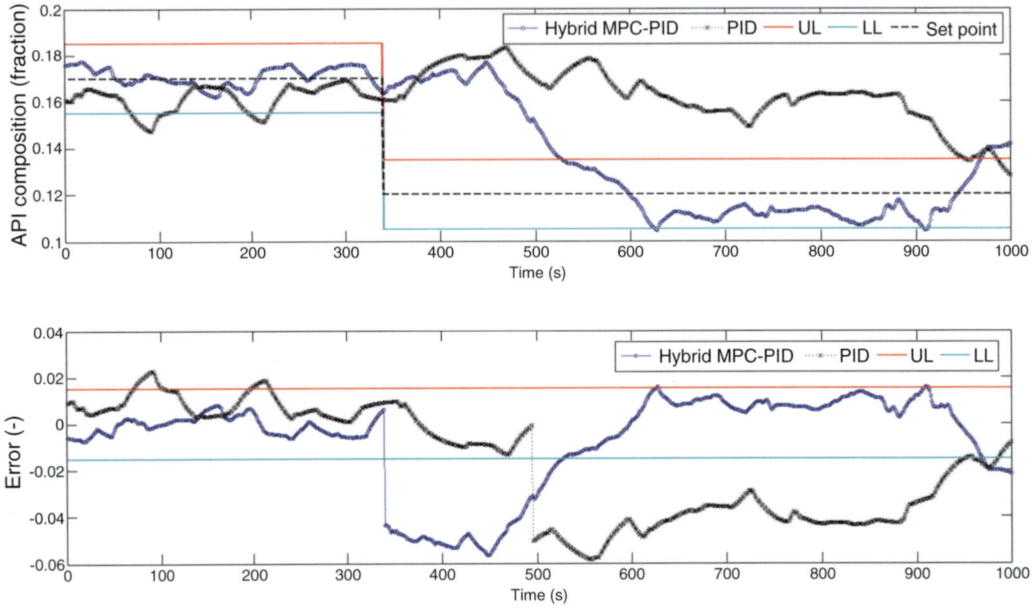

Fig. 11 Comparison of MPC with PID controller

4 Case Study 2: Control of a Continuous Tumble Mixer as Part of an Integrated Continuous Direct Compression Line

4.1 System Description

The continuous production of tablets through direct compression can be achieved through different models, all using automatic feeders, continuous mixers and a tablet press. In this second study, the continuous mixer is a tumble-based one. As can be seen in Fig. 12, two automatic feeders are integrated to the continuous tumble mixer, which connects directly to the tablet press.

Between the mixer and the tablet press there is a conveyor belt, which serves two purposes. One is to move the powders from the mixer to the tablet press and the second, and more important, is to maintain a constant powder density in front of the NIR probe. This NIR is used to predict the blend API concentration, which is used by the control algorithm in the industrial control system to guide the process to maintain the desired quality.

In this continuous tumble mixer powder enters the tumble via the center of the front wall and then escapes through a series of orifices in the radial wall. The back wall of the drum is connected to a variable speed motor, which allows the manipulation of the mixer speed to impact the mixing performance. It has been demonstrated that this continuous tumble mixer can achieve the same level of performance as commercially available continuous mixers.

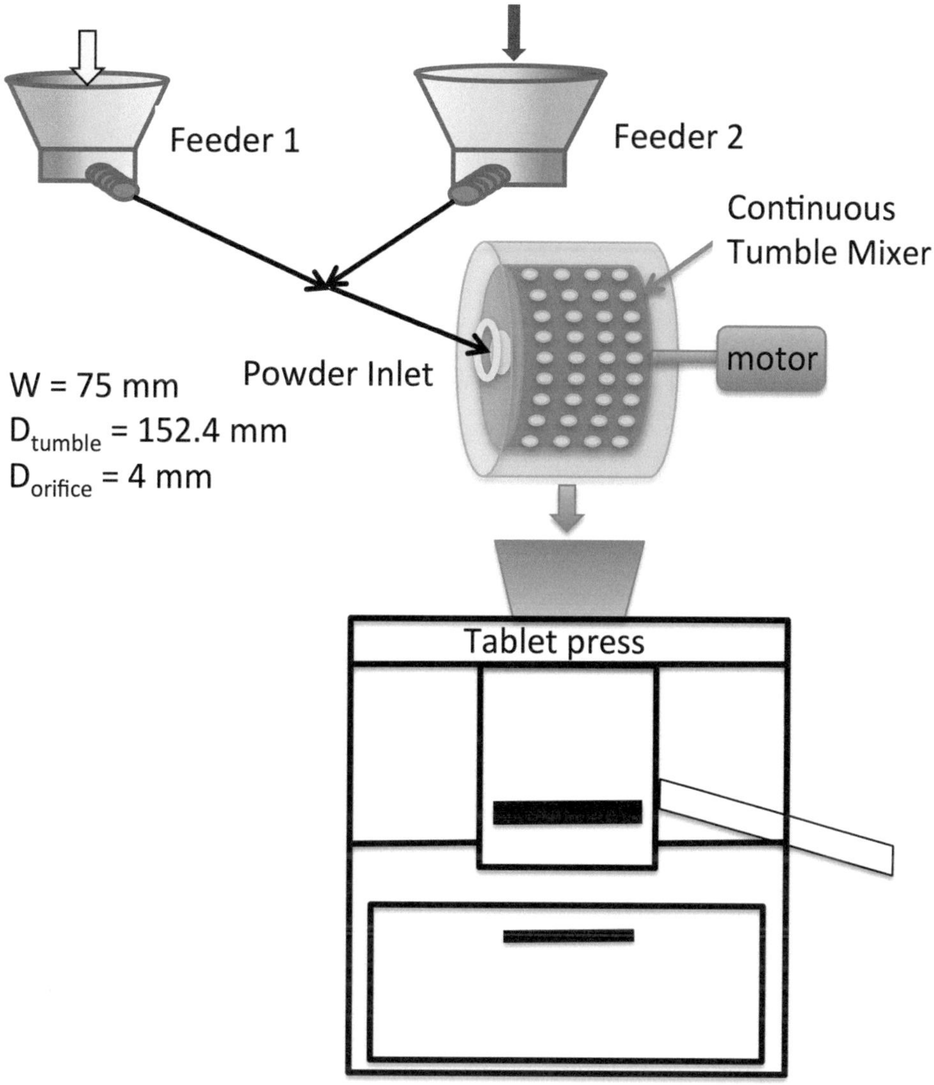

Fig. 12 Schematic of the integrated system for the continuous direct compaction tablet production using tumble mixing

As the tumble rotates, counterclockwise in this case, the powders inside form an avalanche that enables mixing. In general, high rpms exert a large centrifugal force in the powders avoiding an adequate mix. On the other hand, low rpms do not create an adequate avalanche and the powders leave shortly after they enter (low residence time), which produces a low level of mix.

Therefore, the control of the mixer together with the other units is necessary to avoid any inadequate operation. In the tumble mixer, the control variables are the outlet powder concentrations, the total flow rate, and the blend homogeneity. As in the case of the continuous paddle mixer, the outlet concentration is controlled by

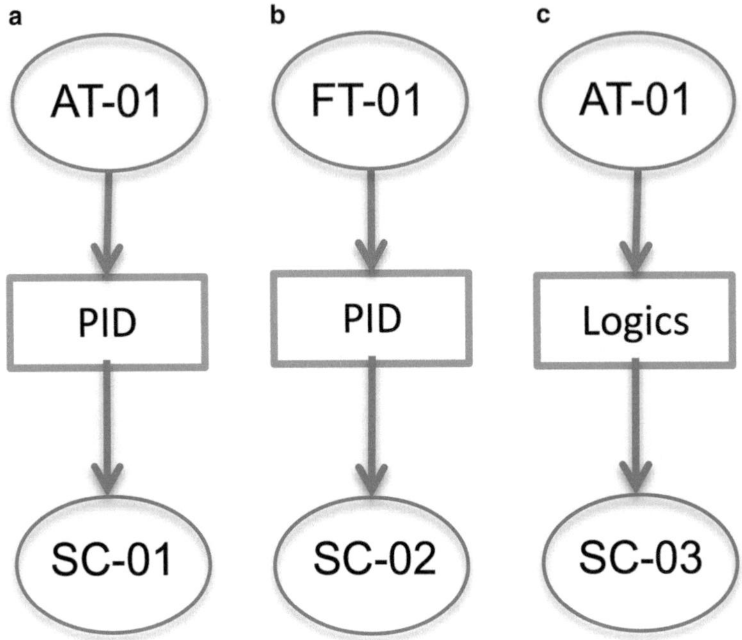

Fig. 13 SAMA drawings of the control loops in the tumble mixer: (**a**) API concentration, (**b**) total flow rate, and (**c**) RSD

manipulating the API inlet flow rate, whilst the total flow rate is controlled by the excipient inlet flow rates. For blend homogeneity, the variable to be manipulated is the mixer rpms.

Figure 13 (SAMA drawings) describes each loop with its corresponding sensor, control element and the control strategy. Notice that the name of each accessory is the same that is used in the P&ID of the system (Fig. 14) and the tags (inputs and outputs) in the control algorithm programmed in the industrial control system. This is critical to avoid any confusion, especially during troubleshooting.

Figure 15 depicts an example of an interface that could be developed at the industrial control system to run the system from within. The key requirement of the interface is that it should be complete and lean. This means it has to have only those elements that are necessary to run the system and to help understand what is going on in the process, and nothing more. As can be seen in the figure, the interface includes two feeders, the tumble mixer and the tablet press. In addition, it includes dialog boxes to punch in any operating conditions and to display the status of the variables for each unit. The additional element in the interface is the name of the units and variables.

In addition to the interface, it is necessary to program the control loops in the controller and connect those to the interface. Figure 16 presents an example of a control loop or module, in this

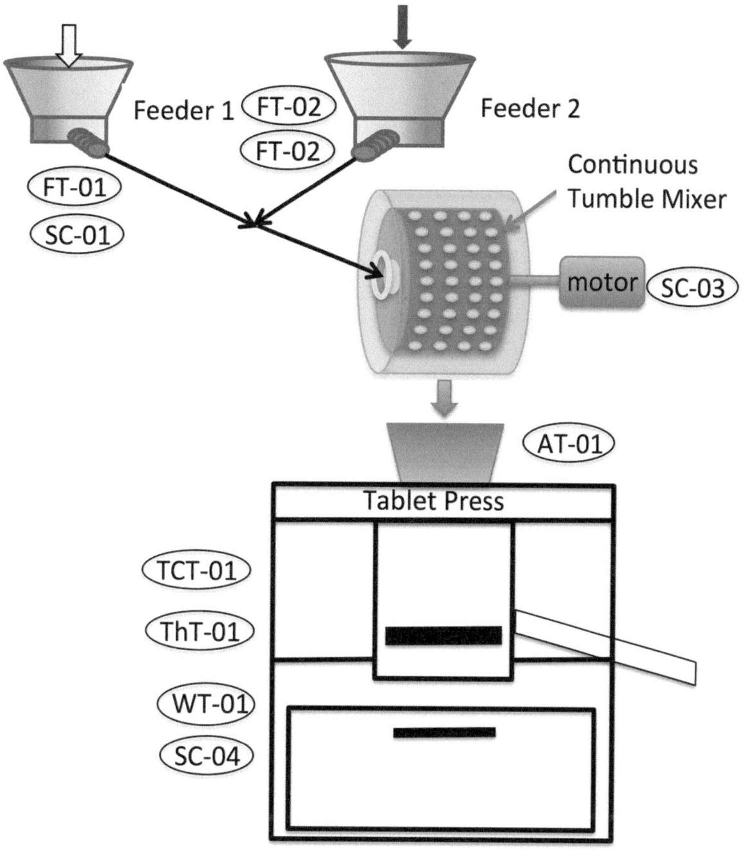

Fig. 14 P&ID of the integrated continuous direct compaction line

case for the API outlet concentration. In this case, the control module has the same components as the SAMA drawing: the sensor (left block, NIR-PRED), the PID control strategy (center block) and the API feeder (right block). The rest of the blocks are to read other parameters from the feeder local controller which can also be displayed in the interface. This control module is connected to a dialog box at the interface that calls a template representing the module every time the box is pressed.

4.2 Sensor Software and Communication Integration

Figure 15 demonstrates the sequence of the processing units and sensors to complete the integrated line. However, there is more integration behind the hardware to achieve the desired control of the continuous manufacturing line. Figure 17 depicts the integration of the signal of the sensor through synTQ (Optimal) and chemometrics software and the signals entering and leaving the control system (DeltaV, Emerson Process Management).

The software synTQ triggers the NIR to collect an NIR spectrum, which is then evaluated using a chemometrics software (for example, SIMCA QP, Umetrics) to predict the API concentration.

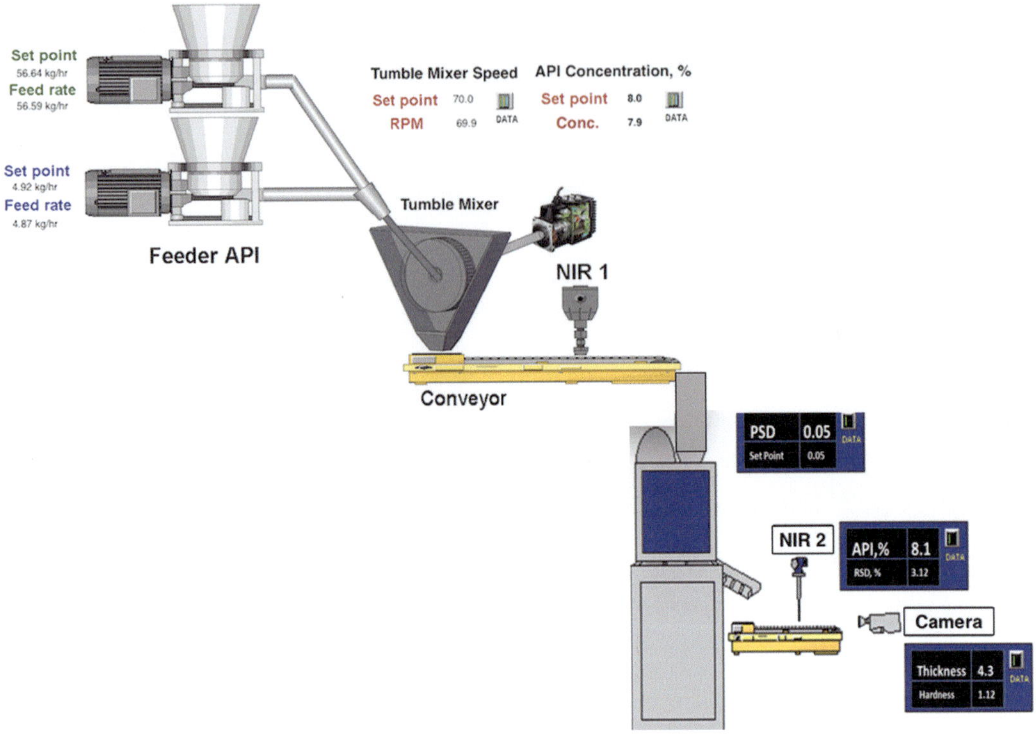

Fig. 15 Example of interface for the integrated direct compaction system in the industrial controller

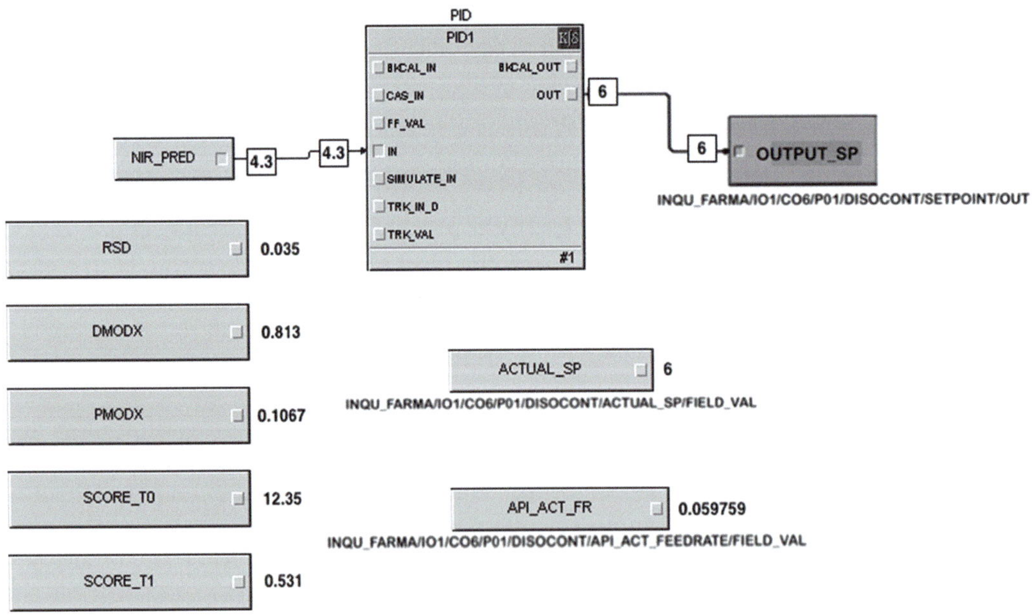

Fig. 16 Control module for the API concentration loop

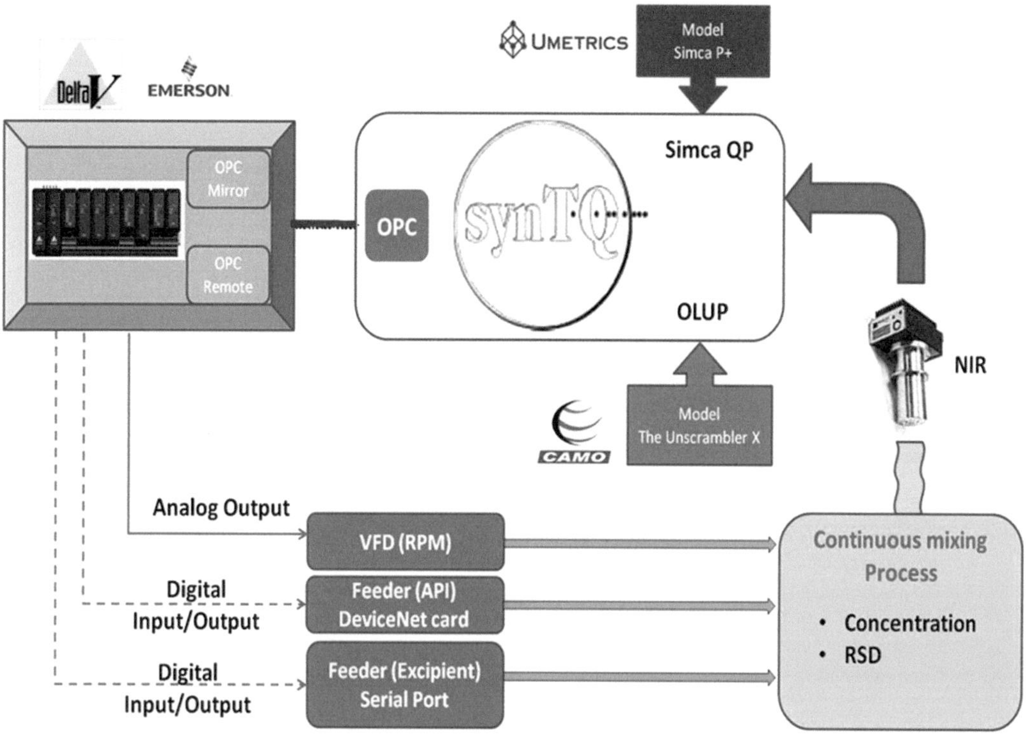

Fig. 17 Integration and communication of software

This concentration is sent to the controller (DeltaV) using OPC communication. Then, the control loop (PID based) computes the necessary changes in the API flow rate, which are communicated to the feeder using the digital communication protocol known as DeviceNet.

The RSD control uses the same integration as the API flow rate; however, it is manipulated via the mixer rotational speed (rpm). The changes of the rpm are based on a correlation rather than the PID control law. In this case, the communication between the controller and the mixer is the typical analog 4–20 mA.

The control of the total flow rate does not use the mentioned integration because the total flow rate is based on the flow transmitters of the feeders. In the case of the excipient feeder, the digital communication is serial.

4.3 Results: API Concentration Control

Figure 18 demonstrates the control performance on the API concentration exiting the tumble mixer using the described system. Initially, the experiment was set to start with a blend composition of 4 % of the API (naproxen sodium). The figure shows that the NIR prediction of the API concentration initially varied at a higher range around the set point. After the system reached some sort of steady state, the set point was increased to 8 % API. Then, the control

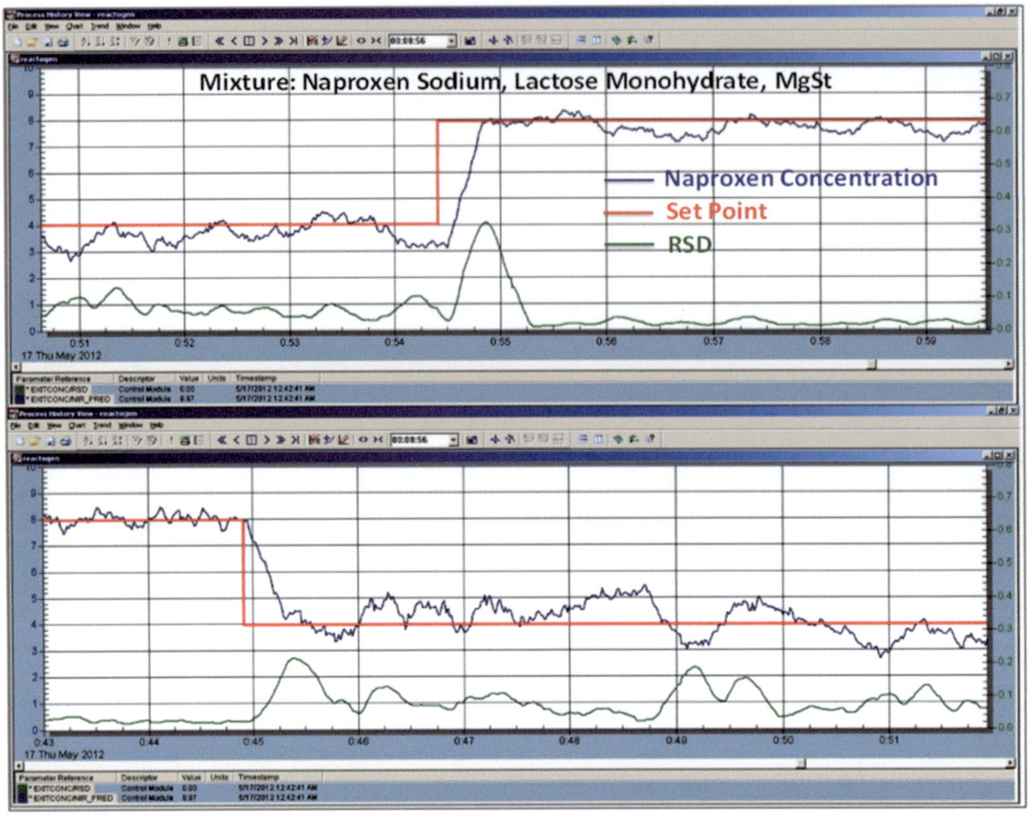

Fig. 18 Control system response to step up and step down in the set point

algorithm was able to drive the mixing to the target, but with a smaller variation in the predicted API concentration. After reaching steady state, the set point was changed once more (step down). The new set point was set at 4 %, which was reached once again with the same variation as in the beginning of the experiment.

Despite the acceptable performance by the control system, the variation is too high by industrial standards. Hence, the next step focused on decreasing the variation in the API concentration prediction by improving the accuracy of the NIR model. This was completed via inclusion of additional spectra acquired at different operating conditions. After the new more accurate model was developed, the old model (used for Fig. 18) was replaced in the control algorithm (inside the synTQ software).

Figure 19 depicts the new API concentration prediction using the new calibration model and the same control system. As can be seen, the variation decreased to acceptable levels while the system was able to drive the API concentration to the target value. A more accurate and robust model means the predicted value is closer to the actual value and therefore, the manipulations in the process

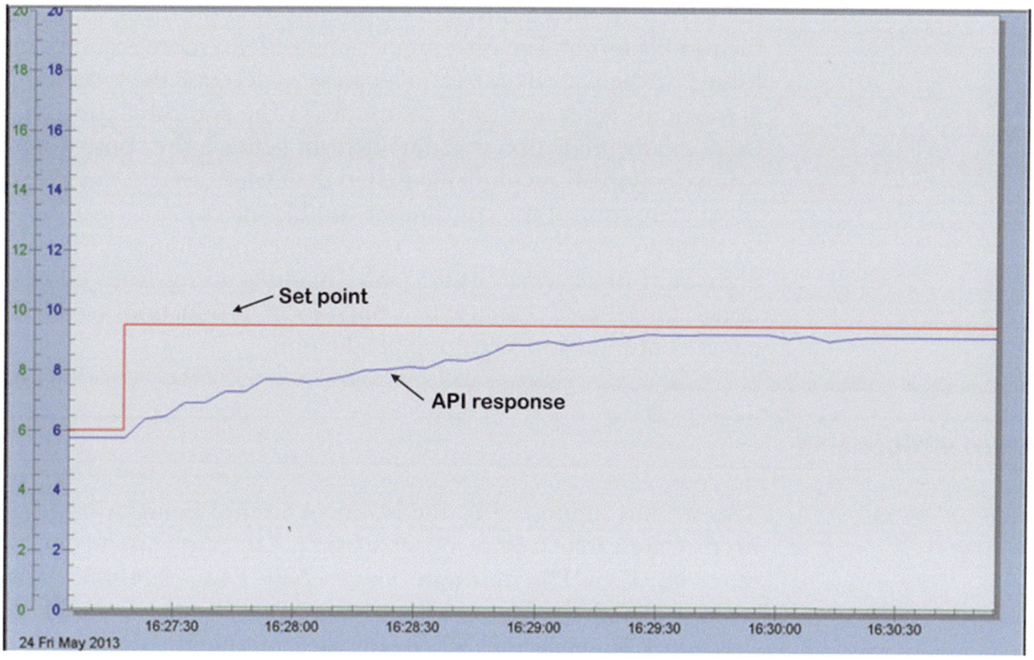

Fig. 19 Control system response with the more robust NIR calibration model

by the control system are closer to the actual needed change. This avoids "chattering" of the control actions, documented in the literature to happen when large errors in measurements or predictions are present.

5 Conclusions

A PID control scheme as well as an advanced model predictive controller (MPC) have been introduced for the direct compaction process. A systematic methodology to design a control system for a direct compaction continuous tablet manufacturing process was described. In Case Study 1, the performance of the control system was evaluated using process flowsheet models. Successful implementation of the control system into a continuous tablet manufacturing pilot plant situated at ERC-SOPS, Rutgers University, Piscataway NJ, USA, demonstrated the validity of the approach. Hybrid MPC-PID control scheme on an API composition control loop was used to close the loop.

In Case Study 2, a direct compaction tablet manufacturing process via a continuous tumble mixer was evaluated. In this study, the pilot plant pertained to the one situated at University of PR—Mayaguez Campus, Mayaguez PR, USA. This implementation followed the same macro-scale approach as for the Case Study 1 pilot plant using a continuous paddle mixer. However, the feeders needed a different communication protocol, as well as the tumble mixer. The NIR prediction needed a more accurate and robust mathematical pretreatment so as to decrease the variation in predictions. After applying the model in the control algorithm, a more steady prediction was obtained. In general, the control algorithm developed and implemented in the DeltaV system was able to adequately control the continuous tumble mixer.

The proposed systematic control framework supports the paradigm shift of pharmaceutical tablet manufacturing from conventional QbT-based batch-wise, open-loop production to QbD-based continuous, closed-loop production.

Acknowledgements

This work is supported by the National Science Foundation Engineering Research Center on Structured Organic Particulate Systems (ERC-SOPS) through Grant NSF-ECC 0540855. The authors would also like to acknowledge Paul Brodbeck (CAI) and Rodolfo J. Romañach (UPRM) for their meaningful discussions.

References

1. PhRMA profile (2012). Washington DC, http://phrma.org/sites/default/files/pdf/PhRMA%20Profile%202013.pdf. Accessed 26 September 2013.
2. FDA (2004). Challenge and opportunity on the critical path to new medical products. U.S. Food and Drug Administration, http://www.fda.gov/downloads/Drugs/ScienceResearch/ResearchAreas/ucm079290.pdf. Accessed 26 September 2013.
3. Singh R, Godfrey A, Gregertsen B, Muller F, Gernaey KV, Gani R, Woodley JM (2013) Systematic substrate adoption methodology (SAM) for future flexible, generic pharmaceutical production processes. Comput Chem Eng 58:344–368
4. Singh R, Sahay A, Muzzio F, Ierapetritou M, Ramachandran R (2013) Systematic framework for onsite design and implementation of the control system in continuous tablet manufacturing process. Comput Chem Eng 66:186–200
5. Singh R, Ierapetritou M, Ramachandran R (2013) System-wide hybrid model predictive control of a continuous pharmaceutical tablet manufacturing process via direct compaction. Eur J Pharm Biopharm 85(3):1164–1182
6. Singh R, Ierapetritou MG, Ramachandran R (2012) An engineering study on the enhanced control and operation of continuous manufacturing of pharmaceutical tablets via roller compaction. Int J Pharm 438(1-2):307–326
7. Charoo NA, Shamsher AAA, Zidan AS, Rahman Z (2012) Quality by design approach for formulation development: a case study of dispersible tablets. Int J Pharm 423:167–178
8. Singh R, Gernaey KV, Gani R (2009) Model-based computer-aided framework for design of process monitoring and analysis systems. Comput Chem Eng 33(1):22–42
9. FDA/CDER (2005) Process Analytical Technology - (PAT) initiative. U.S. Food and Drug Administration, Center for Drug Evaluation and Research. http://www.fda.gov/AboutFDA/CentersOffices/OfficeofMedicalProductsandTobacco/CDER/ucm088828.htm. Accessed 19 June 2012
10. Gnoth S, Jenzsch M, Simutis R, Lübbert A (2007) Process Analytical Technology (PAT): batch-to-batch reproducibility of fermentation processes by robust process operational design and control. J Biotechnol 132(2):180–186
11. FDA (2004) Guidance for industry: PAT-A framework for innovative pharmaceutical manufacturing & quality assurance. http://www.fda.gov/downloads/Drugs/GuidanceComplianceRegulatoryInformation/Guidances/ucm070305.pdf. Accessed 19 June 2012

12. Muzzio F, Singh R, Chaudhury A, Rogers A, Ramachandran R, Ierapetritou MG (2013) PharmTech magazine Europe 37(6): 40–41. http://www.pharmtech.com/pharmtech/Manufacturing/Model-Predictive-Design-Control-and-Optimization/ArticleStandard/Article/detail/814906. Accessed 04 October 2013
13. Boukouvala F, Chaudhury A, Sen M, Zhou R, Mioduszewski L, Ierapetritou M, Ramachandran R (2013) Computer-aided flowsheet simulation of a pharmaceutical tablet manufacturing process incorporating wet granulation. J Pharm Innov 8(1):11–27
14. Boukouvala F, Niotis V, Ramachandran R, Muzzio F, Ierapetritou M (2012) An integrated approach for dynamic flowsheet modeling and sensitivity analysis of a continuous tablet manufacturing process: an integrated approach. Comput Chem Eng 42:30–47
15. Boukouvala F, Ramachandran R, Vanarase A, Muzzio FJ, Ierapetritou M (2011) Computer aided design and analysis of continuous pharmaceutical manufacturing processes. Comput Aid Chem Eng 29:216–220
16. Sen M, Dubey A, Singh R, Ramachandran R (2013). Mathematical development and comparison of a hybrid PBM-DEM description of a continuous powder mixing process. J Powder Technol. http://dx.doi.org/10.1155/2013/843784
17. Sen M, Ramachandran R (2012) A multi-dimensional population balance model approach to continuous powder mixing processes. Adv Powder Technol 24(1):51–59
18. Sen M, Singh R, Vanarase A, John J, Ramachandran R (2012) Multi-dimensional population balance modeling and experimental validation of continuous powder mixing processes. Chem Eng Sci 80:349–360
19. Barrasso D, Ramachandran R (2012) A comparison of model order reduction techniques for a four-dimensional population balance model describing multi-component wet granulation processes. Chem Eng Sci 80:380–392
20. Barrasso D, Walia S, Ramachandran R (2013) Multi-component population balance modeling of continuous granulation processes: a parametric study and comparison with experimental trends. Powder Technol 241:85–97
21. Portillo PM, Vanarase A, Ingram A, Seville JK, Ierapetritou MG, Muzzio FJ (2010) Investigation of the effect of impeller rotation rate, powder flow rate, and cohesion on powder flow behavior in a continuous blender using PEPT. Chem Eng Sci 65:5658–5668
22. Vanarase AU, Alcala M, Rozo J, Muzzio FJ, Romanach RJ (2010) Real-time monitoring of drug concentration in a continuous powder mixing process using NIR spectroscopy. Chem Eng Sci 65:5728–5733
23. Vanarase A, Muzzio FJ (2011) Effect of operating conditions and design parameters in a continuous powder mixer. Powder Technol 208:26–36
24. Vanarase A, Gao Y, Muzzio FJ, Ierapetritou MG (2011) Characterizing continuous powder mixing using residence time distribution. Chem Eng Sci 66(3):417–425
25. Singh R, Gernaey KV, Gani R (2010) ICAS-PAT: a software for design, analysis & validation of PAT systems. Comput Chem Eng 34 (7):1108–1136
26. Hsu S, Reklaitis GV, Venkatasubramanian V (2010) Modeling and control of roller compaction for pharmaceutical manufacturing. Part I: process dynamics and control framework. J Pharm Innov 5:14–23
27. Hsu S, Reklaitis GV, Venkatasubramanian V (2010) Modeling and control of roller compaction for pharmaceutical manufacturing. Part II: control and system design. J Pharm Innov 5: 24–36
28. Ramachandran R, Chaudhury A (2012) Model-based design and control of continuous drum granulation processes. Chem Eng Res Des 90(8):1063–1073
29. Burggraeve A, Tavares da Silva A, Van den Kerkhof T, Hellings M, Vervaet C, Remon JP, Vander Heyden Y, Beer TD (2012) Development of a fluid bed granulation process control strategy based on real-time process and product measurements. Talanta 100:293–302
30. Bardin M, Knight PC, Seville JPK (2004) On control of particle size distribution in granulation using high-shear mixers. Powder Technol 140(3):169–175
31. Sanders CFW, Hounslow MJ, Doyle FJ III (2009) Identification of models for control of wet granulation. Powder Technol 188(3): 255–263
32. Long CE, Polisetty PK, Gatzke EP (2007) Deterministic global optimization for nonlinear model predictive control of hybrid dynamic systems. Int J Robust Nonlin Control 17:1232–1250
33. Gatzke EP, Doyle FJ III (2001) Model predictive control of a granulation system using soft output constraints and prioritized control objectives. Powder Technol 121(2–3): 149–158

34. Pottmann M, Ogunnaike BA, Adetayo AA, Ennis BJ (2000) Model-based control of a granulation system. Powder Technol 108 (2–3):192–201
35. Ramachandran R, Arjunan J, Chaudhury A, Ierapetritou M (2012) Model-based control loop performance assessment of a continuous direct compaction pharmaceutical processes. J Pharm Innov 6(3):249–263
36. Seborg DE, Edgar TF, Mellichamp DA (2004) Process dynamics and control, 2nd edn. John Wiley, New York
37. Ziegler JG, Nichols B (1942) Optimum settings for automatic controllers. Trans ASME 64:759–765
38. Cutler CR, Ramaker BL (1979) Dynamic matrix control—a computer control algorithm. AIChE National Meeting, Houston, TX, April 1979
39. Prett DM, Gillette RD (1980) Optimization and constrained multivariable control of a catalytic cracking unit. AIChE National Meeting, Houston, TX, April 1979
40. Wojsznis W, Gudaz J, Blevins T, Mehta A (2003) Practical approach to tuning MPC. ISA Trans 42:149–162
41. Singh R, Boukouvala F, Jayjock E, Ramachandran R, Ierapetritou M, Muzzio F (2012b) Flexible multipurpose continuous processing of pharmaceutical tablet manufacturing process. GMP news, European Compliance Academic (ECE). http://www.gmpcompliance.org/ecanl_503_0_news_3268_7248_n.html. Accessed 26 Sept 2013.
42. Singh R, Boukouvala F, Jayjock E, Ramachandran R, Ierapetritou M, Muzzio F (2012c) Flexible multipurpose continuous processing. PharmPro Magazine, Pharmaceut Process 27 (6): 22–25
43. Kawakita K, Ludde KH (1971) Some considerations on powder compression equations. Powder Technol 4:61–68
44. Kuentz M, Leuenberger H (2000) A new model for the hardness of a compacted particle system, applied to tablets of pharmaceutical polymers. Powder Technol 111:143–145
45. Kimber JA, Kazarian SG, Stepánek F (2011) Microstructure-based mathematical modelling and spectroscopic imaging of tablet dissolution. Comput Chem Eng 35: 1328–1339
46. Singh R, Gernaey KV, Gani R (2010) An ontological knowledge based system for selection of process monitoring and analysis tools. Comput Chem Eng 34(7):1137–1154
47. Bristol E (1966) On a new measure of interaction for multivariable process control. IEEE Trans Autom Control 11(1):133–134
48. Blevins T, Wojsznis WK, Nixon M (2013) Advanced control foundation: tools, techniques and applications. International Society of Automation, NC, USA. ISBN ISBN: 978-1-937560-55-3
49. Trierweiler JO, Farina LA (2003) RPN tuning strategy for model predictive control. J Process Control 13:591–598
50. Singh R, Sahay A, Karry KM, Muzzio F, Ierapetritou M, Ramachandran R (2014) Implementation of a hybrid MPC-PID control strategy using PAT tools into a direct compaction continuous pharmaceutical tablet manufacturing pilot-plant. Int J Pharm 473: 38–54
51. Singh R, Barrasso D, Chaudhury A, Sen M, Ierapetritou M, Ramachandran R (2014) Closed-loop feedback control of a continuous pharmaceutical tablet manufacturing process via wet granulation. J Pharm Innov 9:16–37

Chapter 8

Mathematical Tools for the Quantitative Definition of a Design Space

Amanda Rogers and Marianthi G. Ierapetritou

Abstract

This chapter focuses on the application of process modeling to determination of design space for pharmaceutical manufacturing processes. The first two sections define design space and related terms from a regulatory perspective and discuss how these apply in practice during pharmaceutical process development. In Sect. 3, a variety of mathematical techniques that can be used to guide design space development are introduced. An emphasis is placed on the use of feasibility analysis and the relationship between process feasibility and design space. Statistical techniques, including latent variable and Bayesian methods, are also discussed in detail. For methods that have been reported in the literature for pharmaceutical manufacturing applications, an overview of the relevant case studies is provided. If methods have not yet been applied to pharmaceutical processes, potential applications for these techniques are indicated. To illustrate the applicability of these concepts to pharmaceutical manufacturing processes, a brief case study is presented in Sect. 4. A discussion of design space verification and issues related to scale-up is provided in Sect. 5. Finally a brief summary of the concepts presented and their potential role in design space development for pharmaceutical processes is given in Sect. 6.

This chapter is intended to provide readers with an understanding of mathematical techniques that can be used during process development to assist in the determination of process design space. An overview of the problem formulation and solution approaches for each method are presented. More detailed mathematical developments for each technique are available in the references provided. Table 1 provides a list of symbols that are used throughout this chapter, while Table 2 describes acronyms used within this chapter.

Key words Design space, Feasibility analysis, Flexibility analysis, Black-box feasibility, Derivative-free optimization, Bayesian reliability, Partial least squares (PLS), Principal component analysis (PCA), Pharmaceutical process development, Process modeling

1 Definition of Design Space

Guidelines for the development and validation of pharmaceutical products have been described International Conference on Harmonisation of Technical Requirements for Registration of Pharmaceuticals in Human Use (ICH). Specifically, ICH Q8 describes standards for pharmaceutical development based on the notion of Quality by Design (QbD). QbD is a process development paradigm

Marianthi G. Ierapetritou and Rohit Ramachandran (eds.), *Process Simulation and Data Modeling in Solid Oral Drug Development and Manufacture*, Methods in Pharmacology and Toxicology, DOI 10.1007/978-1-4939-2996-2_8, © Springer Science+Business Media New York 2016

in which the quality of a pharmaceutical product is ensured through fundamental understanding of how process design and operating variables affect product quality attributes [1–5]. ICH has also developed guidance describing best practices for quality risk assessment (Q9) [6] and the implementation of Pharmaceutical Quality Systems (Q10) [7]. These guidelines include a discussion of process modeling and other mathematical tools that can be used to develop and understand pharmaceutical processes with the goal of achieving quality by design. ICH guidelines emphasize the importance of determining process design space, defined as the "multidimensional parametric space within which acceptable quality product is obtained" [3, 8, 9].

In order to implement a QbD approach it is necessary to have quantitative indicators of product quality. These are referred to as critical quality attributes (CQAs), which are physical, chemical or microbiological properties or characteristics that are controlled directly or indirectly to ensure product quality [5]. CQAs should be correlated in some way with acceptable clinical performance of a drug. The design space must be defined with respect to specific process design and operating parameters, which are referred to as critical process parameters (CPPs). CPPs are formally defined as process inputs which must be controlled within a specific range in order to ensure product quality [4], and generally include process inputs that directly and significantly affect product quality when varied within the range of anticipated operating conditions [5].

The quality by design approach to pharmaceutical process development has important regulatory and financial implications. Under the previously implemented quality by testing approach, the effect of process design and operating parameters on product quality was not always well understood. As a result, conservative quality specifications could be set, potentially causing clinically acceptable materials to be discarded and increasing the cost of manufacturing [5]. A testing-based approach also does not consider how reducing process disturbances could help to increase the operating range for a manufacturing process or change the optimal operating conditions [10]. If the design space for a process is well-understood, it is possible to change process operating conditions within this space without filing changes with the regulatory agencies. Therefore operating conditions can be adjusted to achieve optimal process performance from both a quality and cost perspective. The establishment of design space also clearly connects process operating conditions with control strategies for product quality, facilitating successful dialogue between regulators and industry with respect to justifying potential process modifications [11]. From a regulatory perspective, if a process is operating within its filed design space no further documentation or approval is required [12]. Thus having a larger design space is beneficial in terms of a company's ability to make future changes to a manufacturing process. When operating beyond the filed design space it is necessary to provide additional studies to regulatory agencies prior to implementing changes in the manufacturing process [5, 13–16]. These supplements are time consuming to file and due to their relatively large number can be difficult for regulatory agencies to review. Thus a comprehensive understanding of design space is critical to quality by design and can save both time and money in the development and regulatory control of pharmaceutical products [5].

2 Developing Design Space

In order to develop a design space for a solid dosage form, it is first necessary to know what product quality attributes must be controlled to ensure the desired biological performance with respect to safety and efficacy. The Quality Target Product Profile (QTPP) describes the collection of desired performance characteristics for a drug product based on its intended clinical use. QTPP encompasses the entire drug development and manufacturing process, from the selected dosage form and route of administration to the quality attributes of the drug substance. In the context of process development for oral solid dosage forms, the QTPP is specifically concerned with tablet quality criteria that should be met in order to ensure acceptable clinical performance. The collection of these attributes constitutes a target material profile (TMP), which is comprised of critical quality attributes (CQA) for the drug product [11, 17]. Once the TMP is known, an understanding of the relationship between process inputs, design and operating parameters and product quality attributes must be developed. Finally, it is necessary to identify the design and operating parameters for the drug product manufacturing process that give rise to acceptable product quality as indicated by the TMP [11, 17, 18]. These are the critical process parameters (CPP). The remainder of this section discusses identification of critical quality attributes and critical process parameters for the manufacture of solid dosage forms.

2.1 Identification of Critical Quality Attributes

Critical quality attributes (CQAs) for solid dosage forms often include both chemical and mechanical properties. Indicators of drug loading and uniformity with respect to active pharmaceutical ingredient (API) content such as content uniformity (CU) and relative standard deviation (RSD) in API concentration are generally considered CQAs. Consistency with respect to tablet weight as indicated by RSD is also a potential CQA [5]. In addition, dissolution performance is of interest as an indicator of drug release in vivo [11]. Physical attributes such as tablet strength, tablet hardness and friability may be considered critical, as these physical attributes can be correlated with product performance during packaging and transportation. Tablet hardness may be related to product release in vivo, and therefore is also a potential CQA [19]. Other attributes such as moisture content and tablet coating thickness may also be of interest, particularly if they are related to tablet stability [20].

2.2 Identification of Critical Process Parameters

Critical process parameters (CPPs) can include equipment design, configuration and operating parameters as well as raw material properties for both the active pharmaceutical ingredient (API) and excipient materials. Since the API for the formulation is considered a raw material from the perspective of drug product manufacturing, its physical and chemical attributes are often considered critical process parameters. Thus physical and chemical properties of the API are considered critical quality attributes for the API manufacturing process, which has its own design space intended to ensure these quality specifications are met [11]. Development of design space at the drug substance manufacturing stage is not discussed at length in the current work, which focuses on the manufacture of drug product. However, it is extensively discussed in the literature [21–24]. The remainder of this section focuses on the identification of critical process parameters for the drug product manufacturing process.

2.2.1 Formulation and Manufacturing Route

Basic decisions about formulation and manufacturing route selection are generally made prior to the establishment of a regulatory design space. However, decision variables related to formulation and manufacturing have the potential to impact the feasible operating region for a process. For formulation these decisions include the selection of excipients and lubricants and the relative quantities in which they are to be combined for various tablet strengths [25, 26]. For solid dosage forms, potential manufacturing routes include direct compression, wet granulation and dry granulation. Figure 1 depicts a sequence of unit operations that may be used to manufacture pharmaceutical tablets continuously via each of these methods. Direct compression involves the feeding of materials including API and excipients to a blender via a comill, which delumps the bulk powders. While Fig. 1 depicts a continuous blending operation, blending may also be carried out using batch equipment such as a bin blender, tote blender or V-blender [27]. Lubricant may be added to the API and excipients at the blending stage to facilitate mixing and improve powder flow. Subsequently the lubricated blend is fed to a tablet press, wherein it is compacted into tablets. Dry granulation involves the addition of a roller compaction step following blending, in which the bulk powder is densified into ribbons [28]. Roller compaction is followed by a milling step for size reduction of the ribbons. The dry granulate is then fed to a tablet press for compression into tablets. Alternatively, the blend may be granulated via wet granulation. Wet granulation can be carried out in a variety of equipment, including high shear granulators, fluid bed granulators or twin screw

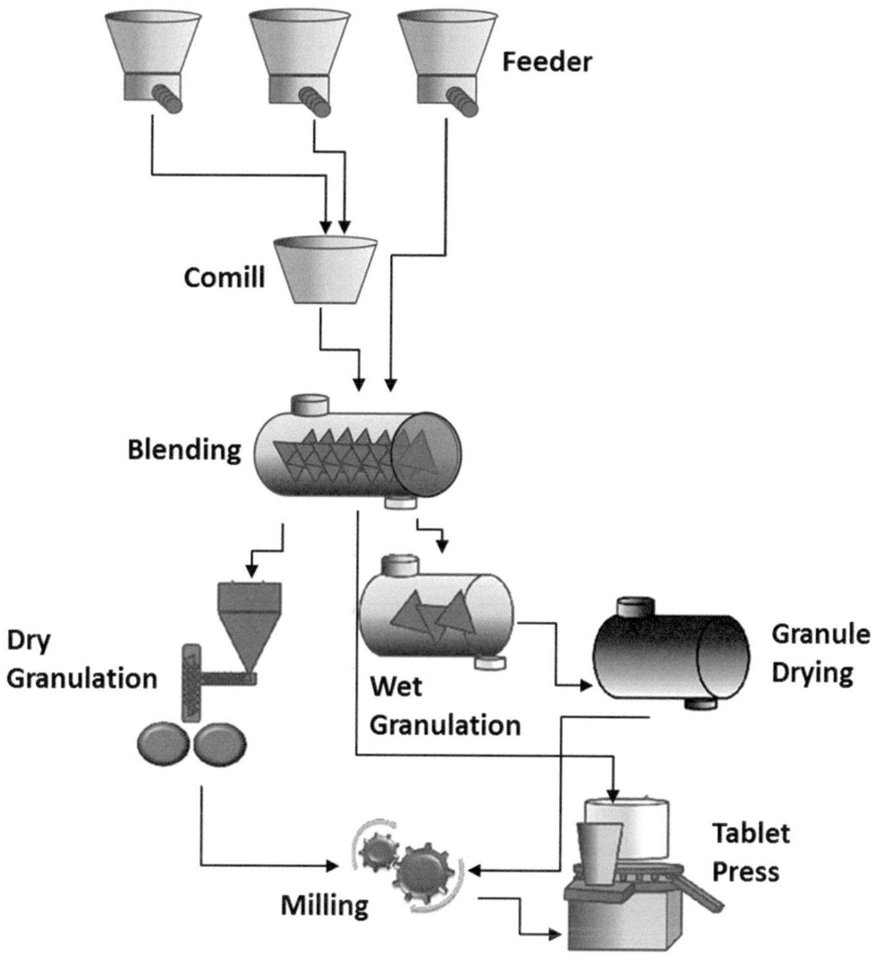

Fig. 1 Sequence of manufacturing steps to produce tablets via continuous direct compression, wet granulation or dry granulation. Though continuous processes are shown here, operations such as blending and wet granulation may also be carried out in batch equipment

granulators [29]. After wet granulation, granules must be dried prior to milling for size reduction and compression into tablets.

It should be noted that while Fig. 1 depicts continuous manufacturing routes, several operations such as blending and wet granulation can also be carried out in batch equipment. In addition, alternate sequences of the processing steps shown in Fig. 1 may also be implemented. It is beyond the scope of this chapter to describe each of these unit operations and the potential sequences in which they may be combined during manufacturing in detail. Further discussion of individual unit operations, the mathematical modeling of these operations, and the development of drug product manufacturing processes in general is provided in the literature [30–33].

Among the aforementioned manufacturing routes, direct compression has the advantage of being straightforward and including fewer unit operations than either wet or dry granulation. However, it is also more sensitive to variability in powder properties and may not be ideal for low drug load formulations, products with tight content uniformity constraints or materials with poor flow properties [33–35]. Granulation tends to reduce variability in composition and improve the flow properties and compactibility of materials prior to tableting [36, 37]. As shown in Fig. 1, granulation can be accomplished using either wet (e.g., continuous fluid bed, high shear agglomeration, twin screw extrusion) [29, 38] or dry (e.g., roller compaction) [28, 39, 40] methods depending on the API and excipient physical and chemical properties. The selection of a formulation and manufacturing method for a specific product will depend in part on API characteristics, as well as on commercial considerations like projected demand for a product. Considering process flexibility when selecting a manufacturing route can also help to ensure robustness of a manufacturing process. Process flexibility is discussed in greater detail in Sect. 3.1.

2.2.2 Equipment Design and Operating Conditions as CPPs

Because decisions about formulation and manufacturing route are generally made in early stages of process development, basic design features such as whether or not to include wet or dry granulation are generally not considered critical process parameters. Instead, the equipment design and operating conditions for the selected manufacturing route may be considered as candidate CPPs. A comprehensive listing of process design and operating parameters for pharmaceutical unit operations is provided by Yu [5]. A summary of process design and operating parameters for the pharmaceutical unit operations shown in Fig. 1 is given in Table 3. The list provided in Table 3 is meant to demonstrate examples of parameters that could be varied during the development of design space. For instance, in a continuous blending process design parameters like the blade configuration and operating parameters like the blade speed could interact to have an influence on blend uniformity. Selecting the appropriate blade configuration could help to enhance process robustness by enhancing mixing performance over the range of possible blade speeds. However, this list is not exhaustive, as some parameters will be equipment and process specific. In any given process a subset of these design and/or operating parameters may be deemed critical based on their potential to effect product quality, as discussed in Sect. 2.2.4.

2.2.3 Raw Material Properties as CPPs

Pharmaceutical products consist of multi-component blends of excipients, active ingredients and lubricants. The physical and chemical properties of these materials have the potential to influence product quality attributes including purity, content and weight uniformity, dissolution performance and stability [41].

Table 1
List of symbols used in this book chapter

Symbol	Definition
θ	Uncertain parameter(s)
T	Parameter range for uncertain parameters—the uncertain region
x	State variable(s)
z	State variable(s)
d	Design variable(s)
g_j	Inequality constraints
h_i	Equality constraints
$\Psi(d,\theta)$	Feasibility function
u	Scalar variable introduced to solve the feasibility problem
$\chi(d)$	Flexibility test function
δ	Scalar introduced to solve the flexibility index problem
$T(\delta)$	Feasible region corresponding to a given value of δ in the flexibility test problem
θ^N	Nominal uncertain parameter value(s) flexibility index problem
$\Delta\theta^{-/+}$	Uncertain parameter deviations for flexibility index problem
F	Flexibility index
θ^C	Critical point for flexibility index problem
kV	Set of vertices for the uncertain parameter range T
λ_j	Lagrange multiplier for constraint g_j
y_j	Binary variable indicating activity of constraint g_j
s_j	Slack variable for constraint g_j
n_z	Number of control variables
Nk	Number of sub regions for split and bound strategy
T_i	Sub region of the uncertain region in the split and bound strategy
F_{1i}, F_{2i}	Sub problems for ith sub region in the split and bound strategy
$j(\theta)$	Joint probability distribution function for the uncertain variables
SF(d),SF(s)	Stochastic flexibility index for design d or state s
E(SF)	Expected stochastic flexibility
$P(s)$	Probability of state s occurring
TFS	Total number of feasible states
$E_{\text{feas}}[I(x)]$	Expected improvement (EI) function
U	Value of feasibility function at a given point x for the EI function

(continued)

Table 1
(continued)

Symbol	Definition
Φ	Distribution function
φ	Standard normal density
s_u	Kriging prediction variance
T,P	Principal component scores and loadings
X,Υ	Predictor (X) and response (Υ) blocks for latent variable methods
W^*	PLS weights
Q	PLS Y-block loadings
$\hat{\Upsilon}(\theta_1, \theta_2)$	Mean predicted response for a quality attribute that is a function of two uncertain variables
$g_k^{\text{path}}, g_k^{\text{point}}$	Path and point constraints for dynamic feasibility analysis
H	Time horizon for dynamic feasibility problem
τ	Lag time for dynamic feasibility with time delay

Table 2
List of acronyms used in this work

Acronym	Definition
ICH	International Conference on Harmonisation of Technical Requirements for Registration of Pharmaceuticals in Human Use
QbD	Quality-by-design
CQA	Critical quality attribute
CPP	Critical process parameter
QTPP	Quality target product profile
CU	Content uniformity
API	Active pharmaceutical ingredient
LP	Linear program
NLP	Nonlinear program
MIP	Mixed-integer program
MINLP	Mixed-integer nonlinear program
KKT conditions	Karush-Kuhn-Tucker necessary conditions for optimality
RSM	Response surface methodology

(continued)

Table 2
(continued)

Acronym	Definition
ANN	Artificial neural networks
HDMR	High dimensional model representation
DOE	Design of experiments
PCA	Principal component analysis
PLS	Partial least squares regression/projection to latent structures
LQR	Linear quadratic regulator
FDA	Food and Drug Administration
EMA	European Medicines Agency

For the active pharmaceutical ingredient, chemical properties like purity and potency must be controlled in order to ensure tablet quality [42]. In addition, physical properties of the active ingredients and excipients can affect flow behavior and compaction performance during tableting. In order for a material property to be defined as a critical process parameter, it must be possible to test it reliably. For this reason, specifications on physical properties are typically limited to bulk powder properties, which are more straightforward to test. These include bulk density, particle size distribution, flow factor, cohesion and compression behavior [31]. The relationship between bulk properties and product quality is not always explicitly known, but the range of properties corresponding to acceptable product quality can be determined experimentally [2, 34, 41].

The proper selection of excipients is an important aspect of pharmaceutical process development. Each active pharmaceutical ingredient has its own set of unique physical and chemical properties that play a role in dictating which excipient should be selected as well as which manufacturing route (e.g., direct compaction, wet or dry granulation) is most effective [2, 43]. Selection of excipients and manufacturing method may require some trial and error, though statistical methods have been developed to aid in this decision-making process [41]. Excipients commonly used in pharmaceutical manufacturing, such as micro crystalline cellulose and lactose, are generally well characterized with respect to their chemical properties. The physical properties of excipients may vary from one manufacturer to another [44, 45]. This is important to consider when selecting vendors for these materials. If the physical properties of an excipient are well controlled by the manufacturer, it may not be necessary describe them as critical process parameters.

Table 3
Pharmaceutical unit operations and relevant design and operating parameters

Pharmaceutical unit operation	Design parameters	Operating parameters	References
Feeding	Feeder tooling	Screw speed	[140, 156, 157]
	Hopper size	Refill strategy	
		Operating mode—gravimetric versus volumetric	
Milling	Mill type (comill, jet mill, hammer mill)	Mill speed	[158–161]
	Screen size/type	Feed rate	
	Blade configuration	Grinding pressure	
	Equipment size	Nozzle pressure	
	Air nozzle arrangement		
Mixing	Mixing, batch		
	Blender geometry (e.g., V blender, double cone blender, bin blender, convective mixer)	Order of addition	[5, 27]
	Blender size (volume)	Fill level	
	Blade configuration (for convective batch mixers)	Rotation speed	
		Blending time	
	Mixing, continuous		
	Blender geometry (length, diameter)	Fill level	[162–165]
	Blade angle	Blade rotation rate	
	Shaft angle		
	Blade width		
Wet granulation	Common to all wet granulation processes		
	Granulator type (e.g., high shear granulator, fluid bed granulator, twin screw extrusion)	Binder fluid temperature	[36, 37, 166]
	Binding solution properties	Binder addition rate and/or binder content	
	Method/location of binder addition	Granulation temperature	
	High shear granulator		
	Impeller configuration	Impeller speed	[29, 167]
	Chopper configuration	Chopper speed	
	Spray nozzle configuration	Pre-binder addition mix time	
		Post-granulation mix time	
	Fluid bed granulator		
	Spray nozzle configuration	Mixing time	[36, 37, 167]
	Filter properties, size	Solids flow rate	
	Shaking intervals	Inlet air flow rate, volume, temperature, and dew point	
		Exhaust air temperature, flow	
	Twin screw extruder		
	Screw geometry	Screw speed	[29, 37, 166–168]
	Number of screws	Solids flow rate	
	Barrel length		

(continued)

Table 3 (continued)

Pharmaceutical unit operation	Design parameters	Operating parameters	References
Dry granulation	Roll configuration Roll diameter Roll surface Powder feed method	Flow rate Roll speed Compaction pressure Roll gap	[40, 147, 169]
Drying	Common to all drying processes	Product temperature limits Drying time	[170]
	Fluidized bed Filter properties Shaking intervals	Inlet air volume, temperature, dew point Exhaust air temperature, flow	[171–173]
	Tray Quantity carts and trays per chamber Quantity of product per tray	Air flow Inlet dew point	[170, 174]
	Vacuum/microwave drying Dryer geometry Dryer size (volume) Microwave potency Energy Supplied	Jacket temperature Condenser temperature Impeller speed Dryer operating pressure	[5, 170, 175]
Compaction (tablet press)	Hopper size (height, aperture) Feedframe type Die and punch size and geometry Die filling method Lubrication method	Turret speed Feed frame rotation rate Punch penetration depth Depth of fill Pre-compression force Compression force	[35, 47, 48, 176, 177]

Lubricants like talc or magnesium stearate often have a disproportionately significant effect on process performance relative to their low concentrations in drug product formulations. For instance, lubricants can significantly improve flow behavior and mixing quality at relatively low concentrations (0.25–10 wt.%) [46] and have been demonstrated to reduce tablet hardness and adversely affect friability and dissolution as well [35, 47].

All of these factors complicate the process of characterizing design space with respect to raw material properties. Hlinak et al. [41] have provided a list of raw material properties and their potential effects on process performance and product quality. This can be used as a starting point for identifying critical process parameters related to material properties. In addition, data-based and statistical methods for determining the contribution of API properties to tablet quality have been demonstrated in the literature [48–50]. Specific methods that can be used for data-based evaluation of design space are discussed in Sect. 3.3.

2.2.4 Selecting CPPs

The designation of a parameter as critical should be related to the extent that it affects product quality and the likelihood that it will deviate from its nominal or acceptable range [11]. Criticality can be evaluated in part via risk assessment, which is an important aspect of pharmaceutical process development in the quality by design framework [6]. Risk assessment categorizes potential adverse events based on probability, severity, detectability and sensitivity [51]. Probability refers to the likelihood of an event occurring. In the context of pharmaceutical manufacturing, the event of interest is a deviation from normal process operation. Process parameters that are not well controlled or have a large degree of inherent variability may be deemed critical, as the likelihood of a deviation from normal operating conditions is greater. Severity describes the extent to which an adverse event will impact product quality. If a parameter deviating from nominal operating conditions has severe consequences, that parameter may be designated as critical. Detectability refers to whether an adverse event can be identified or observed, for instance through real time process monitoring or off-line analytical testing. If inadequate raw material quality is detectable, it is possible to avoid using this material and therefore to mitigate the corresponding product quality consequences. Finally, sensitivity refers to the possible exacerbation or attenuation of deviations in subsequent processing steps such that product quality is affected [52]. Sensitivity analysis can play a role in risk assessment by identifying process parameters that have the greatest potential to impact process performance [10, 53]. Sensitivity metrics for quality attributes are indicators of factor importance with respect to product quality. As such, sensitivity indices can be used to rank variables with respect to their effect on process outcomes [54]. Process inputs and operating parameters to which product quality is highly sensitive can be identified as CPPs [34, 55, 56].

Risk assessment requires process knowledge and understanding in order to be meaningful. It is an ongoing endeavor throughout process development, and as additional process knowledge is gained it is possible to more accurately assess the likelihood or severity of an adverse event. Ultimately risk assessment and sensitivity analysis provide a framework for identifying and justifying the selection of critical process parameters. The process design space can then be established with respect to these CPPs and the corresponding CQAs which they affect [51].

3 Mathematical Methods for Design Space Definition

The notion of design space is related to the concepts of process operability and flexibility. Flexibility tends to be used in the context of process design under uncertainty while operability tends to be used more often in the controls literature, but both the concepts are similar. They describe the ability of a process to generate acceptable product across a range of process design and operating conditions in the presence of potential variability. In other words, they are concerned with feasible operation of a process, given uncertainty in its inputs, design and operating conditions [57]. Sources of variability include variation in input material quality, disturbances in operating conditions, or plant-model mismatch [58]. These concepts can be readily applied to the determination of design space for pharmaceutical processes. Design space is, by definition, the region within which material of acceptable quality can be produced. Thus applying the concept of feasibility or operability analysis to a process constrained by specifications on product quality will result in the identification of a feasible region within which the process design space can be established. It is important to note that the boundaries of feasible operation identified via various methods for feasibility and flexibility analysis provide information regarding the edges of failure for a process. This information can be used to guide the selection of design space to ensure that it lies well within the feasible operating region. However, it is not recommended that the boundaries of the feasible operating region be established as the design space for a process.

In the remaining sections, a number of mathematical formulations and approaches for determining the feasible operating region of a process are presented. Examples of how these methods have been or could be applied to design space evaluation for pharmaceutical manufacturing processes are also provided.

3.1 Feasibility and Flexibility Problems

In feasibility analysis of pharmaceutical manufacturing processes, the boundaries of the feasible region for the process are consistent with edges of the design space [11, 59]. The mathematical concepts of process feasibility and flexibility, which are well established in the process systems engineering literature, [60–62] are ideally suited to describe design space. Several recent publications discuss feasibility analysis of pharmaceutical processes [59, 63]. However, there are significant opportunities to expand the use of feasibility and flexibility analysis within the pharmaceutical industry.

3.1.1 Mathematical Formulation

Rigorous mathematical formulations describing the concepts of process feasibility and flexibility were introduced in the 1980s in the process systems engineering literature [60–62, 64]. Feasibility refers to the limits within which a process is operable, while flexibility quantifies the ability of a process to maintain feasible operation in the presence of inherent process variability or external disturbances [65]. Mathematically, feasibility is defined according to the feasibility function, shown in Eq. (1), where d represents a given process design and z indicates a vector of control variables. θ represents a vector of uncertain variables that are expected to vary on the interval $T = \{\theta | \theta^{L} \leq \theta \leq \theta^{U}\}$ where θ^{L} and θ^{U} are lower

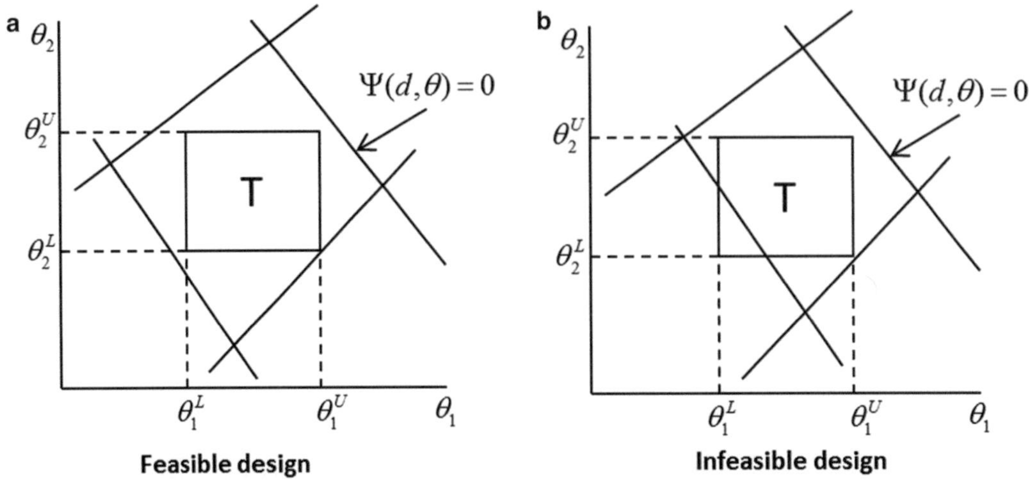

Fig. 2 Geometrical representation of the concept of feasibility. Figure (**a**) depicts a process design that is feasible. Figure (**b**) depicts a design that is infeasible

and upper bounds respectively. Implicit in this problem is also the set of state variables, x, which can be expressed in terms of the design, control and uncertain variables such that $x=x(d,z,\theta)$. The functions g_j are constraints on the process of the form $g_j(d,z,\theta) \leq 0$ which must be met in order for the process to operate successfully [61, 62, 65]. Additionally, a set of equality constraints $h_i(d,z,\theta) = 0$ that constitute a model describing the process physics may be included in the problem formulation.

$$\Psi(d, \theta) = \min_{z} \max_{j \in J} \left\{ g_j(d, z, \theta) \right\} \tag{1}$$

If the feasibility function $\Psi(d,\theta)$ is less than or equal to 0 then the process is feasible for the given design over the range of uncertain parameters considered. If not, then the process is infeasible and cannot be made feasible through adjustment of the control variables. Figure 2 depicts the concept of feasibility for a process that has two uncertain variables (θ_1 and θ_2) and four constraints. The boundaries of the feasible region are shown as the region where the feasibility function (θ_1,θ_2) is equal to 0. Figure 2a shows a process design that is feasible, because the entire uncertain space T is contained within the feasible region. Figure 2b depicts an infeasible process design, where the range of the uncertain variables T exceeds the boundaries of the feasible region.

Problem (1) can be solved through the introduction of a scalar variable u and reformulation of the problem as Eq. (2). This problem can be solved using standard linear (LP) or nonlinear (NLP) programming solvers, depending on the nature of the constraints [57, 65].

$$\begin{aligned} \Psi(d,\theta) &= \min_{z,\,u} u \\ \text{s.t. } g_j(d,z,\theta) &\le u \quad \forall j \in J \end{aligned} \tag{2}$$

In order to determine the extent to which a process is feasible over a range of potential deviations of $\theta \in$ T, one can solve the flexibility test problem. This is formulated as the max-min-max problem (3). In this expression, if $\chi(d) \le 0$, the control variables can be adjusted to achieve feasible process operation over the full range of deviations $\theta \in T$. If not, the process is infeasible for some values of θ [61, 62, 64, 65].

$$\chi(d) = \max_{\theta \in T} \min_{z} \max_{j \in J} \left\{ g_j(d,z,\theta) \right\} = \max_{\theta \in T} \Psi(d,\theta) \tag{3}$$

The flexibility test problem begs the question; to what extent can a process tolerate deviations in θ and remain feasible? This can be determined by solving the flexibility index problem (4). In problem (4), the region $T(\delta)$ describes a variable range for the uncertain parameters. δ is a non-negative scalar value such that if $\delta < 1$, the corresponding feasible region is $T(\delta) \subset T$, if $\delta > 1$ the corresponding feasible region is $T(\delta) \supset T$ and if $\delta = 1$ the corresponding feasible region is $T(\delta) = T$. The quantities $\Delta\theta^-$ and $\Delta\theta^+$ represent deviations from the nominal values of the uncertain parameters in the negative and positive directions, respectively [57, 61, 62, 65].

$$\begin{aligned} &F = \max \delta \\ &\text{s.t.}\, \chi(d) = \max_{\theta \in T} \min_{z} \max_{j \in J} \left\{ g_j(d,z,\theta) \right\} \le 0 \\ &T(\delta) = \left\{ \theta \middle| \theta^N - \delta\Delta\theta^- \le \theta \le \theta^N + \delta\Delta\theta^+ \right\} \\ &\delta \ge 0 \end{aligned} \tag{4}$$

F is the flexibility index. It represents the largest possible rectangle, or in the case of higher dimensional spaces hypercube, that can be inscribed within the multi-dimensional feasible region for the given design. If $F > 1$ the design is actually more flexible than it needs to be to handle the given deviations, while if $F < 1$ the design is not sufficiently flexible to handle these deviations. If $F = 1$ the process is exactly flexible enough to maintain feasible operation over the range of deviations from $\Delta\theta^-$ to $\Delta\theta^+$ [61, 62, 65]. If $F < 1$ its value indicates the fraction of the expected deviation in the parameters that can be tolerated in the given design. The value of θ that corresponds to the solution of the flexibility index problem is referred to as the critical parameter point θ^C. This is the parameter value that limits the flexibility of the design, and the corresponding constraint(s) along which θ^C occurs are the critical constraints that limit flexibility [57]. Relaxing these constraints would thus increase

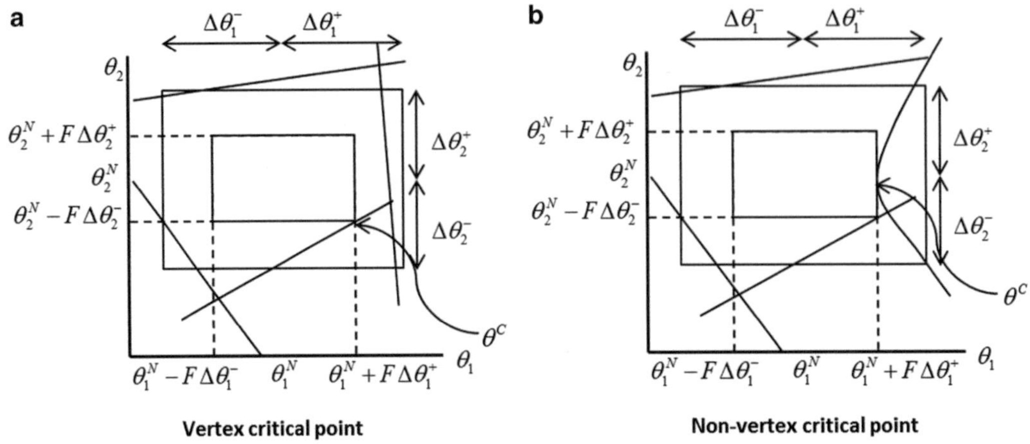

Fig. 3 Geometrical representation of the concept of flexibility and the flexibility index. Figure (**a**) depicts a process design with a vertex critical point. Figure (**b**) depicts a process design with a non-vertex critical point. Figures (**a**, **b**) both depict processes where the original design was not sufficiently flexible to tolerate the full range of possible deviations in θ_1 and θ_2

the size of the feasible region for the process. The critical point may occur at the extreme values of the uncertain parameters space *T*, in which case it is referred to as a vertex critical point. Otherwise it is referred to as a non-vertex critical point. Figure 3 depicts the concept of process flexibility index for a problem with two uncertain parameters, θ_1 and θ_2, and four constraints. In the examples shown, the flexibility (*F*) is less than one, so the process design is not sufficiently flexible to tolerate the full range of possible deviations in θ_1 and θ_2 considered. Figure 3a depicts a problem with a vertex critical point θ^C. In this case the critical point lies at a set of extreme values of the feasible region, corresponding to the upper bound of θ_1 and the lower bound of θ_2. Figure 3b depicts a non-vertex critical point, as θ^C occurs at a point that does not correspond to an upper or lower bound with respect to θ_2.

Several methods have been proposed for the solution of the flexibility index problem (4). In the case where the critical parameter value θ^C corresponds to a vertex of the feasible region, the solution of the flexibility index problem can be simplified [57, 66]. The flexibility test problem (3) can be restated in terms of the vertices of the region *T*, defined as the set of points θ^k for $k \in V$, where *V* is the set of vertices. The flexibility test problem expressed in terms of the vertices of the feasible region is shown in Eq. (5).

$$\chi(d) = \max_{k \in V} \Psi\left(d, \theta^k\right) \tag{5}$$

Problem (3) can be solved by evaluating problem (5) for all $k \in V$. Given that θ^k is a scalar, this can be done using deterministic optimization algorithms in commercially available optimization software. Once problem (5) has been solved for all vertices, $\chi(d)$ can be set to the maximum value of $\Psi(d,\theta^k)$ obtained through evaluation of all $k \in V$ [57, 66]. Similarly, the flexibility index problem can be solved by reformulating problem (4) in terms of the vertices, as shown in Eq. (6).

$$\begin{aligned} \delta^k &= \max_{z,\delta} \delta \\ \text{s.t.}\ & g_j\left(d, z, \theta^k\right) \le 0 \\ \theta^k &= \theta^N + \delta \Delta \theta^k \end{aligned} \tag{6}$$

By solving Eq. (6) for all vertices, a series of parameter ranges $T(\delta^k)$ are obtained. The smallest value of δ^k gives the flexibility index F while the corresponding region $T(F)$ indicates smallest rectangle/hypercube that can be inscribed in the feasible region [57, 66].

The solution to the flexibility test problem may not always occur at a vertex of the feasible region, as shown in Fig. 3b. In this case the flexibility test and flexibility index problems can be solved using an active set strategy, which involves formulating them as mixed-integer optimization problems [64]. The flexibility test problem (3) can be expressed as the two-level optimization problem (7).

$$\begin{aligned} \chi(d) &= \max_{\theta \in T} \Psi(d,\theta) \\ \text{s.t.}\ \Psi(d,\theta) &= \min_{z} \max_{j \in J} g_j(d,z,\theta) \end{aligned} \tag{7}$$

By applying the Karush–Kuhn–Tucker (KKT) necessary conditions for optimality [65, 67] to the feasibility function in Eq. (1), problem (7) can be reformulated as the constrained optimization problem (8).

$$\begin{aligned} \chi(d) &= \max_{\theta \in T} u \\ \text{s.t.}\ & \sum_{j \in J} \lambda_j = 1 \\ & \sum_{j \in J} \lambda_j \frac{\partial g_j(d,z,\theta)}{\partial z} = 0 \\ & \lambda_j \left[g_j(d,z,\theta) - u \right] = 0 \quad j \in J \\ & \lambda_j \ge 0 \quad g_j(d,z,\theta) - u \le 0 \quad j \in J \end{aligned} \tag{8}$$

In this problem, λ_j are the Lagrange multipliers corresponding to the constraint g_j. The value of λ for a particular constraint g indicates whether or not that constraint is active. If $\lambda_j = 0$ then the corresponding constraint g_j is not active. By treating the activity of a constraint as a discrete decision, problem (8) can be reformulated as a mixed integer optimization problem that can be solved using standard mixed integer (MIP) or mixed integer nonlinear programming (MINLP) methods. This involves the introduction of binary variables y_j which take a value of 1 when constraints are active and 0 otherwise. In addition, slack variables s_j are introduced for the constraints that are 0 when the binaries $y_j = 1$. Finally, the parameter n_z indicates the number of control variables (z) in the system. This informs the number of active constraints in the system, which will be $n_z + 1$ subject to certain conditions outlined in a series of papers by Swaney and Grossmann [61, 62]. The resulting MIP/MINLP is shown in Eq. (9).

$$\begin{aligned}
&\chi(d) = \max_{u,\theta,z,\lambda_j,s_j,y_j} \quad u\\
&\text{s.t. } g_j(d,z,\theta) + s_j = u\\
&\sum_{j\in J}\lambda_j = 1\\
&\sum_{j\in J}\lambda_j \frac{\partial g_j(d,z,\theta)}{\partial z} = 0\\
&s_j \le U\left(1 - y_j\right) \quad j\in J\\
&\lambda_j \le y_j \qquad\qquad j\in J\\
&y_j = n_z + 1\\
&\theta^L \le \theta \le \theta^U\\
&\lambda_j, s_j \ge 0 \quad y_j\in\{0,1\} \quad j\in J
\end{aligned} \tag{9}$$

Similarly the flexibility index problem can be reformulated as a mixed-integer problem as shown in Eq. (10).

$$\begin{aligned}
&F = \min_{\delta,\lambda_j,s_j,y_j} \quad \delta\\
&\text{s.t. } \quad g_j(d,z,\theta) + s_j = 0\\
&\sum_{j\in J}\lambda_j = 1\\
&\sum_{j\in J}\lambda_j \frac{\partial g_j(d,z,\theta)}{\partial z} = 0\\
&s_j \le U(1 - y_j) \quad j\in J\\
&\lambda_j \le y_j \qquad\qquad j\in J\\
&y_j = n_z + 1\\
&\theta^N - \delta\Delta\theta^- \le \theta \le \theta^N + \delta\Delta\theta^+\\
&\delta, \lambda_j, s_j \ge 0 \quad y_j\in\{0,1\} \quad j\in J
\end{aligned} \tag{10}$$

Thus the active set strategy can be used to determine the feasible region and the flexibility index through reformulation of the flexibility and feasibility problems as mixed integer optimization problems. These can be solved using commercially available MIP or MINLP solvers.

Ostrovsky et al. [68, 69] have proposed an alternate strategy that involves determining upper and lower bounds to the flexibility index by solving a series of simpler problems. This so-called split and bound strategy includes partitioning the uncertain region (T) into N_k sub regions ($T_1, T_2, \ldots, T_{Nk}$) and defining a series of sub problems F_{1i} and F_{2i}, shown in Eqs. (11) and (12).

$$F_{1i} = \max_{\theta \in T_i} \min_{z \in Z} \max_{j \in J} g_j(d, z, \theta) = \max_{\theta \in T_i} \Psi(d, \theta) \tag{11}$$

$$F_{2i} = \min_{z \in Z} \max_{j \in J} \max_{\theta \in T_i} g_j(d, z, \theta) \tag{12}$$

These problems have the property that $F_{2i} \geq F_{1i}$, so that F_{2i} is an upper bound of $\Psi(d,\theta)$ in the sub region T_i. A tight lower bound is then obtained by evaluating F_{1i} at t^i, the value of the uncertain parameters obtained by solving problem (11). In this way, the sub region that provides the maximum upper bound is obtained. The lower and upper bounds, $F_1{}^{L}$ and $F_1{}^{U}$ are then checked for feasibility. If they are feasible, the difference between the lower and upper bounds is determined. The process of partitioning and identifying bounds is carried out iteratively until the criterion in Eq. (13) is satisfied, where is a user-defined tolerance. A branch-and-bound procedure is used to explore potential process designs to find the one corresponding to the greatest flexibility [70].

$$F_1^{U} - F_1^{L} \leq \varepsilon \tag{13}$$

A full development of the split and bound strategy for solving flexibility index problems can be found in Ostrovsky et al. [68–70].

In order to apply the aforementioned solution strategies, the feasibility function and constraints must satisfy certain conditions [71]. For instance, the vertex strategy is only effective when the critical values of the feasibility problem occur at vertices of the set of uncertain parameters, T, as shown in Fig. 3a [62, 66]. This corresponds to the feasible region being jointly convex in the control variables and the uncertain parameters [70]. The active set strategy requires that the feasibility function $\chi(d)$ be quasi-concave with respect to the uncertain parameters θ and quasi-convex with respect to control variables z for fixed values of the uncertain parameters as well as quasi-concave in the control variables and uncertain parameters [64]. Given the required conditions, several methods have been developed to convexify constraints such that the vertex or active set strategies can be used to solve the feasibility problem. Methodologies for dealing with nonconvex equality or inequality constraints are detailed in Floudas [72] and Floudas et al. [71].

Deterministic formulations of the feasibility and flexibility index problems can be used to solve problems of optimal design, scheduling and operation under uncertainty. These problems are formulated as mixed integer problems, with the selection of design variables represented by integers. Subrahmanyam et al. [73] have applied deterministic flexibility formulations to solve integrated design and scheduling problems for batch processes under market uncertainty. In this case, the uncertain parameters are related to demand for different products that can be produced using a batch plant and feasibility analysis is implemented in the scheduling stage of the combined design and scheduling problems. Vin and Ierapetritou [74] have applied flexibility analysis to quantify robustness for the scheduling of multiproduct batch plants in the presence of demand uncertainty and potential process deviations. These implementations have potential relevance for pharmaceutical applications, which often involve batch processing. Floudas et al. [71] have applied deterministic problem formulations combined with convexification and branch and bound algorithms to conduct flexibility analysis of heat exchanger networks and integrated chemical processing plants that include heat exchangers, pumps and reactors. Similar methods could be applied to feasibility analysis of integrated networks of processing equipment for the manufacture of pharmaceutical products. Finally Linninger and Chakraborty [75] have applied deterministic flexibility methods, among other approaches, to develop robust waste management procedures for pharmaceutical plants given production and scheduling uncertainty.

3.2 Stochastic Feasibility and Flexibility

The Stochastic Flexibility Index (SF) was introduced by Pistikopoulos and Mazzuchi [76] and Straub and Grossmann [77] in the 1990s. This metric describes the probability that a particular design will achieve feasible operation given uncertainties that are described by a probability distribution. The stochastic flexibility problem is formulated with inequality constraints as in Eq. (4), but in this case the uncertain parameters θ are defined by some joint probability distribution function $j(\theta)$. The individual distributions may be independent, but do not have to be. So-called sigma bounds can be applied to individual distributions to constrain sampling to regions where changes in the cumulative probability are nontrivial, (e.g., within $\pm 3\sigma$ of the mean for a normal distribution). These bounds also create a closed feasible region, which simplifies solution of the stochastic flexibility problem [78]. The stochastic flexibility index for a given design can be determined through evaluation of the integral in Eq. (14), which must be evaluated over the feasible region in the θ space. The concept of stochastic

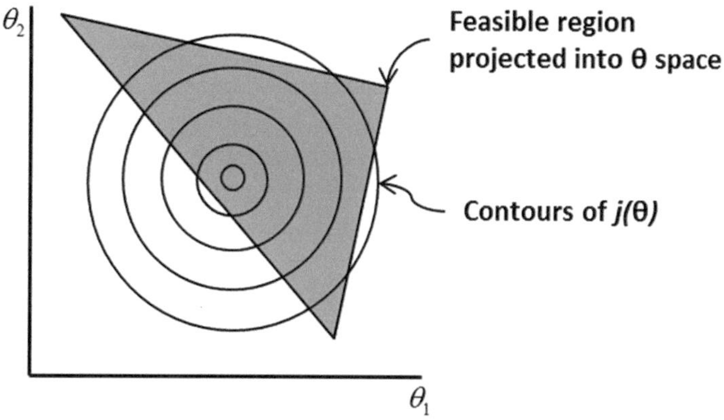

Fig. 4 Geometrical representation of the concept of stochastic flexibility. The stochastic flexibility index SF(*d*) is obtained by integration over the feasible region projected into θ space

flexibility is illustrated in Fig. 4, which shows the contours of the joint distribution function $j(\theta)$, as well as the projection of the feasible region into the space of the uncertain parameters.

$$\mathrm{SF}(d) = \int_{\theta:\Psi(d,\theta)\le 0} j(\theta)d\theta \tag{14}$$

Evaluating the stochastic flexibility index is challenging due to the multiple integration involved in problem (14). For this reason initial solution strategies focused on problems defined by linear constraints [79]. Pistikopoulos and Mazzuchi [76] suggest an approach based on an assumed normal distribution for the joint uncertainty distribution $j(\theta)$. By expressing the feasibility function in terms of the Lagrangian, the stochastic flexibility can be determined for a specific set of inequalities k via a one dimensional integral. These in turn can be used to obtain tight lower and upper bounds on stochastic flexibility index, SF(d). This method is related to the active set strategy for determining the flexibility index, but in this case the active set must be specified a priori [80].

Alternatively, Gaussian quadrature can be used to approximate the feasible region for arbitrary distribution functions [77]. An iterative procedure introduced by Straub and Grossmann [77] involves solving a series of feasibility problems in which the constraints are successively projected into lower dimensional space with respect to the uncertain parameters θ. This inequality reduction scheme allows bounds on the uncertain variables to be computed analytically. A set of quadrature points is then generated within the feasible operating space. The stochastic flexibility index can be numerically approximated using these bounds and the quadrature points [77, 79, 80]. This method also involves identifying active sets of constraints, and for this reason is limited to problems with relatively few constraints [78]. In addition, it cannot be readily extended to consider design variables, and thus is not appropriate for problems of integrated design and optimization [78].

For cases where both discrete and continuous uncertainties are involved, Straub and Grossmann [77] have described an expected stochastic flexibility index $E(\mathrm{SF})$. This metric considers the probability of each state occurring as well as the stochastic flexibility index with respect to the continuous uncertainties corresponding to each state. This is shown in Eq. (15) where s is a state consistent with a specific set of values for the discrete uncertainties, $P(s)$ is the probability of state s occurring, TFS is the total number of feasible states and $\mathrm{SF}(s)$ is the stochastic flexibility index corresponding to state s [77, 79, 81].

$$E(\mathrm{SF}) = \sum_{\mathrm{TFS}} \mathrm{SF}(s)P(s) \tag{15}$$

In practice the expected flexibility $E(\mathrm{SF})$ is found by determining upper and lower bounds on the flexibility index. Individual states, s, are evaluated systematically in such a way that states containing active units are prioritized, as they have a higher probability of occurring. The bounding scheme limits the number of states that are evaluated in determining the expected stochastic flexibility such that it is not necessary to evaluate all possible states [78, 79].

Initially the SF metric was developed specifically for linear process models, but it has been extended to nonlinear systems and nonconvex systems by Straub and Grossmann [78] and Pistikopoulos and Ierapetritou [82]. Straub and Grossmann [78] present a simultaneous approach to determining stochastic flexibility for nonlinear systems, as the previously mentioned inequality reduction scheme cannot be readily extended to nonlinear constraints [77]. In this approach, the determination of bounds and quadrature points is embedded into a single NLP formulation. The bounds and quadrature points are treated as variables for the optimization and are selected in order to maximize the stochastic flexibility metric, SF. This method is applicable to feasible regions that are convex, or at least $1 - d$ convex with respect to each uncertain parameter. For nonconvex regions, this approach can provide an accurate estimation of the feasible region if the stochastic flexibility is large. However, for small SF it is possible that the quadrature points will be outside of the feasible region [78]. Pistikopoulos and Ierapetritou [82] formulate the general problem of optimal design under uncertainty as a two-stage stochastic programming problem. In the design stage, a process design is selected in order to maximize profit. In the operating stage, control variables are selected in order to ensure operational feasibility for all possible realizations of the uncertain parameters. This problem can be solved using an iterative algorithm, in which a design is selected, the feasibility problem is solved for the given design and the expected profit for the selected design and corresponding operating parameters is determined using Gaussian quadrature. This provides a lower bound on the expected profit. Subsequently a new design is obtained by maximizing expected profit within the current feasible region. This provides an upper bound on the expected profit. If the upper bound is greater than the lower bound within some tolerance, this design is used as a starting point for the next iteration. This iterative procedure continues until the current upper bound is less than or equal to the lower bound on the expected profit. The result of this problem is that a profit-maximizing, feasible design is selected. This problem formulation is limited to feasible regions that are convex or $1 - d$ convex with respect to each uncertain parameter. In addition, for a large number of uncertain parameters the solution of the problem may become computationally expensive [82].

The concept of stochastic flexibility has been extended to dynamic systems by Bansal et al. [83]. The proposed methodology is applied to linear dynamic systems in which the uncertainty is time-invariant, but the process model may be described by both differential and algebraic equations (DAE). The proposed solution method involves an active set strategy to determine the limiting cases for feasibility and the corresponding critical vertex directions for the uncertain parameters. This methodology can be applied to problems of optimal design under uncertainty for dynamic operation. Dynamic feasibility and flexibility analysis is described in greater detail in Sect. 3.5.

In pharmaceutical manufacturing applications, the concept of stochastic flexibility is useful in determining the feasible operating region for a process with input uncertainty that is characterized by a probability distribution. This type of uncertainty can arise from measurement variability in testing raw material quality attributes or due to variability in the quality of raw materials provided by a vendor. It may also be useful to consider stochastic flexibility in applications where real-time process monitoring is employed as part of a process control strategy. The uncertainty in PAT measurements can be characterized by a distribution around the true value, which can be explicitly incorporated into feasibility analysis. Accounting for the probability distributions associated with measurements can guide the selection of the operating limits for a process in such a way that the design space is conservative relative to the expected measurement variability.

3.3 Sampling-Based Methods for Process Feasibility Analysis

The previously discussed formulations of the feasibility test and flexibility index problems involve the use of deterministic optimization solvers. These require explicitly known constraints and may make assumptions regarding the convexity of the feasible region [65]. For instance, many of the aforementioned methods require that the feasible region be convex or $1 - d$ quasi-convex in the uncertain variables [64, 70, 78, 82]. In addition, deterministic solvers sometimes experience difficulty when dealing with large or complex optimization problems. Sampling-based methods can address some of these challenges by developing reduced-order representations for the feasible region. These reduced-order representations are developed using input-response data sampled experimentally from a process or obtained through simulation of a complex process model. In surrogate-based methods, the feasible region is approximated as a response-surface. When using operational envelopes, the shape of the feasible region is approximated and the bounds on process feasibility are determined through successive over and under approximation.

3.3.1 Surrogate-Based Methods

Recent developments in surrogate-based optimization have enabled the solution of optimization problems described by complex process models and problems with black-box constraints. Surrogate-based methods do not require inferences regarding the convexity of the feasible region [63, 84]. These approaches facilitate the use of complex or expensive-to-evaluate process models because surrogate functions can provide a reduced-order approximation of the complex process model.

Surrogate-based methods for design space evaluation involve the construction of a response surface representing the value of the feasibility function with respect to critical process parameters and/or uncertain model parameters [12, 48–50]. A response surface is a functional representation of a system output with respect to input variables that is meant to approximate the true variable-response relationship, which may not be explicitly known. Input-response data is collected by sampling the process or process model through random sampling, a design of experiments, or directed sampling in a specific region of interest [85]. The methodology selected for developing a response surface from input-output data is a critical aspect of surrogate-based strategies [85–87]. Techniques for response surface modeling include response surface methodology (RSM) [86], Kriging [88–90], artificial neural networks (ANN) [91, 92], and high dimensional model representation (HDMR) [93–95]. Response surface techniques for modeling pharmaceutical processes have been compared in several previous works [59, 87, 96]. In these, Kriging has been found to outperform other methods in terms of prediction accuracy. Predicting flow rate variability in a loss-in-weight feeder using Kriging gave lower prediction error than response surface methodology with a second-order function approximation [87, 97]. The design space for loss-in-weight feeders and for a continuous mixer was also determined with greater accuracy using Kriging as compared with other methods [59]. A particular advantage of Kriging is that an estimate of the prediction uncertainty is obtained for each test point. This information can be used to direct sampling to regions where additional data would be most beneficial in terms of improving the prediction accuracy of the surrogate model [84]. Kriging is also capable of modeling highly nonlinear and complex input-response relationships because it is interpolating in nature and does not assume a closed form model [96]. In all cases, the reduced-order modeling methodology should be selected in such a way that it is consistent with the intended application. For some studies a closed-form representation of the process model may be required. In this case response surface methodology might be more appropriate. HDMR can be used for applications in variance-based global sensitivity analysis. ANN and Kriging can model nonlinear systems accurately, but have the drawback of requiring potentially large training or fitting sets [92, 96].

The general algorithm for surrogate-based feasibility analysis involves sampling the uncertainty space, evaluating the process or model, determining the feasibility function value for each realization of the uncertain parameters, building a response surface for the feasibility function, and ultimately using this to identify the boundaries of the feasible region. Development of the response surface can be iterative in nature, with additional samples collected to improve the representation of the feasible region at each iteration. Additional sampling can be directed through the use of an expected improvement function, like that described in Boukouvala et al. [63, 98]. For feasibility analysis, an expected improvement function of the form shown in Eq. (16) can be used. This function directs additional sampling towards feasible points and points near the boundary of the feasible region. The first term in Eq. (16) indicates the probability of the feasibility function value (U) to be less than 0 while the second term indicates the probability of U to be greater than 0. When the objective is to locate the boundary of the feasible region, the second term alone can be used.

$$\begin{aligned} E_{\text{feas}}[I(x)] &= E_{\text{feas}}\{\max(0 - U), 0\} \\ &= -U\Phi\left(\frac{-U}{s_u}\right) + s_u\varphi\left(\frac{-U}{s_u}\right) \end{aligned} \tag{16}$$

The terms Φ and φ in Eq. (16) refer to the distribution function and the standard normal density, respectively. The term s_u in Eq. (16) is the surrogate model prediction variance. If Kriging is used to develop the response surface, prediction variance can be determined explicitly at each test point. An advantage of using Kriging approximations for surrogate-based feasibility analysis is that a prediction uncertainty surface can be generated concurrently with the response surface. The prediction variance can be used as an input to the expected improvement function. This facilitates the implementation of an adaptive sampling strategy which can be used to develop an accurate surrogate representation without exhaustively sampling the input space for the process. In addition, it has been demonstrated that a Kriging-based approach to feasibility analysis can be used to represent complex, nonconvex, or even disjoint feasible regions [63].

Given the inherent complexity of solids-based process modeling, surrogate-based approaches are well suited to pharmaceutical feasibility and flexibility analysis. Boukouvala et al. [59] have used surrogate-based feasibility analysis to explore the design space of several pharmaceutically relevant operations. The feasible operating space for a continuous solids mixer was identified with respect to powder flow rate and blender blade rotation rate, where relative standard deviation (RSD) of the API concentration at the exit of the mixer was considered a quality attribute. Both all-forward and alternating blade configurations were considered, yielding an integrated design and optimization problem. In this work, surrogate models based on RSM, HDMR and Kriging were developed. Kriging was found to most accurately capture the nonlinearity of the design space with respect to blender blade rotation rate. A similar approach was used to evaluate the design space of a loss-in-weight feeder. In this case the design space was evaluated with respect to RSD of the feeder mass flowrate and the flow index of the powder and the screw speed for the feeder were treated as uncertain parameters. A number of design parameters, including screw size and configuration, were also considered in this work. In both cases, the established design space was experimentally validated. In these case studies, Kriging outperformed RSM and HDMR in terms of accurately representing input-response relationships and therefore determining design space correctly. Boukouvala and Ierapetritou [63] have also demonstrated the use of a surrogate-based approach with an adaptive sampling strategy to determine the feasible region for a roller compaction process with respect to constraints on the ribbon thickness and ribbon density. In this case the sources of uncertainty were variability in the nip angle, an estimated model parameter, and fluctuations in the entering powder density [99]. The feasible region identified using a surrogate-based approach was found to agree with the feasible space obtained analytically through solution of the original complex process model.

3.3.2 Operational Envelopes for Feasibility Analysis

Another sampling-based approach to feasibility analysis involves treating the feasible region as a shape and approximating the features of this shape through successive over and/or underestimation techniques. Initial approaches to identifying feasible regions using operational envelopes involved simple shapes, like hyper rectangles [100].

More detailed approximation of feasible region shape was motivated in part by the fact that deterministic approaches to solving the feasibility problem may underestimate feasibility for complex and nonconvex feasible regions [101]. Goyal and Ierapetritou [102] present a Simplicial Approximation Approach [103] to determine the feasible region, which can help to address the issue of underestimating process feasibility. This procedure can be used to locate the boundary (∂R) of an n-dimensional feasible region (R) by determining the convex hull of a set of sample points consisting of m samples where $m \geq n + 1$. The simplicial convex hull provides a lower bound on the feasible region. A convex polytope is then constructed around the convex hull which yields an upper bound for the feasible region. A feasibility metric is introduced based on the ratio between the volume of the convex hull and that of the outer approximation. The feasible region is determined by iteratively solving the aforementioned problems until the volume of the convex hull matches that of the convex polytope within some tolerance. This methodology has been used to determine the feasible region and the flexibility index for a number of case studies, including the determination of operating envelopes for an air separation plant [102].

Banerjee and Ierapetritou [104] have suggested the use of a shape reconstruction technique similar to that of Goyal and Ierapetritou [102] to determine feasibility for nonconvex systems. The method is based on that of Goyal and Ierapetritou [102], but involves the use of shape reconstruction techniques to more accurately represent the feasible region. This method has been demonstrated to accurately represent feasible regions with complex shapes, including nonconvex and disjoint feasible regions [104].

Sampling-based methods for feasibility analysis have numerous potential applications in design space determination for pharmaceutical manufacturing processes. These methods provide a means of directly representing the feasibility function value in terms of process inputs, design and operating parameters. These reduced-order representations can be used to identify the boundaries of the feasible region without necessitating optimization over a complex process model. In addition, sampling-based methods can be used to identify nonconvex or even disjoint feasible regions.

3.4 Statistical Methods for Evaluating Process Feasibility

Statistical techniques such as Bayesian analysis, principal component analysis and multivariate regression can also be used to describe and analyze process design space. This section provides a brief discussion of latent variable and Bayesian methods, and describes their application to pharmaceutical process design space.

3.4.1 Latent Variable Methods

Latent variable methods like principal component analysis (PCA) and partial least squares/projection to latent structures regression (PLS) involve the projection of input and response data into a reduced space on a set of orthogonal axes. In this reduced space, the principal component scores (T) represent the transformed variables while the loadings (P) represent the axes upon which the transformed variables can be viewed [105]. The objective of PCA is to find scores and loadings that represent the most variance in a dataset (X) subject to certain orthogonality conditions [106]. This transformation is shown in Eq. (17), where is the error introduced by expressing X in the reduced space.

$$X = TP' + \varepsilon \tag{17}$$

Partial least squares (PLS) regression determines a transformation that explains variance in both the input (X) and response (Υ) data while retaining directions in the input space which are useful in predicting the responses of interest [107]. If the response data is high dimensional, it may be projected into a latent space as well. The response data can then be regressed onto the reduced state data (T). The transformation and regression equations for PLS are shown in Eq. (18). More detailed mathematical developments of both PCA and PLS are available in Hoskuldsson [108] and Burnham et al. [107].

$$\begin{aligned} X &= TP' + \varepsilon_X \\ \Upsilon &= TQ' + \varepsilon_\Upsilon \\ T &= XW^* \end{aligned} \tag{18}$$

Latent variable methods are often used to model the relationships between raw material properties and process performance or product quality in pharmaceutical process development [109, 110]. For these applications the interpretability of the model parameters is helpful in terms of developing process understanding [105, 111–113]. Latent variables are useful in defining specifications on raw material properties, as this information is often multivariate in nature. Reduced variables (e.g., PCA scores) can facilitate the definition of a low-dimensional and well-conditioned design space from material properties measurements that can be very high-dimensional in nature [112–114]. This avoids the issue of setting specifications on each material property individually and facilitates the development of a joint design space with respect to multiple material properties. Joint design spaces are of interest because they may provide a better and less restrictive representation of the design space than specifications based on individual properties [115]. The design space for pharmaceutical processes with respect to raw material properties can be defined in terms of principal component

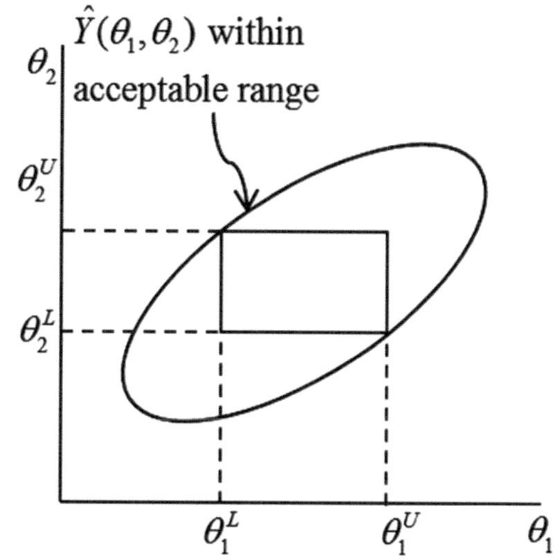

Fig. 5 Representation of a multivariate feasible space with respect to two uncertain variables. This figure illustrates the difference between univariate and multivariate specifications with respect to uncertain parameters. $\hat{Y}(\theta_1, \theta_2)$ represents the predicted mean response for a product quality attribute. This quality attribute is within an acceptable range for all combinations of uncertain parameter values falling within the ellipse. However, if univariate specifications are applied to θ_1 and θ_2 the feasible operating region will be restricted to the rectangle inscribed within the larger ellipse

scores obtained via either PCA or PLS [115, 116]. Figure 5 indicates the potential advantage of considering multivariate specifications, like those based on latent variables, rather than univariate specifications. In Fig. 5, the ellipse represents the region within which the predicted mean response for a product quality attribute, $\hat{Y}(\theta_1, \theta_2)$ is acceptable. If univariate specifications are set with respect to θ_1 and θ_2, the feasible operating region for the process will be restricted to the rectangle defined by the lower and upper bounds on these variables. However, if multivariate specifications are used instead, the feasible region can be expanded to incorporate the full ellipse corresponding to acceptable quality attributes.

Duchesne and MacGregor [115] have demonstrated this approach for a polymer film manufacturing process, in which the design space with respect to raw material (polyolefin pellet) properties was expressed in terms of two principal components obtained from a PLS model relating film properties to raw material properties. The resulting design space is lower dimensional and more interpretable than one expressed in terms of the ten material properties measured for the polyolefin pellets. García-Muñoz [114] proposes an extension to this approach for cases where most of the data is collected at a scale other than the proposed manufacturing scale, as is often the case in pharmaceutical process development. The author suggests estimating the hypothetical operating conditions at the manufacturing scale that are necessary to mimic the effect of scaling on quality. This can be accomplished using an optimization approach as discussed in García-Muñoz et al. [117, 118].

MacGregor and Bruwer [112] have proposed a framework for defining the joint design space with respect to material properties and process variables for a chemical process and using this information to establish effective feed-forward and feedback control strategies. The method involves performing successive design of experiments (DOE), first to determine the design space with respect to individual raw materials and then to determine the intersection of those design spaces. This is accomplished using principal properties as determined from PCA or PLS. Subsequently a DOE is conducted within the raw material design space. In this DOE the process variables are manipulated according to the desired process control scheme. Finally a DOE is performed in the combined design space of the raw materials and the process variables to identify the joint design space [112].

Huang et al. [111] have described a hybrid approach to design space definition that involves the development of response surfaces for critical quality attributes as a function of a subset of process design and operating parameters. This is followed by a principal component analysis to analyze all available process data for relationships among variables. Finally PLS is used to determine the relationship between all material properties, design variables, operating parameters and the quality attributes of interest. This includes even factors that are not explicitly varied in the original DOE. The combined response surface and multivariate analysis approach was demonstrated to enhance process understanding and facilitate the development of control strategies for ensuring product quality in a pharmaceutical manufacturing process [111].

García-Muñoz et al. [113] have extended the use of latent variable approaches to develop control strategies that can actually expand the design space for a process with respect to material properties. This work incorporates control strategies developed using the inverse of latent variable regression models, which is extensively discussed in Jaeckle and MacGregor [119]. A feed-forward control strategy is developed such that process operating parameters are adjusted to account for variability in the physical properties of input materials. As a result, the design space of a process with respect to uncertainty in raw material properties may be increased. Rather than demonstrating that a fixed process can tolerate all possible combinations of material properties, the authors suggest that control variables be adjusted such that acceptable product quality can be achieved for a wider range of raw material properties [113].

3.4.2 Bayesian Methods

Bayesian methods for design space development are based on the optimization of multiple response surfaces. Multiple response surface methods involve building response surfaces for each quality attribute, examining the regions within which these response surfaces overlap and identifying the input parameter space within which acceptable values for quality attributes are achieved [12]. The most straightforward way to address the issue of multiple response surfaces is the use of overlapping mean responses, which determines the space within which the mean value of the responses (quality attributes) are acceptable [120]. A more elegant approach involves introducing a desirability function, which transforms individual responses to a [0,1] value. Each quality attribute is assigned its own desirability metric in the range [0,1] and these are averaged to define a desirability function value. Individual metrics can be weighted in this average to emphasize certain responses. The desirability function is then optimized to identify process input values that provide acceptable quality attributes [12, 121].

Determining the design space using the aforementioned multiple response surface approach could result in an artificially small design space. In addition, this method does not consider correlations among responses or prediction uncertainty for the process outputs [12]. Introducing a quadratic loss function [122–124] can address the issue of response correlation, but still does not address prediction uncertainty. Peterson [125] introduces a Bayesian reliability approach that can account for response uncertainty in the solution of multiple response surface problems. A posterior predictive distribution of the multivariate process response is used to determine the probability that a future response will be within the range of acceptable values [12, 125]. This approach has been applied to the study of a pharmaceutical mixture composition for a controlled release formulation. In this case the design space was determined with respect to the particle size and glass transition temperature of controlled release particles as a function of mixture composition.

A Bayesian reliability approach has also been demonstrated effective at optimizing the probability of achieving good performance with respect to peak shape, signal to noise ratio and run time for a high performance liquid chromatography (HPLC) process [125]. This optimization approach can be extended to determine design space reliability in the presence of process noise [12]. Using a Bayesian approach based on the posterior predictive distribution, it is possible to determine the probability that a given process response or set of responses will fall within an acceptable region for a specific set of uncertain process parameters and noisy process data. The acceptable region is defined based on a specific reliability value, R [12]. An advantage of this methodology is its flexibility. It can incorporate mixed effect regression models, account for uncertainty in model form as well as in model parameters and can incorporate prior information, such as data collected on a pilot scale. This methodology has been applied to design space development for a synthetic chemical reaction, where the factors considered include temperature, pressure, catalyst loading and reaction time and quality attributes of interest are product purity, residual starting material and impurity levels. A separate model was developed for each response, generating a multivariate response surface. A posterior prediction approach was used to locate the joint region where the probability of meeting all quality specifications would be greatest with respect to all process operating parameters. In addition, a reduced design space is developed for a specific process reliability value. Within this reduced space the probability of all quality constraints being met is at least some minimum reliability value, R [12].

3.5 Dynamic Feasibility and Flexibility Analysis

It is important to consider dynamics in analyzing design space of pharmaceutical processes, as pharmaceutical manufacturing is often carried out in batch mode. Batch processes are inherently time-dependent and it is therefore possible that their design space might vary in time. Flexibility analysis for dynamic systems is fundamentally different than that for steady state systems [126]. Initial attempts at addressing dynamic flexibility focused on the notion of dynamic resilience, introduced by Morari [60] and discussed in Holt and Morari [127, 128] and Skogestad and Morari [129]. These methods require that process models be linearized around some steady state conditions and can describe resiliency with respect to a limited set of factors [126]. Other work has focused on the successful or optimal operation of a dynamic process in the presence of uncertainty, without accounting for the ability of control variables to enhance flexibility [130, 131].

Dimitriadis and Pistikopoulos [126] introduced formulations for the dynamic feasibility and flexibility index problems based on the steady state formulation of Swaney and Grossmann [61, 62]. This approach can be used to determine process feasibility given time-varying uncertainty and to determine a dynamic flexibility index for processes described by differential and algebraic equations. Before introducing this formulation it is important to define two different types of constraints that may be imposed on a dynamic process. The first are path constraints, which must be satisfied during the entire time that the process is operating as show in Eq. (19). The second are point constraints, which must be satisfied at specific time points during the operation as shown in Eq. (20). Common examples of point constraints are those imposed at the beginning or end of a process, like final conversion in a batch reactor [126].

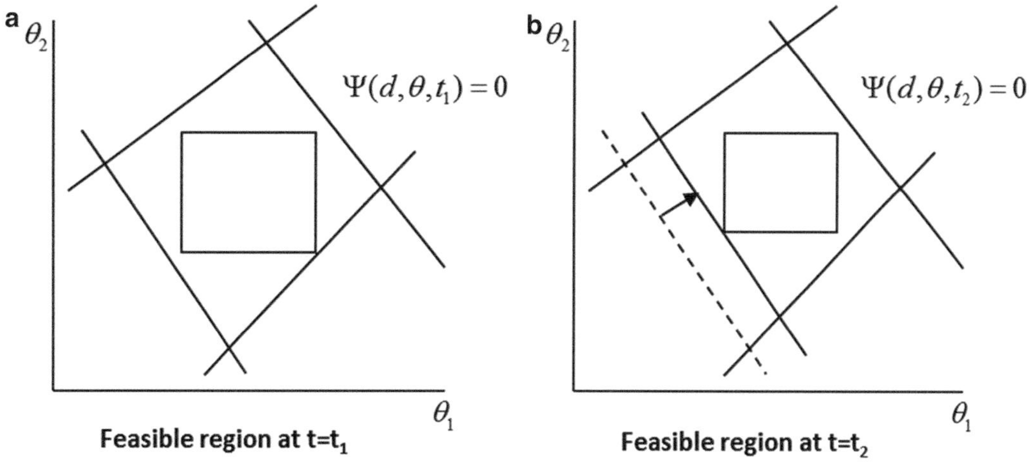

Fig. 6 Geometrical representation of dynamic feasibility. Figure (**a**) shows the feasible region for a process with two uncertain variables (θ_1 and θ_2) and four constraints at some initial time $t = t_1$. Figure (**b**) shows the feasible region for the same process at some later point in time $t = t_2$. Between $t = t_1$ and $t = t_2$, one of the constraints shifts from its initial position along the dashed line to its final position, shown as a solid line in (**b**). As a result, the feasible region at $t = t_2$ is smaller than the feasible region at $t = t_1$

$$g^{\text{path}}(d, x(t), z(t), \theta(t), t) \leq 0 \tag{19}$$

$$g_k^{\text{point}}\left(d, x\left(t^k\right), z\left(t^k\right), \theta\left(t^k\right), t^k\right) \leq 0 \quad k = 1, 2, \ldots, \text{NP} \tag{20}$$

Given that the constraints in Eqs. (19) and (20) are time-dependent, it is clear that the feasible region and therefore the design space for the process will vary over time. This is depicted in Fig. 6, which shows the change in feasibility for a dynamic process over time. Figure 6a shows the feasible region for a process at some initial time $t = t_1$. Figure 6b illustrates the feasible region at some later point in time $t = t_2$. In Fig. 6b it can be seen that one of the process constraints, which is defined by a differential equation, shifts during the time elapsed between t_1 and t_2. As a result, the feasible operating region at $t = t_2$ is smaller than that at $t = t_1$.

In order to solve the dynamic feasibility problem it is necessary to establish a time horizon (H) over which the process must remain feasible. Then the feasibility problem can be formulated as a dynamic version of the steady state problem defined by Halemane and Grossmann [66], as shown in (21).

$$\begin{aligned}
&\chi(d) = \max_{\theta(t)\in T(t)} \min_{z(t)\in Z(t)} \max_{j\in J,\, t\in[0,H]} g_j(d, x(t), z(t), \theta(t), t) \\
&\text{s.t. } h(d, \dot{x}(t), x(t), z(t), \theta(t), t) = 0; \quad x(0) = x^0 \\
&T(t) = \{\theta(t) | \theta^L(t) \leq \theta(t) \leq \theta^U(t)\} \\
&Z(t) = \{z(t) | z^L(t) \leq z(t) \leq z^U(t)\}
\end{aligned} \tag{21}$$

If $\chi(d) \leq 0$ in Eq. (21) then there is at least one dynamic control strategy characterized by $z(t)$ that can satisfy the path and point constraints $g_j(d,x(t),z(t),\theta(t),t)$ for all possible profiles of the uncertain variables $\theta(t)$ throughout the time period $t \in [0,H]$. The dynamic feasibility problem in Eq. (21) can be reformulated as the bi-level optimization problem (22) using the feasibility function $\Psi(d,\theta(t))$. The dynamic flexibility index problem can be expressed similarly based on deviations from the nominal time-varying profile of the uncertain parameters $\theta^N(t)$, as shown in Eq. (23).

$$\begin{aligned}
&\chi(d) = \max_{\theta(t)\in T(t)} \Psi(d, \theta(t)) \\
&\text{s.t. } \Psi(d, \theta(t)) = \min_{u,\, z(t)\in Z(t),\, t\in[0,H]} u \\
&\text{s.t. } h(d, \dot{x}(t), x(t), z(t), \theta(t), t) = 0; \quad x(0) = x^0 \\
&g(d, x(t), z(t), \theta(t), t) \leq u \\
&T(t) = \{\theta(t) | \theta^L(t) \leq \theta(t) \leq \theta^U(t)\} \\
&Z(t) = \{z(t) | z^L(t) \leq z(t) \leq z^U(t)\}
\end{aligned} \tag{22}$$

$$\begin{aligned}
&DF(d) = \max_{\delta,\, z(t)\in Z(t),\, t\in[0,H]} \delta \\
&\text{s.t. } \chi(d) = \max_{\theta(t)\in T(\delta,t)} \min_{z(t)\in Z(t)} \max_{j\in J,\, t\in[0,H]} g_j(d, x(t), z(t), \theta(t), t) \leq 0 \\
&\text{s.t. } h(d, \dot{x}(t), x(t), z(t), \theta(t), t) = 0; \quad x(0) = x^0 \\
&T(\delta, t) = \{\theta(t) | \theta^N(t) - \delta\Delta\theta^-(t) \leq \theta(t) \leq \theta^N(t) + \delta\Delta\theta^+(t)\} \\
&\delta \geq 0 \\
&Z(t) = \{z(t) | z^L(t) \leq z(t) \leq z^U(t)\}
\end{aligned} \tag{23}$$

Problems (22) and (23) are two-stage, semi-infinite dynamic optimization problems. These are potentially difficult to solve, as the dynamic profiles $z(t)$ and $\theta(t)$ represent an infinite number of decision variables.

Dimitriadis and Pistikopoulos [126] have shown that the dynamic feasibility test and flexibility index problems can be significantly simplified if the uncertainty profile $\theta(t)$ and the critical deviation $\Delta\theta^C(t)$, from the nominal uncertainty profile $\theta^N(t)$ are known. In this case, the dynamic feasibility and flexibility problems can be formulated as standard dynamic optimization problems, as shown in Eqs. (24) and (25).

$$\begin{aligned}
\chi(d) = & \min_{u,\, z(t)\in Z(t),\, t\in[0,H]} u \\
& \text{s.t. } h(d,\dot{x}(t),x(t),z(t),\theta(t),t) = 0; \quad x(0) = x^0 \\
& g(d,x(t),z(t),\theta(t),t) \le u \\
& Z(t) = \left\{ z(t) \middle| z^L(t) \le z(t) \le z^U(t) \right\}
\end{aligned} \tag{24}$$

$$\begin{aligned}
DF(d) = & \max_{\delta,\, z(t)\in Z(t),\, t\in[0,H]} \delta \\
& \text{s.t. } h(d,\dot{x}(t),x(t),z(t),\theta(t),t) = 0; \quad x(0) = x^0 \\
& g(d,x(t),z(t),\theta(t),t) \le 0 \\
& \theta(t) = \theta^N(t) + \delta\Delta\theta^c(t) \\
& \delta \ge 0 \\
& Z(t) = \left\{ z(t) \middle| z^{\mathrm{L}}(t) \le z(t) \le z^{\mathrm{U}}(t) \right\}
\end{aligned} \tag{25}$$

For the problem formulations given in Eqs. (22)–(25) it is necessary to implement a solution algorithm that accounts for both differential and algebraic equations (DAE). One way to do this is to embed a differential equation solver into the optimization algorithm and solve the differential equations for each control profile and/or set of decision variables selected by the optimizer. However, this approach would require the solution of all differential equations at each iteration, which can be time-consuming. Alternatively, some or all of the differential equations can be converted into algebraic equations. This can be accomplished via control parameterization or full discretization. In the former approach, the vector of controls is parameterized based on a set of finite parameters, resulting in a partitioning of the time horizon (H) into a subset of control intervals that may vary in length [132, 133]. The resulting problem contains both differential and algebraic equations (DAE). In the latter approach, also known as direct transcription, all of the equations in the dynamic feasibility and/or flexibility problem are discretized with respect to time. Thus the system consists of only algebraic equations. Both algorithms generate nonlinear programming (NLP) problems, but control parameterization is considered a sequential method while discretization is a simultaneous method [134].

Control parameterization involves the use of piecewise polynomials to represent the control variables. The polynomial coefficients become variables in the optimization problem and are determined using an NLP solver. A DAE solver can then be used to solve the system of equations for the given set of control parameters at each iteration. Subsequently the gradients of the objective function and constraints with respect to the polynomial coefficients and parameters can be determined from the sensitivity equations or by integration of the adjoint equations [126, 134, 135]. One advantage of control parameterization is that it can be readily implemented using commercially available solvers for systems of differential and algebraic equations (DAE). In addition, the NLP solver only has to handle a relatively small optimization problem with respect to the control parameters. However, because of the sequential nature of first determining control coefficients followed by integration of the DAE system, this approach can be time consuming for large problems. In addition, control parameterization is sensitive to open-loop instability and typically handles path constraints only approximately [134–136].

Full discretization involves dividing the time horizon into finite elements, where the state and control trajectories are approximated using polynomials on these elements [136]. Different polynomial representations, including power series, sums of orthogonal polynomials, or Lagrange forms, may be selected depending on the discretization used [134]. A common discretization method is orthogonal collocation on finite elements [126], though strategies involving other collocations or backward difference discretization may also be used [136–138]. The model equations are satisfied at all collocation points within the finite elements. Constraints can be enforced at these points as well [134, 136]. Continuity of the state profiles is enforced at the boundaries of the elements and the length of the elements can be treated as an additional optimization variable [126]. Simultaneous approaches can be used to accurately solve problems with path constraints and unstable modes. In addition, state and control variable constraints can be enforced at the same level of discretization, providing advantages in the solution of large-scale problems [134]. A disadvantage is that discretization generates very large NLPs, which require specialized solution approaches [135]. Full discretization can be readily applied to boundary value problems and optimal control problems but is generally not well suited to initial value problems [134, 137].

Problems (24)–(25) can be solved directly using the aforementioned strategies. However, problems (22)–(23) are bi-level optimization problems and require additional manipulation to be solved. The differential equations can be converted to algebraic equations using orthogonal collocation on finite elements. Once the system has been expressed in terms of finite elements, the feasibility problem can be solved using an active set strategy similar to that proposed by Grossmann and Floudas [64] for steady state feasibility analysis. The KKT optimality conditions for discretized system are enumerated, and slack variables are incorporated to account for a subset of constraints being active. The bi-level dynamic feasibility problem is divided into an inner and outer optimization, where the inner problem is used to determine the optimal values of the control variables and locations of the collocation points for a given uncertainty vector, while the global solution provides the uncertainty vector that corresponds to a limiting case for feasible operation given the best possible control action. Both problems are mixed integer nonlinear programs (MINLP), which can be solved using commercially available optimization software [65]. The resulting uncertainty and control vectors are parameterized according to the scheme used in the discretization of the original problem. The dynamic flexibility index problem can also be expressed as an MINLP using orthogonal collocation coupled with an active set strategy.

Dimitriadis and Pistikopoulos [126] have demonstrated the use of control parameterization to determine feasibility and the flexibility index for a jacketed batch reactor system in which an exothermic first order reaction is being carried out. The problem is constrained with respect to the desired final conversion, maximum allowable operating temperature throughout the process, and maximum allowable temperature at the end of the operation. The cooling water flow rate in the reactor jacket is the control variable. The heat of reaction is uncertain, but the uncertainty is parametric and therefore constant with respect to time. For this reason, formulations (24)–(25) can be applied. Flexibility analysis can be used to determine the extent to which uncertainty in the heat of reaction can be tolerated. Using dynamic feasibility analysis, the cooling water flowrate profile corresponding to feasible operation at the critical point with respect to heat of reaction can also be obtained.

Mohideen et al. [139] have applied the dynamic feasibility formulation to identify the feasible region and optimal control scheme for operation of a distillation column subject to path constraints on product quality and operational safety with uncertainty in the feed stream flow rate and composition. Constraints are also added to describe the operation of PI multiloop controllers for the process, where the control structure and the controller tuning parameters are variables to be optimized. In this case, an objective function is added to minimize the total cost associated with the process and the total cost is minimized subject to process feasibility. The resulting problem is a mixed-integer stochastic nonlinear optimal control problem, which the authors propose a decomposition algorithm to solve. Orthogonal collocation is used to convert differential equations to algebraic equations and the problem is divided into two sub problems that are solved iteratively. The first is a multiperiod design and control problem that provides the optimal design and control scheme for a given approximate uncertainty structure. Subsequently a dynamic feasibility analysis is applied to the design and control scheme to determine if it is feasible everywhere. The two problems are solved iteratively until the design and control structure provided by the first sub problem are feasible across the time horizon of interest. The result is a design and control structure that ensures feasibility while minimizing total cost of operation.

Although the aforementioned examples are not specific to the pharmaceutical industry, they demonstrate the potential for dynamic feasibility analysis to contribute to pharmaceutical process development. In both cases, dynamic feasibility analysis is used to achieve optimal control trajectories for processes with time-varying uncertainty. The same methods can be applied to pharmaceutical manufacturing processes, where feasible operation may be time-dependent in both batch and continuous operation. In continuous pharmaceutical manufacturing processes, time-varying uncertainty can arise from variability in powder feeding operations. This may include periodic fluctuations in feed rate that are inherent to the equipment being used and the materials being fed. In addition, transient disturbances in feed rate may be observed during feeder refilling operations [140].

Batch processes, which are often encountered in pharmaceutical manufacturing, are inherently time-dependent and thus require the use of dynamic strategies for feasibility analysis. In batch systems, initial conditions can affect both process performance and operability.[131] For this reason, it is important to account for initial conditions in estimating process feasibility and flexibility. Zhou et al. [141] propose a method for incorporating initial conditions into dynamic feasibility analysis of batch processes by treating initial conditions as control variables via a point constraint. This involves replacing the statement $x(0) = x^0$ in Eqs. (21)–(25) with the constraint given in Eq. (26). In this statement, the initial condition is treated as a control variable with upper and lower bounds rather than as a fixed point, facilitating flexibility analysis with respect to variability in initial conditions and optimization of initial conditions to enhance flexibility.

$$x(0) = z_x = \left\{ z_{x(i)} \middle| z^{\mathrm{L}}_{x(i)} \leq z_{x(i)} \leq z^{\mathrm{U}}_{x(i)} \right\} \tag{26}$$

In order to solve the dynamic flexibility analysis problem incorporating Eq. (26), a decomposition algorithm similar to that used in Luus and Hennessy [142] can be applied. The algorithm is iterative and begins with guessing an initial value for the flexibility index. The differential equations are discretized in the time horizon. The algorithm then proceeds by successive optimization of the control variables and tests of feasibility with respect to path constraints and point constraints. When the algorithm converges, a set of control vectors and initial conditions have been found that satisfy all path and point constraints for the problem and correspond to the edge of the feasible region. In addition, a dynamic flexibility index for the system is determined. The proposed algorithm has been applied to dynamic flexibility analysis of two batch processes, a simple case study of a first order irreversible reaction and a highly exothermic polymerization process. In both cases, initial conditions and control trajectories corresponding to maximum flexibility were identified [141]. This methodology could be applied to batch processes in pharmaceutical manufacturing to identify optimal initial conditions.

In some dynamic systems, time delays may be introduced due to material transit between units or material recycle streams. Time delays can affect process operability and controllability, so if significant time delays are present in a process it is important to account for them when evaluating feasibility and flexibility. Huang et al. [143] have introduced a modified finite element collocation method that can address time delays in dynamic flexibility analysis. In this approach, the dynamic flexibility index problem is reformulated to incorporate a new time variable $t' = t - \tau$, where τ is the lag time for the system. In order to account for the effect of lag time on the dynamic control of the system, the linear quadratic regulator (LQR) technique is used. The general form of the control optimal solution for the dynamic time-delay system is incorporated directly into the flexibility index problem, resulting in the formulation in Eq. (27). The term $z(x,y(\cdot))$ represents the control optimal solution for a system with a time delay as obtained from the LQR technique, where N is a symmetric positive definite matrix, B is a constant matrix, P is a symmetric positive matrix, $D(s)$ is a solution to the Generalized Riccati Equations and $y(-\tau) = x(t - \tau)$. Additional details pertaining to the LQR technique can be found in the literature [143–145]. The flexibility index problem can be formulated similarly, as shown in Eq. (28). Note that in these formulations it is assumed that the variation of the uncertainty vector is known, as is the critical deviation from the nominal uncertainty profile [126].

$$
\begin{aligned}
&\chi(d) = \min_{u,\, z,\, t\in[0,H]} u \\
&\text{s.t. } h(d, \dot{x}(t), x(t), x(t-\tau), z(t), \theta(t), t) = 0; \quad x(0) = x^0 \\
&g(d, x(t), x(t-\tau), z(t), \theta(t), t) \leq u \\
&z(x, y(\tilde{n})) = -N^{-1}B' \left[Px + \int_{-\tau}^{0} D(s)y(s)ds \right]
\end{aligned}
\tag{27}
$$

$$
\begin{aligned}
&DF(d) = \max_{\delta, z, t \in [0,H]} u \\
&\text{s.t.}\, h(d, \dot{x}(t), x(t), x(t-\tau), z(t), \theta(t), t) = 0; \quad x(0) = x^0 \\
&g(d, x(t), x(t-\tau), z(t), \theta(t), t) \leq 0 \\
&\theta(t) = \theta^N(t) + \delta \Delta \theta^c(t) \\
&\delta \geq 0 \\
&z(x, y(\cdot)) = -N^{-1} B' \left[Px + \int_{-\tau}^{0} D(s) y(s) ds \right]
\end{aligned}
\tag{28}
$$

The problems (27) and (28) can be solved using a finite element collocation method to find the optimal solution for the state variables and the LQR method to find the optimal solution for the control variables. Huang et al. [143] have used this approach to determine the flexibility index for several processes in which a first order, irreversible reaction is carried out. In the first case, the system consists of two reactors in series, where a time delay is introduced by a recycle loop of unreacted material from the second reactor back to the first reactor. Uncertainty arises from variability in the concentration of the inlet stream to the first reactor. In the second case, a reactor–separator system in which unreacted starting material is recycled back to the reactor after separation is considered. In this instance uncertainty is introduced through the heat transfer coefficient for the system, as the process is exothermic and must be maintained below a specified temperature for safety reasons. In both instances, the modified finite element collocation method can be successfully used to determine the flexibility index for the system given the time delay introduced by the recycle loops. In pharmaceutical manufacturing processes, time delays can be introduced through residence time within certain unit operations, the transit of powders from one unit operation to another, or through the requisite relaxation time for tablets prior to undergoing coating. Thus feasibility and flexibility analysis methods that can incorporate the effect of time delays are useful for the evaluation of process design space. This is particularly important for continuous pharmaceutical manufacturing processes, where time delays in one unit operation can have effects that propagate to subsequent units.

4 Case Study: Feasibility Analysis of a Roller Compaction Process

This section illustrates the concept of feasibility analysis as applied to a pharmaceutical manufacturing process using a roller compaction case study. Roller compaction is used in dry granulation, and involves the compaction of powder blends into ribbons of densified material which are subsequently milled to form granules. In a roller compactor, powder is fed to a system of counter rotating rolls. When it initially enters the rolls, the powder is considered to be in the *slip* region. This region is characterized by the powder particles slipping at the surface of the rolls. The rolls convey the material forward and as the material approaches the point where the rolls are closest together the powder in contact with the rolls begins to move at the same speed as the rolls and substantial pressure is exerted on the powder. This is the *nip* region, in which the powder is compressed to form a ribbon. The nip angle, α, defines the transition from the slip to the nip region. The ribbon continues to be conveyed forward and as the roll gap increases again the powder enters the *release* region, where it is released from the rolls [146].

The roller compaction process can be mathematically modeled using Johanson's theory of rolling granular solids [99, 147]. Bindhumadhavan et al. [146] have demonstrated that for pharmaceutical powders this theory can accurately describe pressure profiles in the *nip* region of a roller compaction process, where the ribbon is formed. The roller compaction model used in this example is based on a modified Johanson's model developed by Hsu et al. [148] This model considers the effect of powder feed rate on the roll gap, which is assumed constant in the original Johanson's model. This process model has also been used to demonstrate a surrogate-based feasibility analysis strategy by Boukouvala et al. [63].

In this case study, the objective is to identify the feasible operating region for a roller compaction process with respect to constraints on two quality attributes of the ribbon: ribbon thickness (h_0) and ribbon density (ρ_{exit}). The uncertain variables considered are the inlet angle, θ_{in}, which is an estimated model parameter, and the density of the powder entering the system, ρ_{in}. The control variables, which can be manipulated to try to maintain feasible operation, are the applied hydraulic pressure on the rolls (P_h), the roll speed (ω) and the powder feed speed (u_{in}). The formulation of the feasibility problem is given in Eqs. (29)–(36). Equation 29 is the formulation of the feasibility objective, as shown in Eq. (2). Equations (30)–(32) describe the process model, which are equality constraints for the feasibility analysis problem. Equation 30 describes the relationship between the roll gap and the feed speed. Equation 31 describes the relationship between roll pressure and stress in the *nip* region, which is related to ribbon density via the empirical correlation given in Eq. (32). Further development of the mathematical model is available in Hsu et al. [148]. The constraints on the critical quality attributes are shown in Eqs. (33) and (34), for the ribbon thickness and the ribbon density respectively. The bounds given in (35) describe the limitations on the control variables. Finally the expressions in (36) describe the range of

Table 4
Parameter definitions and values used in the roller compaction case study

Parameter	Symbol	Value/range	Units
Roll radius	R	0.125	m
Roll width	W	0.05	m
Compression parameter	K	4.97	
Compression parameter	C_1	7.5×10^{-8}	$Pa/(kg/m^3)^{4.97}$
Compact surface area	A	0.01	m^2
Effective angle of friction	δ	0.7069	rad
Nip angle	α	0.173	rad
Angular position	θ	NA	rad
Hydraulic pressure (roll pressure)	Ph	0.9–1.1	MPa
Angular velocity of rolls	ω	4–6	rpm
Powder feed speed	u_{in}	2.27×10^{-4}–4.27×10^{-4}	m/s
Inlet powder density	ρ_{in}	200–400	kg/m^3
Inlet angle	θ_{in}	0.3–0.5	rad
Ribbon thickness	h_0	1.7×10^{-3}–1.9×10^{-3}	m
Ribbon density	ρ_{exit}	850–950	kg/m^3

uncertain parameters. This is equivalent to the region T in the feasibility test problem formulation. The specific process parameters used for this case study can be found in Table 4.

$$\min_{P_h, u_{in}, \rho_{exit}, h_0, u} \quad u \qquad \text{s.t.} \tag{29}$$

$$\rho_{in} \cos\theta_{in}\left(1 + \frac{h_0}{R} - \cos\theta_{in}\right)\left(\frac{u_{in}}{\omega h_0}\right) = \rho_{exit} \tag{30}$$

$$P_h = \frac{W}{A}\frac{\sigma_{exit} R}{1 + \sin\delta}\int_0^{\alpha}\left[\frac{h_0/R}{(1 + h_0/R - \cos\theta)\cos\theta}\right]^K \cos\theta d\theta \tag{31}$$

$$\sigma_{exit} = C_1 \rho_{exit}{}^K \tag{32}$$

$$\begin{aligned} h_0 - h_0^{\mathrm{U}} &\le u \\ h_0^{\mathrm{L}} - h_0 &\le u \end{aligned} \tag{33}$$

$$\begin{aligned} \rho_{\mathrm{exit}} - \rho_{\mathrm{exit}}^{\mathrm{U}} &\le u \\ \rho_{\mathrm{exit}}^{\mathrm{L}} - \rho_{\mathrm{exit}} &\le u \end{aligned} \tag{34}$$

$$\begin{aligned} & P_h \in \left[P_h^{\min}, P_h^{\max}\right] \\ & \omega \in \left[\omega^{\min}, \omega^{\max}\right] \\ & u_{\mathrm{in}} \in \left[u_{\mathrm{in}}^{\min}, u_{\mathrm{in}}^{\max}\right] \end{aligned} \tag{35}$$

$$\begin{aligned} & \theta_{\mathrm{in}} \in \left[\theta_{\mathrm{in}}^{\min}, \theta_{\mathrm{in}}^{\max}\right] \\ & \rho_{\mathrm{in}} \in \left[\rho_{\mathrm{in}}^{\min}, \rho_{\mathrm{in}}^{\max}\right] \end{aligned} \tag{36}$$

By solving the feasibility problem described by (29)–(36), the goal is to determine:

- Whether the roller compaction process can provide ribbons of acceptable density and thickness over the full range of possible inlet densities (ρ_{in}) to the process given the uncertainty associated with estimating the model parameter θ_{in}.
- If the roller compaction process cannot achieve the desired product quality, what combinations of ρ_{in} and θ_{in} result in off-specification product?
- Which quality attributes are not within specification for the combinations of ρ_{in} and θ_{in} identified as infeasible?
- What potential control strategies can be employed to guarantee acceptable ribbon density and thickness?

In this case the feasibility problem is solved analytically using a sampling-based approach similar to those described in Sect. 3.3.2. The feasibility test problem is evaluated for a series of points within the uncertain region. The resulting feasibility function values are collected and the convex hull of those points with feasibility values less than or equal to 0 is taken. In this case the problem is convex with a feasible region described by linear constraints, so the convex hull provides an adequate approximation of the feasible region provided a sufficient number of samples are collected. To ensure adequate sampling near the boundary of the feasible region, an expected improvement function based on the second term in Eq. (16) is employed. The feasible operating region for the roller compaction process with respect to the two uncertain variables is shown in Fig. 7. The polygon indicates the boundaries of the feasible region with respect to ρ_{in} and θ_{in}. The points scattered

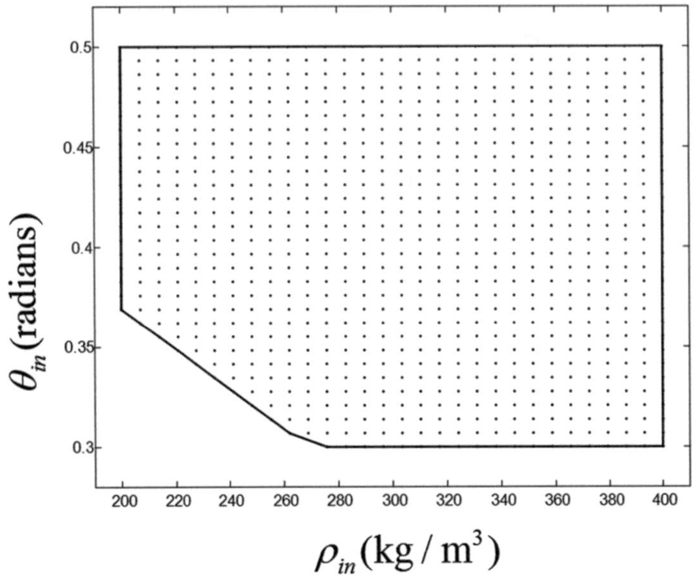

Fig. 7 Feasible region for the roller compaction case study. The polygon represents the feasible region with respect to the uncertain parameters, the inlet powder density and the inlet angle. The points scattered within the feasible region represent combinations of the uncertain parameters which result in acceptable product quality with respect to the constraints on the ribbon thickness and density. The boundaries of the feasible region are given by the convex hull of these points

within this polygon indicate combinations of the uncertain parameters that resulted in acceptable product quality with respect to the ribbon thickness and ribbon density. In this case it is clear that the process is infeasible when ρ_{in} is less than 260 kg/m^3 and the value of θ_{in} is below 0.37 radians. By examining the feasibility with respect to each of the constraints, it is possible to determine which quality attributes are off-specification for the infeasible points identified in Fig. 7. In Fig. 8, the feasible region with respect to each constraint is shown separately. It is clear from Fig. 8a that the ribbon thickness meets the specifications shown in Eq. (33) for all realizations of the uncertain parameters in the region defined by (36). However, Fig. 8b reveals that the ribbon density does not meet the specifications in Eq. (34). In fact, the ribbon density is below its lower bound for combinations of ρ_{in} and θ_{in} below 260 kg/m^3 and 0.37 radians respectively. Note that if the inlet angle (θ_{in}) were greater than 0.37 radians, the ribbon density would be within specification for the full range of possible powder densities entering the process. Improving the estimation of the model parameter θ_{in}, perhaps through the collection of additional process data, could reduce the variability associated with this parameter. This could increase the size of the feasible region with

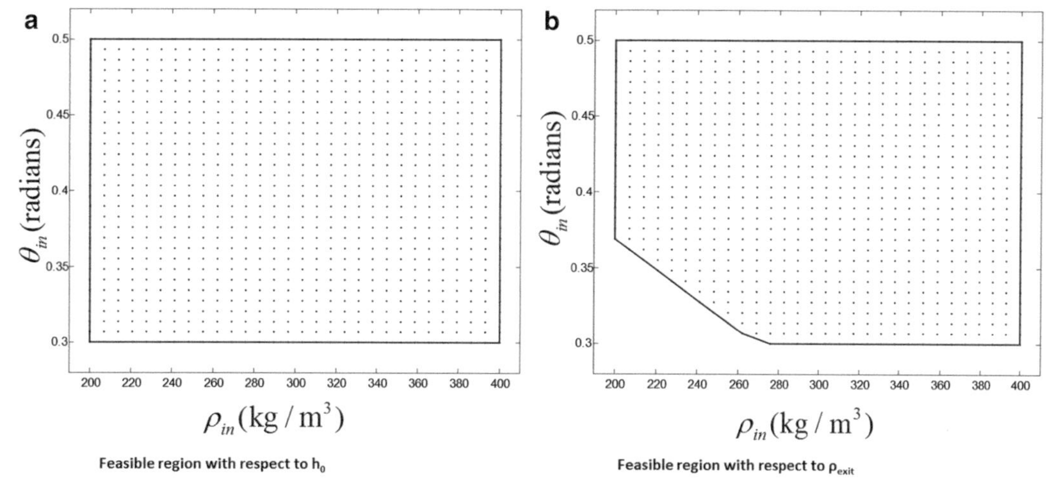

Fig. 8 Feasible region with respect to each of the quality attributes for the roller compaction case study. Figure (**a**) shows the feasible region with respect to the ribbon thickness, while (**b**) shows the feasible region with respect to the ribbon density. Figure (**a**) indicates that all combinations of the uncertain parameters yield acceptable ribbon thickness. Figure (**b**) indicates that some combinations of the uncertain parameters result in unacceptable ribbon density

respect to the inlet density. Alternatively, if the density of the powder entering the process could be controlled between 270 and 400 kg/m^3 then the estimation error associated with θ_{in} could be tolerated without limiting process feasibility. Based on the results of this feasibility analysis, the following conclusions can be drawn.

- The roller compaction process *cannot* provide ribbons of acceptable density and thickness over the full range of possible powder inlet densities (ρ_{in}) to the process given the uncertainty associated with estimating the model parameter θ_{in}.
- The desired product quality is not achieved when ρ_{in} is less than 260 kg/m^3 and θ_{in} is less than 0.37 radians.
- The ribbon density (ρ_{exit}) does not meet its specifications for the previously identified combinations of ρ_{in} and θ_{in}.
- Improving the estimation of θ_{in} or reducing the amount of variability in the density of the powder entering the process (ρ_{in}) could help to expand the feasible operating region.

5 Verification of Design Space

Before discussing the verification of process design space, it is important to reiterate that the information obtained from feasibility and flexibility analysis is useful in providing guidance to establish design space. However, the feasible operating region identified using these methods should not be considered the process design space. Instead, feasibility analysis should be viewed as a method for identifying the edges of failure for a process, with the design space being set well within the feasible region. Flexibility analysis can also help to identify factors that limit process feasibility and therefore suggest potential strategies for expanding design space. In this way flexibility analysis can be used to ensure that a manufacturing process is developed with the objective of maximizing design space in mind.

Recommendations for pharmaceutical process validation, including validation of design space, are described in ICH guidance documents Q8, Q9, and Q10 [1, 6, 7]. Specifically ICHQ8 indicates that the use of design spaces developed at small or pilot scale at a commercial manufacturing scale must be justified [1]. Concerns about scalability have prompted regulators, including the US FDA and the European Medicines Agency (EMA) to request verification of process design space at commercial scale. While it is clearly important to demonstrate the adequacy of the developed manufacturing process and corresponding design space for commercial production, extensive design space exploration during process validation may be impractical, time-consuming and costly [149]. Thus in order to address regulatory concerns, strategies to demonstrate understanding of scale effects on design space throughout process development are needed [5].

First, it is important to note that not all process parameters are scale-dependent. For instance, formulation composition and the required raw material quality attributes are not necessarily a function of scale [5, 149]. Continuous processes such as continuous blending, roller compaction, twin-screw wet granulation and tablet compaction can scale readily through increased throughput or parallelization rather than an increase in size. This can mitigate issues of scale dependence in developing and verifying design space for continuous processes [2, 150]. For parameters that are scale-dependent, it may be possible to use straightforward scaling relations to extend information collected at lab or pilot scale to expected performance at commercial scale. For instance, tip speed, number of blade passes or revolutions, or power per unit volume can be used to scale up blending processes [149, 151]. Scaling relations are also available for high-shear and fluid-bed wet granulation [152]. Often unit operations do not comply with standard scaling rules, and in these instances it is important to take advantage of data collected throughout process development. Valuable process data can be collected during the manufacture of supplies for clinical trials, especially during the later stages of process development when the manufacturing process is well established [149].

Mechanistic and semiempirical models can also aid in the development of design space, especially for processes that are relatively scale independent, like homogenous chemical reactions. In such cases, models developed using data collected on a laboratory scale can be used to estimate process performance over a range of conditions and thereby guide the selection of design space. The edges of the corresponding design space can be verified on a laboratory scale [153, 154]. For scale-dependent phenomena, acceptable performance on a laboratory scale may not be sufficient to verify the design space. However, if it can be demonstrated that the process can be controlled well within the design space at a commercial scale it may not be necessary to demonstrate the edges of the design space on-scale. In this case demonstrating effective control of the process at a commercial scale, thereby ensuring operation well within the design space, may be sufficient. This can be accomplished during traditional process validation, which is required prior to regulatory approval [5].

6 Summary

This chapter provides an overview of mathematical methods that can be used in the development of design space for pharmaceutical processes. These methods include flexibility and feasibility analysis, stochastic feasibility analysis, sampling-based approaches, statistical approaches, and dynamic feasibility and flexibility analysis. Of the methods discussed, statistical methods like latent variable and Bayesian approaches have been applied most extensively in pharmaceutical applications to date. However, advances in process modeling and derivative free optimization have created new opportunities to apply mathematical techniques like feasibility and flexibility analysis to pharmaceutical processes. Feasibility and flexibility formulations have long been used in other chemical process industries to solve problems of integrated design and optimization, but have only recently been applied to pharmaceutical processes. Feasibility analysis can help to identify the operating regions that provide acceptable product quality and can inform quality risk management activities such as control strategy development and design space evaluation. The use of rigorous mathematical approaches to address process development challenges remains limited in the pharmaceutical industry. Process systems engineering concepts such as feasibility and flexibility analysis can contribute to the design of manufacturing processes that are robust and efficient. This is critical in the current regulatory and economic environment, in which stricter regulations coupled with increased competition are driving forces for innovation both in the introduction of new molecular entities and in the development of reliable and expedient manufacturing processes [155].

Acknowledgements

The authors would like to acknowledge financial support from Bristol-Myers Squibb as well as from the Engineering Research Center for Structured Organic Particulate Systems at Rutgers University (NSF-0504497, NSF-ECC 0540855).

References

1. ICH (2005) ICH Q8 pharmaceutical development
2. McKenzie P, Kiang S, Tom J, Rubin E, Futran M (2006) Can pharmaceutical process development become high tech? AIChE J. 52 (12)
3. ICH Harmonised Tripartite Guideline: Pharmaceutical Development Q8(R2) Current Step 4 Version (2009) International Conference on Harmonisation of Technical Requirements for Registration of Pharmaceuticals for Human Use
4. Guidance for Industry (2006) Q8 pharmaceutical development, USA
5. Yu LX (2008) Pharmaceutical quality by design: product and process development, understanding, and control. Pharm Res 25(4):781–791. doi:10.1007/s11095-007-9511-1
6. ICH (2005) ICH Q9 quality risk management
7. ICH (2005) ICH Q10 pharmaceutical quality systems
8. Chatterjee S (2008) Overview of models used in design space determination: a regulatory perspective. Paper presented at the AIChE annual meeting, Philadelphia, PA
9. Chatterjee S (2012) Design space considerations. In: AAPS annual meeting, Chicago, Il
10. Degerman M, Westerberg K, Nilsson B (2009) A model-based approach to determine the design space of preparative chromatography. Chem Eng Technol 32(8):1195–1202. doi:10.1002/ceat.200900102
11. Leopore J, Spavins J (2008) PQLI design space. J Pharm Innov 3(2):79–87
12. Peterson JJ (2008) A Bayesian approach to the ICH Q8 definition of design space. J Biopharm Stat 18(5):959–975. doi:10.1080/10543400802278197
13. Guidance for Industry (1995) Immediate release solid oral dosage forms scale-up and postapproval changes: chemistry, manufacturing, and controls, in vitro dissolution testing, and in vivo bioequivalence documentation, USA
14. Guidance for Industry (1997) Modified release solid oral dosage forms scale-up and postapproval changes: chemistry, manufacturing, and controls, in vitro dissolution testing, and in vivo bioequivalence documentation, USA
15. Guidance for Industry (1997) Nonsterile semisolid dosage forms scale-up and postapproval changes: chemistry, manufacturing, and controls, in vitro dissolution testing, and in vivo bioequivalence documentation, USA
16. Guidance for Industry (2004) Changes to an approved NDA or ANDA, USA
17. Charoo NA, Shamsher AA, Zidan AS, Rahman Z (2012) Quality by design approach for formulation development: a case study of dispersible tablets. Int J Pharm 423(2):167–178. doi:10.1016/j.ijpharm.2011.12.024
18. Kenett R, Kenett D (2008) Quality by design applications in biosimilar pharmaceutical products. Accredit Qual Assur 13(12):681–690. doi:10.1007/s00769-008-0459-6
19. Lionberger RA, Lee SL, Lee L, Raw A, Yu LX (2008) Quality by design: concepts for ANDAs. AAPS J 10(2):268–276. doi:10.1208/s12248-008-9026-7
20. Airaksinen S, Karjalainen M, Shevchenko A, Westermarck S, Leppanen E, Rantanen J, Yliruusi J (2005) Role of water in the physical stability of solid dosage formulations. J Pharm Sci 94(10):2147–2165. doi:10.1002/jps.20411
21. Wu H, White M, Khan MA (2011) Quality-by-Design (QbD): an integrated process analytical technology (PAT) approach for a dynamic pharmaceutical co-precipitation process characterization and process design space development. Int J Pharm 405(1-2):63–78. doi:10.1016/j.ijpharm.2010.11.045
22. am Ende D, Bronk KS, Mustakis J, O'Connor G, Santa Maria CL, Nosal R, Watson TNJ (2007) API quality by design example from the torcetrapib manufacturing process. J Pharm Innov 2:71–86
23. Cimarosti Z, Bravo F, Castoldi D, Tinazzi F, Provera S, Perboni A, Papini D, Westerduin P

(2010) Application of the QbD principles in the development of the casopitant mesylate manufacturing process. Process research studies for the definition of the control strategy of some drug substance-CQAs for stages 2a, 2b, and 2c. Org Process Res Dev 14(4):805–814. doi:10.1021/Op1000622

24. Dach R, Song JHJ, Roschangar F, Samstag W, Senanayake CH (2012) The eight criteria defining a good chemical manufacturing process. Org Process Res Dev 16 (11):1697–1706. doi:10.1021/Op300144g
25. Adeyeye MC (2008) Drug-excipient interaction occurrences during solid dosage form development. Drugs Pharm Sci 178:357–436
26. Campisi B, Chicco D, Vojnovic D, Phan-Tan-Luu R (1998) Experimental design for a pharmaceutical formulation: optimisation and robustness. J Pharm Biomed Anal 18(1-2):57–65
27. Muzzio FJ, Alexander A, Goodridge C, Shen E, Shinbrot T (2004) Solids Mixing. In: Paul EL, Atiemo-Obeng VA, Kresta SM (eds) Handbook of Industrial mixing: science and practice. Wiley, Hoboken, NJ, pp 887–985
28. Faldu B, Sharma A, Sharma A, Chauhan CS (2012) Roller compaction: imperative process for tablet manufacturing: a review. Int J Pharm Res Dev 4(10):40–47
29. Lee KT, Ingram A, Rowson NA (2013) Comparison of granule properties produced using twin screw extruder and high shear mixer: a step towards understanding the mechanism of twin screw wet granulation. Powder Technol 238:91–98
30. Rogers AJ, Hashemi A, Ierapetritou MG (2013) Modeling of particulate processes for the continuous manufacture of solid-based pharmaceutical dosage forms. Processes 1 (2):67–127
31. Ng KM (2002) Design and development of solids processes – a process systems engineering perspective. Powder Technol 126 (3):205–210
32. Strong J (2009) Scale-up of pharmaceutical manufacturing operations of solid dosage forms. In: Qiu Y, Chen Y, Zhang GZ, Liu L, Porter WR (eds) Developing solid oral dosage forms: pharmaceutical theory and practice. Academic Press, New York, NY, USA, pp 615–636. doi:10.1016/B978-0-444-53242-8.00027-8, An Imprint of Elsevier
33. Järvinen MA, Paaso J, Paavola M, Leivisk K, Juuti M, Muzzio F, Järvinen K (2013) Continuous direct tablet compression: effects of impeller rotation rate, total feed rate and drug content on the tablet properties and drug release. Drug Dev Ind Pharm 39 (11):1802–1808
34. Boukouvala F, Niotis V, Ramachandran R, Muzzio FJ, Ierapetritou MG (2012) An integrated approach for dynamic flowsheet modeling and sensitivity analysis of a continuous tablet manufacturing process. Comput Chem Eng 42:30–47. doi:10.1016/j.compchemeng.2012.02.015
35. Gentis ND, Betz G (2012) Compressibility of binary powder formulations: investigation and evaluation with compaction equations. J Pharm Sci 101(2):777–793. doi:10.1002/jps.22794
36. Faure A, York P, Rowe RC (2001) Process control and scale-up of pharmaceutical wet granulation processes: a review. Eur J Pharm Biopharm 52:269–277
37. Vervaet C, Remon JP (2005) Continuous granulation in the pharmaceutical industry. Chem Eng Sci 60:3949–3957
38. Iveson SM, Litster JD, Hapgood K, Ennis BJ (2001) Nucleation, growth and breakage phenomena in agitated wet granulation processes: a review. Powder Technol 117:3–39
39. Kleinebudde P (2004) Roll compaction/dry granulation: pharmaceutical applications. Eur J Pharm Biopharm 58:317–326
40. Lecompte T, Doremus P, Thomas G, Perier-Camby L (2005) Dry granulation of organic powders – dependence of pressure 2D-distribution on different process parameters. Chem Eng Sci 60:3933–3940
41. Hlinak AJ, Kuriyan K, Morris KR, Reklaitis GV, Basu PK (2006) Understanding critical material properties for solid dosage form design. J Pharm Innov 1(1):12–17
42. Jornitz MW (2008) Vendor qualification and validation. In: Agalloco J, Carleton FJ (eds) Validation of pharmaceutical processes, 3rd edn. Informa Healthcare, New York, NY, USA, p 529
43. Zhang X, Lionberger RA, Davit BM, Yu LX (2011) Utility of physiologically based absorption modeling in implementing quality by design in drug development. AAPS J 13(1):59–71. doi:10.1208/s12248-010-9250-9
44. Jivraj M, Martini LG, Thomson CM (2000) An overview of the different excipients useful for the direct compression of tablets. Pharm Sci Technol Today 3(2):58–63
45. Pifferi G, Santoro P, Pedrani M (1999) Quality and functionality of excipients. Farmaco 54 (1–2):1–14

46. Morin G, Briens L (2013) The effect of lubricants on powder flowability for pharmaceutical application. AAPS PharmSciTech 14 (3):1158–1168
47. Wang J, Wen H, Desai D (2010) Lubrication in tablet formulations. Eur J Pharm Biopharm 75(1):1–15. doi:10.1016/j.ejpb.2010.01.007
48. Onuki Y, Kawai S, Arai H, Maeda J, Takagaki K, Takayama K (2012) Contribution of the physicochemical properties of active pharmaceutical ingredients to tablet properties identified by ensemble artificial neural networks and Kohonen's self-organizing maps. J Pharm Sci 101(7):2372–2381. doi:10.1002/jps.23134
49. Herting MG, Kleinebudde P (2007) Roll compaction/dry granulation: effect of raw material particle size on granule and tablet properties. Int J Pharm 338(1-2):110–118. doi:10.1016/j.ijpharm.2007.01.035
50. Kushner J, Langdon BA, Hiller JI, Carlson GT (2011) Examining the impact of excipient material property variation on drug product quality attributes: a quality-by-design study for a roller compacted, immediate release tablet. J Pharm Sci 100(6):2222–2239. doi:10.1002/jps.22455
51. Nosal R, Schultz T (2008) PQLI definition of criticality. J Pharm Innov 3(2):69–78
52. Garcia T, Cook G, Nozal R (2008) PQLI key topics: criticality, design space, and control strategy. J Pharm Innov 3(2):60–68
53. Saltelli A, Chan K, Scott EM (2000) Sensivivity analysis. Wiley, Chichester
54. Saltelli A, Annoni P, Azzini I, Campolongo F, Ratto M, Tarantola S (2010) Variance based sensitivity analysis of model output. Design and estimator for the total sensitivity index. Comput Phys Commun 181(2):259–270. doi:10.1016/j.cpc.2009.09.018
55. Rogers AJ, Inamdar C, Ierapetritou MG (2013) An integrated approach to simulation of pharmaceutical processes for solid drug manufacture. Ind Eng Chem Res 131015 $32#102838009. doi: 10.1021/ie401344a
56. Kikuchi S, Takayama K (2009) Reliability assessment for the optimal formulations of pharmaceutical products predicted by a nonlinear response surface method. Int J Pharm 374(1-2):5–11. doi:10.1016/j.ijpharm.2009.02.016
57. Grossmann IE, Calfa B, Garcia-Herreros P (2014) Evolution of concepts and models for quantifying resiliency and flexibility of chemical processes. Comput Chem Eng. doi:10.1016/j.compchemeng.2013.12.013
58. Lima FV, Jia Z, Lerapetritou M, Georgakis C (2010) Similarities and differences between the concepts of operability and flexibility: the steady-state case. AIChE J 56(3):702–716. doi:10.1002/Aic.12021
59. Boukouvala F, Muzzio FJ, Ierapetritou MG (2010) Design space of pharmaceutical processes using data-driven-based methods. J Pharm Innov 5(3):119–137. doi:10.1007/s12247-010-9086-y
60. Morari M (1983) Flexibility and resiliency of process systems. Comput Chem Eng 7 (4):423–437
61. Swaney RE, Grossmann IE (1985) An index for operational flexibility in chemical process design. Part II: computational algorithms. AIChE J 31(4):631–641
62. Swaney RE, Grossmann IE (1985) An index for operational flexibility in chemical process design. Part II: computational algorithms. AIChE J 31(4):631–641
63. Boukouvala F, Ierapetritou MG (2012) Feasibility analysis of black-box processes using an adaptive sampling Kriging-based method. Comput Chem Eng 36:358–368. doi:10.1016/j.compchemeng.2011.06.005
64. Grossmann IE, Floudas CA (1987) Active constraint strategy for flexibility analysis in chemical processes. Comput Chem Eng 11 (6):675–693
65. Biegler LT, Grossmann IE, Westerberg AW (1997) Systematic methods of chemical process design. Prentice Hall, Upper Saddle River, NJ
66. Halemane KP, Grossmann IE (1983) Optimal process design under uncertainty. AIChE J 49:425
67. Kuhn HW, Tucker AW (1950) Nonlinear programming. In: Proceedings of the second Berkeley symposium on mathematical statistics and probability. University of California Press, Berkeley, CA, pp 481–492
68. Ostrovsky GM, Volin YM, Barit EI, Senyavin MM (1994) Flexibility analysis and optimization of chemical plants with uncertain parameters. Comput Chem Eng 18:755–767
69. Ostrovsky GM, Achenie LEK, Wang Y, Volin YM (2002) A unique approach for solving sub-problems in flexibility analysis. Chem Eng Comm 189(1):125–149
70. Ostrovsky GM, Achenie LEK, Wang Y (2000) A new algorithm for computing process flexibility. Ind Eng Chem Res 39:2368–2377

71. Floudas CA, Gümüs ZH, Ierapetritou MG (2001) Global optimization in design under uncertainty: feasibility test and flexibility index problems. Ind Eng Chem Res 40:4267–4282
72. Floudas CA (2000) Nonconvex optimization and its applications. In: Floudas CA (ed) Deterministic global optimization: theory, methods and applications, vol 37. Kluwer Academic Publishers, Dordrecht, The Netherlands
73. Subrahmanyam S, Peknyt JF, Reklaitis GV (1994) Design of batch chemical-plants under market uncertainty. Ind Eng Chem Res 33(11):2688–2701. doi:10.1021/Ie00035a019
74. Vin JP, Ierapetritou MG (2001) Robust short-term scheduling for multiproduct batch plants under demand uncertainty. Ind Eng Chem Res 40:4543–4554
75. Linninger AA, Chakraborty A (2001) Pharmaceutical waste management under uncertainty. Comput Chem Eng 25 (4–6):675–681. doi:10.1016/S0098-1354(01)00668-8
76. Pistikopoulos EN, Mazzuchi TA (1990) A novel flexibility analysis approach for processes with stochastic parameters. Comput Chem Eng 14(9):991–1000
77. Straub DA, Grossmann IE (1990) Integrated stochastic metric of flexibility for systems with discrete state and continuous parameter uncertainties. Comput Chem Eng 14 (9):967–985
78. Straub DA, Grossmann IE (1993) Design optimization of stochastic flexibility. Comput Chem Eng 17(4):339–354
79. Grossmann IE, Straub DA (1996) Recent developments in the evaluation and optimization of flexible chemical processes. In: Reklaitis GV, Sunol AK, Rippin DWT, Hortacsu O (eds) Batch processing systems engineering fundamentals and applications for chemical engineering, vol 143, Fundamentals and applications for chemical engineering. Springer in Collaboration with NATO Scientific Affairs Division, Heidelberg, Berlin, pp 495–516
80. Bansal V, Perkins JD, Pistikopoulos EN (2000) Flexibility analysis and design of linear systems by parametric programming. AIChE J 46(2):335–354. doi:10.1002/aic.690460212
81. Bansal V, Perkins JD, Pistikopoulos EN (2002) Flexibility analysis and design using a parametric programming framework. AIChE J 48(12):2851–2868
82. Pistikopoulos EN, Ierapetritou MG (1995) Novel-approach for optimal process design under uncertainty. Comput Chem Eng 19 (10):1089–1110. doi:10.1016/0098-1354(94)00093-4
83. Bansal V, Perkins JD, Pistikopoulos EN (1998) Flexibility analysis and design of dynamic processes with stochastic parameters. Comput Chem Eng 22(Suppl):S817–S820. doi:10.1016/S0098-1354(98)00156-2
84. Boukouvala F, Ierapetritou M (2012) Simulation-based derivative-free optimization for computationally expensive function. In: AIChE annual meeting, Pittsburgh, PA
85. Myers RH, Montgomery DC (2002) Response surface methodology process and product optimization using designed experiments. Wiley, New York
86. Box GEP, Wilson KB (1951) On the experimental attainment of optimum conditions. J Roy Stat Soc B 13(1):1–35
87. Jia ZY, Davis E, Muzzio FJ, Ierapetritou MG (2009) Predictive modeling for pharmaceutical processes using Kriging and response surface. J Pharm Innov 4(4):174–186. doi:10.1007/s12247-009-9070-6
88. Calder CA, Cressie N (2009) Kriging and variogram models. In: Kitchin R, Thrift N (eds) International encyclopedia of human geography, vol 1. Elsevier, Oxford, pp 49–55
89. Kleijnen JPC (2009) Kriging metamodeling in simulation: a review. Eur J Oper Res 192 (3):707–716. doi:10.1016/j.ejor.2007.10.013
90. Metheron G (1963) Principles of geostatistics. Econ Geol 58(8)
91. Agatonovic-Kustrin S, Beresford R (2000) Basic concepts of artificial neural network (ANN) modeling and its application in pharmaceutical research. J Pharm Biomed Anal 22 (5):717–727
92. Basheer IA, Hajmeer M (2000) Artificial neural networks: fundamentals, computing, design, and application. J Microbiol Methods 43(1):3–31
93. Li G, Rosenthal C, Rabitz H (2001) High dimensional model representations. J Phys Chem A 105(33)
94. Li GY, Wang SW, Rabitz H (2002) Practical approaches to construct RS-HDMR component functions. J Phys Chem A 106 (37):8721–8733. doi:10.1021/Jp014567t

95. Li G, Rabitz H, Hu J, Chen Z, Ju Y (2008) Regularized random-sampling high dimensional model representation (RS-HDMR). J Math Chem 43(3):1207–1232
96. Boukouvala F, Muzzio FJ, Ierapetritou MG (2011) Dynamic data-driven modeling of pharmaceutical processes. Ind Eng Chem Res 50(11):6743–6754. doi:10.1021/Ie102305a
97. Jia Z, Davis E, Muzzio FJ, Ierapetritou MG (2009) Predictive modeling for pharmaceutical processes using Kriging and response surface. J Pharm Innov
98. Boukouvala F, Ierapetritou MG (2014) Derivative-free optimization for expensive constrained problems using a novel expected improvement objective function. AIChE J 60 (7):2462–2474
99. Johanson JR (1965) A rolling theory for granular solids. ASME J Appl Mech E32 (4):842–848
100. Samsatli NJ, Papageorgiou LG, Shah N (1999) Batch process design and operating using operational envelopes. Comput Chem Eng Suppl 5887–5890
101. Ierapetritou MG (2001) A new approach for quantifying process feasibility: convex and one dimensional quasi-convex regions. AIChE J 47(6):1407–1417
102. Goyal V, Ierapetritou MG (2002) Determination of operability limits using simplicial approximation. AIChE J 48(12):2902–2909
103. Director SW, Hachtel GD (1977) The simplicial approximation approach to design centering. IEEE Trans Circ Syst 24(7)
104. Banarjee I, Ierapetritou MG (2005) Feasibility evaluation of nonconvex systems using shape reconstruction techniques. Ind Eng Chem Res 44:3638–3647
105. Wold S, Sjostrom M, Eriksson L (2001) PLS-regression: a basic tool of chemometrics. Chemometr Intell Lab 58(2):109–130. doi:10.1016/S0169-7439(01)00155-1
106. López-Negrete de la Fuente R, García-Muñoz S, Biegler LT (2010) An efficient nonlinear programming strategy for PCA models with incomplete data sets. J Chemometr 24(6):301–311
107. Burnham AJ, MacGregor JF, Viveros R (1999) Latent variable multivariate regression modeling. Chemometr Intell Lab 48 (2):167–180. doi:10.1016/S0169-7439(99)00018-0
108. Hoskuldsson A (1988) PLS regression methods. J Chemometr 2:211
109. García-Muñoz S, Polizzi M (2012) WSPLS – a new approach towards mixture modeling and accelerated product development. Chemometr Intell Lab 15:116–121
110. García-Muñoz S, MacGregor JF, Kourti T (2005) Product transfer between sites using Joint-Y PLS. Chemometr Intell Lab 79:101–114
111. Huang J, Kaul G, Cai C, Chatlapalli R, Hernandez-Abad P, Gosh K, Nagi A (2009) Quality by design case study: an integrated multivariate approach to drug product and process development. Int J Pharm 382:23–32
112. MacGregor JF, Bruwer M-J (2008) A framework for the development of design and control spaces. J Pharm Innov 3:15–22
113. García-Muñoz S, Dolph S, Ward HWI (2010) Handling uncertainty in the establishment of a design space for the manufacture of a pharmaceutical product. Comput Chem Eng 34:1098–1107
114. García-Muñoz S (2009) Establishing multivariate specifications for incoming materials using data from multiple scales. Chemometr Intell Lab 98(1):51–57
115. Duchesne C, Macgregor JF (2004) Establishing multivariate specification regions for incoming materials. J Qual Technol 36 (1):78–94
116. Nasr M (2007) The FDA's initiative on pharmaceutical quality for the 21-st century: emphasizing quality by design. In: IFPAC, Baltimore, MD
117. Garcia-Munoz S, MacGregor JF, Kourti T, Apruzzece F, Champagne M (2006) Optimization of batch operating policies. Part I. Handling multiple solutions. Ind Eng Chem Res 45(23):7856–7866
118. García-Muñoz S, MacGregor JF, Neogi D, Latshaw BE, Mehta S (2008) Optimization of batch operating policies. Part II. Incorporating process constraints and industrial applications. Ind Eng Chem Res 47(12): 4202–4208
119. Jaeckle CM, MacGregor JF (1998) Product design through multivariate statistical analysis of process data. AIChE J 44(5):1105–1118. doi:10.1002/aic.690440509
120. Montgomery DC, Bettencourt VMJ (1977) Multiple response surface methods in computer simulation. Simulation 29:113–121

121. Park KS, Kim KJ (2005) Optimizing multi-response surface problems: how to use multi-objective optimization techniques. IIE Trans 37(6):523–532. doi:10.1080/07408170590928992
122. Khuri AI, Conlon M (1981) Simultaneous optimization of multiple responses represented by polynomal regression function. Technometrics 23:363–375
123. Pignatiello JJ (1993) Strategies for robust multiresponse quality engineering. IIE Trans 25(3):5–15. doi:10.1080/07408179308964286
124. Vining GG (1998) A compromise approach to multi-response optimization. J Qual Technol 30:309–313
125. Peterson JJ (2004) A posterior predictive approach to multiple response surface optimization. J Qual Technol 36(2):139–153
126. Dimitriadis VD, Pistikopoulos EN (1995) Flexibility analysis of dynamic-systems. Ind Eng Chem Res 34(12):4451–4462. doi:10.1021/Ie00039a036
127. Holt BR, Morari M (1985) Design of resilient processing plants V – the effect of deadtime on dynamic resilience. Chem Eng Sci 40:1229
128. Holt BR, Morari M (1985) Design of resilient processing plants VI – the effect of right half plane zeros on dynamic resilience. Chem Eng Sci 40:59
129. Skogestad S, Morari M (1987) Design of resilient processing plants IX – the effect of model uncertainty on dynamic resilience. Chem Eng Sci 42:1765
130. Soroush M, Kravaris C (1993) Optimal-design and operation of batch reactors. 1. Theoretical framework. Ind Eng Chem Res 32(5):866–881. doi:10.1021/Ie00017a015
131. Soroush M, Kravaris C (1993) Optimal-design and operation of batch reactors. 2. A case-study. Ind Eng Chem Res 32 (5):882–893. doi:10.1021/Ie00017a016
132. Vassiliadis VS, Sargent RWH, Pantelides CC (1994) Solution of a class of multistage dynamic optimization problems. 1. Problems without path constraints. Ind Eng Chem Res 33(9):2111–2122. doi:10.1021/Ie00033 a014
133. Vassiliadis VS, Sargent RWH, Pantelides CC (1994) Solution of a class of multistage dynamic optimization problems. 2. Problems with path constraints. Ind Eng Chem Res 33 (9):2123–2133. doi:10.1021/Ie00033a015
134. Biegler LT (2007) An overview of simultaneous strategies for dynamic optimization. Chem Eng Process 46(11):1043–1053. doi:10.1016/j.cep.2006.06.021
135. Biegler LT, Cervantes AM, Wächter A (2002) Advances in simultaneous strategies for dynamic process optimization. Chem Eng Sci 57(4):575–593, Doi: Pii S0009-2509 (01)00376-1
136. Kameswaran S, Biegler LT (2006) Simultaneous dynamic optimization strategies: recent advances and challenges. Comput Chem Eng 30:1560–1575
137. Betts JT (2001) Practical methods for optimal control using nonlinear programming. Advances in design and control. SIAM, Philadelphia, PA
138. Bahri PA, Bandoni JA, Romagnoli JA (1997) Integrated flexibility and controllability analysis in design of chemical processes. AIChE J 43(4):997–1015. doi:10.1002/aic.690430415
139. Mohideen MJ, Perkins JD, Pistikopoulos EN (1996) Optimal design of dynamic systems under uncertainty. AIChE J 42(8):2251–2272
140. Engisch WE, Muzzio FJ (2012) Method for characterization of loss-in-weight feeder equipment. Powder Technol 228:395–403. doi:10.1016/j.powtec.2012.05.058
141. Zhou H, Li X, Qian Y, Kraslawski A (2009) Optimizing the initial conditions to improve the dynamic flexibility of batch processes. Ind Eng Chem Res 48:6321–6326
142. Luus R, Hennessy D (1999) Optimization of fed-batch reactors by the Luus-Jaakola optimization procedures. Ind Eng Chem Res 38:1948–1955
143. Huang W, Li X, Yang S, Qian Y (2011) Dynamic flexibility analysis of chemical reaction systems with time delay: Using a modified finite element collocation method. Chem Eng Res Des 89:1938–1946
144. Uchida K, Shimemura E, Kubo T, Abe M (1988) The linear-quadratic optimal control approach to feedback control design for systems with delay. Automatica 24:773–780
145. Kim AV, Han SH, Kwon WH, Pimenov VG (1998) Explicit numerical methods and LQR control algorithms for time-delay systems. In: International conference on electrical engineering, Korea, pp 21–25
146. Bindhumadhavan G, Seville JPK, Adams MJ, Greenwood RW, Fitzpatrick S (2005) Roll compaction of a pharmaceutical excipient: experimental validation of rolling theory for granular solids. Chem Eng Sci 60:3891–3897

147. Reynolds G, Ingale R, Roberts R, Kothari S, Gururagan B (2010) Practical application of roller compaction process modeling. Comput Chem Eng 34:1049–1057
148. Hsu SH, Reklaitis GV, Venkatasubramanian V (2010) Modeling and control of roller compaction for pharmaceutical manufacturing. Part I: process dynamics and control framework. J Pharm Innov 5:14–23
149. Garcia T, McCurdy V, Watson TNJ, Ende MA, Butterell P, Vukovinsky K, Chueh A, Coffman J, Cooper S, Schuemmelfeder B (2012) Verification of design spaces developed at subscale. J Pharm Innov 7(1):13–18. doi:10.1007/s12247-012-9123-0
150. Plumb K (2005) Continuous processing in the pharmaceutical industry – changing the mind set. Chem Eng Res Des 83 (A6):730–738. doi:10.1205/Cherd.04359
151. Brone D, Alexander A, Muzzio FJ (1998) Quantitative characterization of mixing of dry powders in V-blenders. AIChE J 44 (2):217–278
152. Ahmend SU, Katdare A, Naini V, Wadgaonkar D (2014) Scale-up, technology transfer, and process performance qualification. In: Shargel L, Kanfer I (eds) Generic drug product development: solid oral dosage forms, vol 129, 2nd edn, Drugs and the pharmaceutical sciences. CRC Press Taylor & Francis Group, Boca Raton, FL
153. Hallow DM, Mudryk BM, Braem AD, Tabora JE, Lyngberg OK, Bergum JS, Rossano LT, Tummala S (2010) An example of utilizing mechanistic and empirical modeling in quality by design. J Pharm Innov 5(4):193–203. doi:10.1007/s12247-010-9094-y
154. Burt JL, Braem AD, Ramirez A, Mudryk B, Rossano L, Tummala S (2011) Model-guided design space development for a drug substance manufacturing process. J Pharm Innov 6(3):181–192. doi:10.1007/s12247-011-9109-3
155. Shah N (2004) Pharmaceutical supply chains: key issues and strategies for optimisation. Comput Chem Eng 28(6-7):929–941. doi:10.1016/j.compchemeng.2003.09.022
156. Gao Y, Muzzio FJ, Ierapetritou MG (2011) Characterization of feeder effects on continuous solid mixing using Fourier series analysis. AIChE J 57(5):1144–1153
157. Yang S, Evans JRG (2007) Metering and dispensing of powder; the quest for new solid freeforming techniques. Powder Technol 178:56–72
158. Nakach M, Authelin J-R, Chamayou A, Dodds J (2004) Comparison of various milling technologies for grinding pharmaceutical powders. Int J Miner Process 74S: S173–S181
159. Reynolds GK (2010) Modelling of pharmaceutical granule size reduction in a conical screen mill. Chem Eng J 164:383–392
160. Verheezen JJAM, van der Voort Maarschalk K, Faassen F, Vromans H (2004) Milling of agglomerates in an impact mill. Int J Pharm 278:165–172
161. Vendola TA, Hancock BC (2008) The effect of mill type on two dry-granulated placebo formulations. Pharm Technol. 32(11)
162. Pernenkil L, Cooney CL (2006) A review on the continuous blending of powders. Chem Eng Sci 61(2):720–742. doi:10.1016/j.ces.2005.06.016
163. Portillo PM, Ierapetritou MG, Muzzio FJ (2008) Characterization of continuous convective powder mixing processes. Powder Technol 182(3):368–378. doi:10.1016/j.powtec.2007.06.024
164. Marikh K, Berthiaux H, Gatumel C, Mizonov V, Barantseva E (2008) Influence of stirrer type on mixture homogeneity in continuous powder mixing: a model case and a pharmaceutical case. Chem Eng Res Des 86 (9A):1027–1037. doi:10.1016/j.cherd.2008.04.001
165. Vanarase AU, Muzzio FJ (2011) Effect of operating conditions and design parameters in a continuous powder mixer. Powder Technol 208(1):26–36. doi:10.1016/j.powtec.2010.11.038
166. Dhenge RM, Cartwright JJ, Hounslow MJ, Salman AD (2012) Twin screw wet granulation: effects of properties of granulation liquid. Powder Technol 229:126–136
167. Betz G, Junker-Burgin P, Leuenberger H (2003) Batch and continuous processing in the production of pharmaceutical granules. Pharmaceut Dev Tech 8(3):289–297
168. Vercruysse J, Córdoba Díaz D, Peeters E, Fonteyne M, Delaet U, Van Assche I, De Beer T (2012) Continuous twin screw granulation: influence of process variables on granule and tablet quality. Eur J Pharm Biopharm 82:205–211
169. Yu S, Gururajan B, Reynolds G, Roberts R, Adams MJ, Wu CY (2012) A comparative study of roll compaction of free-flowing and cohesive pharmaceutical powders. Int J Pharm 428:39–47
170. Kudra T, Mujumdar AS (2009) Advanced drying technologies, 2nd edn. Taylor & Francis Group, Boca Raton, FL

171. Paltzer S (2007) Drying of wet agglomerates in a continuous fluid bed: influence of residence time, air temperature and airflowrate on the drying kinetics and the amount of oversize particles. Chem Eng Sci 62:463
172. Mortier ST, De Beer T, Gernaey KV, Vercruysse J, Fonteyne M, Remon JP, Vervaet C, Nopens I (2012) Mechanistic modelling of the drying behaviour of single pharmaceutical granules. Eur J Pharm Biopharm 80 (3):682–689. doi:10.1016/j.ejpb.2011.12.010
173. Hovmand S (1995) Fluidized bed drying. In: Mujumdar AS (ed) Handbook of industrial drying, 2nd edn. Marcel Dekker, Inc., New York, NY
174. Carstensen JT, Zoglio MA (1982) Tray drying of pharmaceutical wet granulations. J Pharm Sci 71(1):35–39
175. McLoughlin CM, McMinn WAM, Magee TRA (2003) Microwave-vacuum drying of pharmaceutical powders. Dry Technol 21(9): 1719–1733. doi:10.1081/Drt-120025505
176. Patel S, Kaushal AM, Bansal AK (2007) Effect of particle size and compression force on compaction behavior and derived mathematical parameters of compressibility. Pharm Res 24(1):111–124. doi:10.1007/s11095-006-9129-8
177. Jackson S, Sinka IC, Cocks AC (2007) The effect of suction during die fill on a rotary tablet press. Eur J Pharm Biopharm 65(2):253–256. doi:10.1016/j.ejpb.2006.10.008

Chapter 9

Optimization Methodologies for the Production of Pharmaceutical Products

M. Sebastian Escotet-Espinoza, Amanda Rogers, and Marianthi G. Ierapetritou

Abstract

Finding the most beneficial conditions during the development of a given product or process is among one of the top goals for both scientist and engineers across all industries. In the pharmaceutical industry, as global competition increases and there is a higher demand for accessible quality products, it is important to focus on the improvement of product development and manufacturing. Optimization methodologies can greatly aid the production of pharmaceutical products by providing a systematic framework to process improvement. In this review, general concepts regarding the implementation of optimization methods are introduced along with examples of their application in pharmaceutical manufacturing process design and formulation development. An overview of optimization methodologies used for the improvement of batch and continuous pharmaceutical manufacturing is presented. Challenges in the application of optimization methods in pharmaceutical manufacturing are discussed along with a future outlook of the field and its place in pharmaceutical process and product design. Overall the review points to optimization as a critical component in the design of improved and effective pharmaceutical products, in alignment with the common goals of both regulatory agencies and industry.

Key words Optimization, Manufacturing, Formulation, Surrogate-based, Direct search methods

1 Introduction

The pharmaceutical regulatory environment has recognized the need to implement a systematic approach to drug product development where quality is *built* into the process [1]. Quality-by-Design (QbD), a US FDA initiative, promotes the use of engineering tools such as predictive models and optimization techniques to formulate products based on scientific understanding of the process. Product quality is achieved through design of robust processes that are modeled, validated and optimized using knowledge of process principles [2, 3]. Modeling and validation establish a predictive framework using experimental data and physical principles to create predictive mathematical representations of the system. In the

Marianthi G. Ierapetritou and Rohit Ramachandran (eds.), *Process Simulation and Data Modeling in Solid Oral Drug Development and Manufacture*, Methods in Pharmacology and Toxicology,
DOI 10.1007/978-1-4939-2996-2_9, © Springer Science+Business Media New York 2016

process design phase the goal mathematical models is to evaluate the impact operations, equipment and inputs have on product attributes. Such mathematical tools further allow for system study through evaluation of process scenarios, some which have not been experimentally performed [4]. The predictive ability also provides a framework for process control and optimization, where accurate predictions of the system are required in order to provide the best problem solution [5, 6].

Optimization methods are essential tools for: (1) experimental design; (2) analysis and design of processes; (3) planning, operations research and scheduling; and (4) process control. In the pharmaceutical literature these methods have been used for experimental and computational design problems surrounding lead drug design and product development. Experimental optimization problems involve development of procedures to test a given hypothesis using the least amount of resources. In these studies the goal is to reduce the experimental load while maximizing the value of data collected. In the pharmaceutical literature these methods have been used extensively under the design of experiments (DoE) framework [7]. Computational optimization problems use repeated evaluations of mathematical models to determine input effects on the process outcomes and their best combination to reach a desired output [8]. Such methodology provides great benefit to the design of pharmaceutical drug products by further reducing the number of experiments and substituting them with in silico evaluations. A growing powder modeling technology literature has been developing in the past few years given the improved performance of modeling capabilities, which has led to further implementation of optimization methods. The purpose of this chapter is to provide a brief review of the optimization methods used in design of solid-oral dosage drug products. A short description of the research performed by each journal article reviewed is provided to orient readers in their search for further information about optimization of drug manufacturing processes.

Due to the extensive applications and discussion of experimental optimization methods in the literature they are not discussed in this chapter. Readers are recommended to consult with other sources focusing on the design of experiments methodologies such as the books and reviews by Montgomery [9], Singh et al. [10] and Gareth et al. [11].

2 Optimization Methodologies

Optimization methodologies used in the pharmaceutical modeling literature can be categorized into three groups: direct search, gradient-based and surrogate-based methods. Each group uses different approaches to reach local or global optimum conditions and is used depending on the optimization problem goals. Below, we review some of the methodologies most commonly used in the pharmaceutical manufacturing literature and provide some references on their application.

2.1 Heuristic or Direct Search Methods

Direct search methods are a class of optimization techniques that do not require information from mathematical derivatives to reach an optimum solution. Derived from heuristic methods, these optimization techniques iteratively evaluate a design function to determine input–output effects and obtain optimum values. Various algorithms to improve the sampling give rise to a wide variety of direct search methods, largely described in the optimization literature [6, 11, 12]. The termination criteria by which these methods evaluate design functions are not based on stationary points nor gradients, but the user criteria. This termination criteria make direct search methods more likely to favor global rather than local optimal solutions, however, no rigorous convergence properties to global optimums have been discovered [12].

Even though robust and reliable for optimization, these methods also tend to scale poorly with increasing number of variables and constraints. Direct search methods have been recommended for simple unconstrained problems where the bounds and number of variables will not interfere with the method's sampling strategy [6, 12]. Their main advantage, however, is the ability to be applied in conjunction with more complex optimization techniques to support the attainment of an optimum solution. Some of the most commonly used methods include random search, particle swarm theory, Nelder-Mead simplex lattice, simulated annealing, evolutionary operations and genetic algorithm.

2.1.1 Pharmaceutical Optimization Applications

In the interest of providing the reader with a concise overview of the implementation of direct search methods, the following applications primarily focus on three major areas of solid dose preparation: drug product formulation, drug delivery, and manufacturing processes.

Product Formulation

Product development involves the design of formulations (recipes) in which active ingredients (i.e., APIs) are combined with pharmacologically inactive excipients to form a product with desired quality attributes. As many drug product attributes are dependent on their composition, this process requires optimal selection of both the type and quantity of materials involved. Excipients in particular play an important role in the absorption, stability and bioavailability of drugs in the body. Direct search methods have been applied in formulation development and optimization of capsule and tablet drug products, specifically aiming at the selection of excipients [13–15]. Salazar de Saavedra et al. [15] developed an optimization strategy for design of oral dose dry powder pharmaceutical formulations with little experimental data using a combination of genetic algorithms and simplex lattice methods. In this research, the authors studied blend flow and dissolution properties of API and excipient powders using small laboratory scale experiments to then formulate an optimal product combination using non-experimental methods. The optimum formulation was found theorizing multiple combinations of blends using the genetic algorithm as a case generating method. Properties of several different compound combinations were modeled in order to create a large set of formulations with a broad range of properties. It is important to note that some of modeled combinations would have not been able to be generated experimentally due to material limitations. Constraints were placed on the formulation properties in order to discard blends whose predicted dissolution and flow properties were outside of the desired ranges. A simplex method was used to find an optimal formulation on the regression response surface obtained from acceptable blends developed by the genetic algorithm sequence. This method allowed for the generation of effective and feasible manufacturing capsule formulations using relatively small sets of experimental data.

Worakul et al. [16] used simulated annealing to optimize child liquid-based acetaminophen formulations using solid drug product date. The purpose of the study was to develop a customizable formulation of acetaminophen syrups based on the availability and taste of available co-solvents. The implementation of simulated annealing in the formulation protocol resulted in the selection of an optimal formulation in less computational time than the traditional simplex search method. Rajab et al. [17] studied optimal effervescent floating tablets (EFT) composition in order to target a set of product attributes (i.e., specific buoyancy, hardness and floating time). Using a hybrid simplex lattice and multivariate regression approach the authors were able to predict and optimize the EFT properties and develop a first principles approach to the type of tablet matrix erosion. Patel, et al. [18] performed a similar protocol by investigating the effects and optimal formulations of melt granulation floating tablet products with minimum experimentation. Plumb et al. [19] used a version of simulated annealing, known as guided evolutionary simulated annealing (GESA), in combination with an artificial neural network (ANN) to determine which coating components were needed in their formulation based on its effects on in vitro dissolution, film opacity and crack velocity. More than six excipient concentrations, found to be optimum from previous experimental literature sets, were optimized in the formulation. The result of this method resulted an undesirably large number of optimal formulations for the coating; a results that was attributed to multiple local optimums and the lack of inputs provided by the researchers [19, 20]. Vaghani et al. [21] optimized the formulation and release profile of oral bioadhesives using a simplex methodology. The authors used the simplex method primarily for improving formulation tablet properties; however, they also expanded its use by coupling it with response surface methodologies and desirability function methods to improve the release profile of the drug in vitro and in vivo.

Drug Delivery Systems

The development and optimization of many drug delivery systems (DDS) has been described in the literature using direct search methods. Most of the work for optimization of DDS focuses on sustained release oral formulations, yet some research points to their use in nontraditional methods of drug delivery such as transdermal devices. Zhang et al. [22] successfully formulated an optimized mathematical model for the design and manufacturing of sustained-release pellets using particle swarm theory (PSO) as the optimization algorithm. Abraham et al. [23] designed extended release oral drug products using other modeling techniques being optimized with PSO. The random search algorithm was used to elucidate the shape of functions being modeled without the need of experimentation. This method was shown to reduce the number of experiments needed to develop valuable datasets and generate best potential routes for experimentation. Nauman et al. [24] used a random search method to determine the optimal drug distribution in transdermal delivery devices. The optimal drug distribution was chosen to be one with a maximum delivery in order to maximize its reservoir exhaustion. The method was given a certain number of feasibility constraints, generating a practical output, which could be used for drug delivery device manufacturing and other real-world applications.

Multi-particulate controlled release pellets for oral drug delivery have been designed and optimized using simplex lattice methods by Bhattacharyya et al. [25]. The authors characterized and modeled the drug product release kinetics by using a simplex method accounting for physiochemical properties such as the entrapment efficiency, particle size and surface topography. The release profile was optimized using blends of microparticles, previously characterized by the simplex methodology. Studies by Prajapatia et al. [26] used simplex methodology to develop an optimized gastroretentive oral dose drug delivery system with matrix-forming properties. The authors used the simplex lattice design to optimize formulation parameters such as concentration of excipients and relative amounts of drug per unit dose. Varshosaz et al. [27] developed a prediction algorithm for the release of antihypertensive drug compounds from nanoparticle-based formulations using a genetic algorithm (GA) approach. GA methodology was implemented to optimize nanoparticle formulation regarding size and number of particles to enhance in vivo absorption. The method was proven to be successful in the optimum selection of both formulation and manufacturing parameters while reducing the number of experimental trials. Barmpalexis et al. [28] developed a variation of the genetic algorithm method, known as genetic programming, for the formulation of pharmaceutical products. In this research, the efficiency of the genetic algorithm derivative was better than that of established neural network methods. The improvement was provided by the method for weighting model coefficients. In a later study, Barmpalexis et al. [29] specifically used genetic programming methods for the design and formulation of zero-order release matrix controlled release drug products. The results from this study suggest the use of genetic programming techniques has increased efficiency in optimization of pharmaceutical formulations where reduced experimental datasets are available.

Manufacturing Processes

Examples of optimization in manufacturing processes have been focused on finding the best process settings using the methodologies described in this section. Ranging from the selection of process conditions to the arrangement of manufacturing plants, direct search methods have played a role in finding the most beneficial process settings. Patel et al. [30] optimized the formulation and the critical process parameters used in tablet products for colon therapies using simplex lattice methods. In the research, the authors established optimal operational conditions of the coater, providing a framework for the design of other coating processes. Coating spray rate and batch time were studied and optimized to reduce final product variability and process.

Genetic algorithm (GA) was used to determine optimal sample datasets used for the control of unit operations. Qu et al. [31] used the GA method in conjunction with radial basis functions (RBF) to select optimal network imbedded functions used to understand nondestructive near-infrared (NIR) data from powder blend processes. The proposed algorithm automatically determined the appropriate network structure more efficiently than single RBF methods. Optimization of process parameters affecting chromatographic separation selectivity and efficiency of vitamins in pharmaceutical formulations was also accomplished using GA methods [32]. Investigators established an exponential unit model, including key process parameters such as surfactant concentration, temperature and pH. The process parameters were minimized to improve chromatographic responses between all analytes, simultaneously. The method was shown to greaten differences in resolution and analysis time between samples and pointed to optimal operational parameters for the separation of each compound. Chaudhury et al. [33] used PSO methods to find multiple constant parameter values used in modeling high shear batch granulation processes. The researchers solved the multiobjective optimization problem by implementing a constraint framework: while one of the objective functions is being minimized, other objective functions are incorporated in the form of a constraint. This is done to dictate a feasible domain for the search algorithm and avoid results that may not lay within the feasible space. Muteki et al. [34] used iterative optimization technology (IOT) to predict the composition of a mixture without calibration standards. The IOT effectiveness was demonstrated through its implementation on four online and offline pharmaceutical processes involving two unit operations and mixtures. The unit operations that were part of the proof included bin-blenders and a rotary press feed frame.

Along the lines of plant optimization, direct search methods have used to generate plant arrangements and scheduling patterns that improve the efficiency of manufacturing complexes. Multiple products are usually manufactured in one pharmaceutical plant with limited resources and time. Effective plant operation is based on the selection and scheduling of protocols for the manufacturing of such products. Batch scheduling is a fundamental practice in the pharmaceutical industry as it provides a clear methodology for the maximization of plant throughput and the minimization of costly down time. Most of the methods available for plant scheduling are available using neural computing techniques in combination with optimization techniques such as simulated annealing (SA). Raaymakers et al. [35] used SA to schedule 15 unit operations (instances) in the no-wait restriction environment of a multipurpose pharmaceutical batch plant manufacturing API. The method was shown to be more efficient at scheduling the instances than other simpler dispatching tools used in the industry. Graells et al. [36] provides a more in depth detail on the methodology involving SA on the process of scheduling batch chemical processes, including those in the pharmaceutical industry specifically.

The layout design of pharmaceutical plants, as any important process parameter, requires optimization of several objective costs and manufacturing functions such as throughput rate, accessibility, operational and maintenance cost. Previous methods for developing optimal plant design in the pharmaceutical industry used empirical or single objective function optimization to determine the distance between units or their adjacency. Hamamoto et al. [37] implemented genetic algorithm methods as a multiobjective optimization approach for pharmaceutical batch-based plant design. The framework used an embedded simulation model that allows the user to select and rank relevant objective functions that will be optimized for the plant layout. Multiple linear equations for scheduling, running and production of individual batch unit operations are modeled and optimized using genetic algorithm results (i.e., crossovers). The model was validated using the design of various solid dose plants and shown to significantly outperform similar empirical algorithms used for design of pharmaceutical plants.

Table 1 summarizes the application of such approaches to pharmaceutical manufacturing literature.

Table 1
Direct search optimization methods

Method	Description	Pharmaceutical applications	Refs.
Random Search (RS)/Particle Swarm Theory (PSO)	Stochastic method targeting discrete input problem variables. The methodology approaches optimality by randomly testing data points near the vicinity of a sample value. Optimality is found after multiple iterations around candidate solutions. Highly hybridized with other optimization methods.	Drug discovery; Formulation; Combination therapies; Model data sampling; Data generation and model testing	[22–24, 38–46]
Nelder-Mead Simplex Lattice (SL)	Geometric figure or directionality development optimization technique. Based on testing candidate solutions and assembling the outputs to establish next sample point direction to obtain the optimal solution.	Formulation; Process and unit operation parameters	[13, 14, 17, 18, 21, 25, 26, 30, 47–53]
Genetic Algorithm (GA)	Non-local heuristic method using generated inputs. Hybridization of input parameters to generate novel output solutions. Input parameters are crossed, generating whole new inputs. Sample testing is performed on the discrete crossed inputs.	Formulation; Process and unit operation parameters; Drug release profile; Manufacturing plant arrangement; Model optimization	[15, 27–29, 31, 32, 37, 54–64]
Evolutionary Operations (EVOP)	Process optimization technique which evaluates the function in a single variable design space to determine the impact of such variable on the output. Based on a statistical approach to measure the evolution of the variable in the process.	Process optimization, formulation development, design of experiments application	[65]
Simulated Annealing (SA)	Parameter variation approach based on choosing extreme inputs and targeting a median value to reach optimality. Process end-point parameters are varied slightly until an optimum is found.	Data extraction models; Process and unit operation parameters; Formulations; Batch scheduling operations; Model optimization	[16, 19, 35, 36, 65–76]
Iterative Optimization Technology (IOT)	Iterative evaluation to reduce the dimensionality of data components without the use of calibration sets	Process parameter estimation; Process analytical technology development	[34]

2.2 Gradient-Based Methods

Extensively used for solving linear and nonlinear programming optimization problems, these methods use information from derivative-based tools in order to reach an optimum solution [6]. Among some of the derivative tools are the Hessian matrix and gradient-vector. They are used to compute the slopes and convexity of functions to determine the direction needed for finding an optimum. Movement towards an optimum is achieved by iteratively moving along the function using slope and concavity information. While obtaining direction information from previously performed steps, movements are sized based on individual method techniques. The main difference between the various gradient methods is based on the mathematical functions used to find the point from which a next step is made [77]. They are used extensively for solving many linear and nonlinear programming problems.

Gradient methods can be applied to a range of constrained or unconstrained problems. Unconstrained optimization focuses on determining local or global optima by finding the points of zero-slope. This process involves using derivative tools and it is applied to a range of multidimensional problems. Unconstrained optimization is applied to optimization problems such as finding equilibrium conditions, fitting an equation to a dataset using least squares regression (i.e., regression analysis), and establishing saturation points [78]. Constrained optimization establishes that all problem variables have certain restrictions, which arise from different conditions. These constraints, which can be either inequalities or equalities, establish problem bounds, allocating the optimum condition within a given range. When function constrains are linear functions of the variables then the problem is considered to be a linear programming problem. However, when one of the constraints or the objective functions is not a linear function, the problem is nonlinear in nature.

Most problems in engineering involve some type of constrained optimization given the ranges of operation and process condition. For example, certain process variables in a system (i.e., pressure, volume) cannot have negative values or their sum is a number greater than zero. Such constraint would then be integrated to the problem as a constraint. These are important as other optimum process conditions that are not attained due to applied constraints. In some cases the presence of constraints makes it difficult to apply certain mathematical optimization tools used in unconstrained methods [78].

2.2.1 Pharmaceutical Applications

Gradient-based methods are widely utilized to optimize engineering and science problems. In the pharmaceutical industry the methods have been used consistently as part of the software packages available for analysis and design. Most cases use gradient methods in combination with surrogate and/or direct search methods to improve the optimized solutions. The following sections describe the use of these methods in the pharmaceutical engineering literature applied to solid and oral dose formulations.

Product Formulation

The steepest descent method was used in the calculation of optimal drug doses in clinical trails and the formulation of biostatic referential data for product design [79, 80]. Dawoodbhai et al. [81] optimized the formulation of direct compression tablet products containing talc using SQP methodologies. The tablet dissolution and disintegration time in vitro was optimized from experimentally measurable manufacturing parameters such as friability, hardness, compression and ejection force. All of these parameters were determined within feasible manufacturing values, which generated regions that served as the constraints for the SQP optimization. The characteristics and results of the optimized tablet formulation were analyzed by principal component analysis and described using response surface modeling [81].

Drug Delivery Systems

Li et al. [82] proposed the use of Newton's method for the optimization of effective drug delivery diffusion coefficients for cylindrical devices. The methodology implemented by Li et al. involved the discretization of multiple transient mathematical equations describing Fickian and convective-diffusion out of a cylinder to an external finite volume. The discrete diffusion equations were then optimized using the Newton's method in combination with a least squares multivariate method. The result of this study was the successful determination of effective drug delivery diffusion coefficients. Ghaffari et al. [83] used Newtonian and Quasi-Newton methods for the development of training sets for neural network modeling techniques used for the selection of optimal formulation compounds in extended release drugs.

Manufacturing Processes

Multiple articles referring to the optimization of unit operation conditions and parameters in the literature use gradient based method optimization. Optimization of parameters controlling simulated moving bed (SMB) processes was achieved by Toumi et al. using SQP methods [84]. SMB model operating regimes were optimized for parameters such as column length, particle porosity and particle diameter using the nonlinear SQP method. Toulouse et al. [85] used SQP optimization methods for determining the best operating conditions of batch chemical reactors taking into account safety and yield constraints. This methodology aimed at establishing the time evolutions of temperature, feed rate and fill level in order to reach maximal efficiency of chemical and pharmaceutical reactions using adequate energy inputs. Makrydaki et al. [86] performed a similar study in which SQP optimization was used alongside with Quasi-Newton and linear search methods to optimize the performance of asymmetric catalytic hydrogenation batch reactors. The authors introduce a new methodology for the development of experimental methods known as Design and Dynamic Experiments (DoDE) to acquire data for computational models using minimal resources. Data collected from this method was formulated into a mathematical model optimized using SQP methods programmed in the optimization toolbox of MATLAB™.

SQP was used in flowsheet optimization of continuous pharmaceutical manufacturing operations by Sen et al. [87]. Sen et al. used SQP in order to optimize a purification-processing operation comprised of a crystallizer, filter, dryer and blender unit modeled using gPROMS™ as the flowsheeting software. The optimized principal variables used to evaluate the model were filtration, crystal growth and nucleation rates. Sensitivity analysis using steepest descent methodology was performed by Sen et al. The analysis was performed to evaluate the effects of individual parameters on process outputs.

The Broyden-Fletcher-Goldfarb-Shanno (BFGS) method is a popular QN method reported in the pharmaceutical engineering literature [83, 86]. Information acquired from previous calculations is used for the selection of the new direction, making the process more computationally expensive yet faster reaching the end optimum. Kim et al. [88] proposed a new mathematical model for scheduling the manufacturing of pharmaceutical intermediates in a real plant. The model's objectives using BFGS methods were to minimize the total operation time, subject to mass balance constraints and process boundary conditions. In the research, the optimization was achieved by employing equipment pre-classification and sequential two-stage optimization techniques. The overall result yielded the formulation of such methodology into a new software package for scheduling pharmaceutical intermediate production.

The problem of flexible pharmaceutical manufacturing capacity is modeled and optimized using linear programming techniques by Chambers et al. [89]. The authors evaluated the flexibility offered by pharmaceutical facilities in terms of their ability to tune production rates and number of products being manufactured at a site. Researchers described flexibility as both an advantage and disadvantage that needs to be modeled. One of the major disadvantages described by the authors regarding flexible capabilities included the large number of options it generates, which may delay decision-making processes due to uncertainty and the inability to promptly find an optimum [89]. A solution to this problem was obtained using linear programming techniques alongside with dynamic optimization methods. The overall result of Chambers et al. indicated that plants with flexible capacity are not necessarily an optimum condition when compared to dedicated facilities for pharmaceutical production [89].

The gradient-based methods used in the pharmaceutical industry are summarized in Table 2 below and are divided into two categories: unconstrained and constrained methods.

Table 2
Gradient based optimization method

Method	Description	Pharmaceutical applications	Refs.
Unconstrained optimization			
(Steepest) Descent or Ascent Method	A line-search technique uses movements on the gradient in order to approach the optimum. Method starts at an initial point then calculate the direction of the gradient. Gradient vector is used to calculate the slope of the nearby points and the biggest slope calculated is selected as the direction of travel.	Formulation; Determination of manufacturing parameters; Hybrid method training; Model optimization	[6, 78–80, 83, 90–97]
Conjugate Gradient (GC)	Method determines the line path in order to establish the optimum. The number and size of the steps are reduced and enlarged, respectively in order to attain a local optimum along the line. The local optimum along the line is then considered the new point from which a new gradient is calculated until a global optimum is reached.	Classification of drugs; Formulation; Bioavailability parameter optimization; Hybrid method training; Model optimization	[96, 98–105]
Newton's Method (NM)	Method uses the gradient vector to derive the objective function into a second-order Taylor series expansion. The Hessian matrix that derives from the expansion allows for the determination of convexity, which aids in finding of global or local optimums more quickly. Method is able to find the optimum for a broader range of functions than the steepest descent methods.	Unit operation and process parameters; Feasible solution space; Pharmacokinetic parameters; Drug delivery optimization	[82, 83, 97, 105–111]
Modified Quasi-Newton's Methods (QN)	Compendium of less computationally expensive modifications of Newton's method. Hessian matrix calculations are done with less computation because it is computed as the inverse. The Hessian approximation allows for simple multiplication with the gradient instead of solving the linear system of equations derived in the Newton's method.	Formulation; Drug development; Manufacturing models	[81, 82, 86, 106–109, 111]

Levenberg–Marquardt Algorithm (LMA)	Used for solving nonlinear least square problems, this method interpolates between the Newton's method and the gradient descent. Popular curve fitting algorithm used to find mainly local optimum values.	Model optimization; Formulation; Drug development	[83]
Constrained optimization			
Sequential and Mixed-Integer Linear Programming (SLP and MILP)	Transformation of complex nonlinear constrained problems into simpler linear problems by generating first-order approximations of the objective and constraint functions. The approximation is done using appropriate algorithms around a specific design point, which are then tested using a new design point.	Supply chain operations; Drug molecule characterization and selection	[88, 89, 112]
Sequential and Mixed-Integer Quadratic Programming (SQP and MIQP)	Method used to simplify the constraint functions by approximating them to solvable function. The method uses quadratic approximations of the objective function. Higher approximation order increases the computational cost of the model compared to LP methods, yet the accuracy and breath of implementation is greatly increased.	Formulations; Process conditions; Pharmaceutical reactions scale-up and operation	[81, 84–87]

2.3 Surrogate-Based Optimization

Often referred to as metamodel or emulator-based optimization, this methodology derives expressions of complex and computationally expensive mathematical simulations using lower dimensional space modeling techniques. High-fidelity and computationally expensive engineering models are not easily optimizable because, using conventional optimization methods, functions need to be evaluated multiple times to obtain an optimal solution [113, 114]. Lower dimension models are built from expensive simulation data to quantify a desired response in terms of operating parameters via sampling-based interpolatory regression-based models [115]. Simulations are evaluated using some type of experimental design optimization methodology and the data is then collected to build a surrogate approximation. These surrogate-based models are then evaluated using a conventional optimization tool given their reduced computational expense compared to simulations. Conventional optimization methods work as sub-optimization tools, employed once the surrogate models have established a relatively accurate prediction of the simulation model [116, 117]. If insufficient data is collected to build an effective metamodel, the model can be further improved by adding more data points. The process criterion for adding more data points is often done using an expected improvement (EI) function [118, 119]. A general algorithm for surrogate-based optimization is shown below (Fig. 1) to illustrate the process.

This hybrid implementation of surrogate and conventional methods greatly reduces the computational cost of optimization and provides a feasible solution for process complex pharmaceutical processes [113, 116, 117]. Furthermore, the implementation of the two methods is very valuable as the design efficiency and optimum can be approached without the need of modifying conventional methods and generating more complex approximations of derivatives. Control strategies can also benefit from using the metamodels as they provide fast evaluation of the system in order to provide a prompt corrective action.

In Table 3, we summarize the most commonly used methods for surrogate-based optimization along with their advantages and disadvantages for modeling and optimization problems.

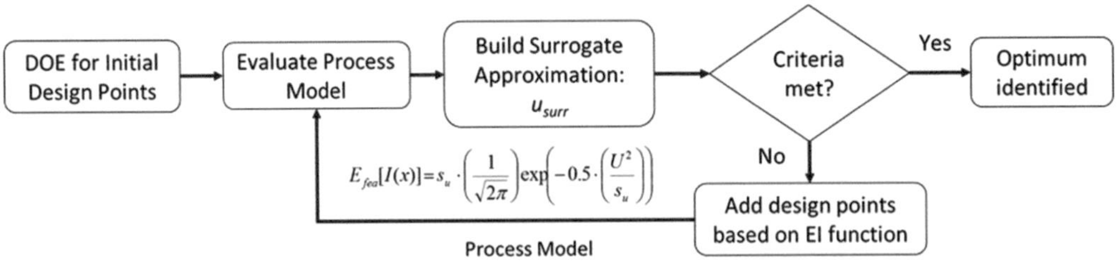

Fig. 1 Algorithm for surrogate-based optimization used by Boukouvala et al. [119]

Table 3
Surrogate-based optimization methods

Model	Description	Advantages	Disadvantages	Ref.
Kriging	Weighted inverse interpolation method fitting variogram and regression coefficients. Establishes equations in a data dimensional set.	Better modeling of nonlinear processes; Provides the largest number of correlation models; Lower prediction error	Requires large datasets to model; Needs previous simulation knowledge; Requires larger number of parameters to be fitted	[115, 119–122]
Response Surface Methodology (RSM)	Method of fitting data points using polynomial equations to create a variable-response relationship.	Lowest dimensional space of all the ROM methods; Uses the least amount of data to build models	Fits the equation to a polynomial equation; Data requires a high degree of pretreatment	[115, 123–126]
High Dimensionality Model Representation (HDMR)	Model fits component function coefficients to a truncated analytical basis function.	Perform global sensitivity analysis; Relatively small number of parameters need to be fitted	Truncated approach may lead to model failure; Relatively more difficult to use for optimization	[127–129]
Artificial Neural Network (ANN)	Input–output model using data clusters to transform inputs using weighted transformation linear or nonlinear equations.	It is a widely accepted and described model; Feedback and feed forward data processing; Learning from datasets	Multiple parameter fitting due to layering process; Requires the use of training sets for modeling	[130–134]
Multivariate Methods (MM)	Compendium of mathematical and statistical tools used to establish relationships between experimental variables. Polynomial and statistical equations are often developed from the relationships.	Reduce the dimensionality of datasets; Establish relationships between variables; Can be used to understand process robustness.	Model creation requires experimental data; Incorrect calculation of variance affects loadings	[135–152]

2.3.1 Pharmaceutical Applications

Product Formulation

Surrogate methods have been extensively used for optimization of drug formulations, from solid oral dose development to hydrogels and implants. For product formulation, the optimization literature primarily focuses on artificial neural networks (ANNs), response surface methods (RSM) and multivariate methods [153]. This use of surrogate models is in accordance with FDA QbD guidance, which suggests the use of design space enabling tools (i.e., design of experiments (DoE) and statistical models) for product design [5, 138, 154, 155].

Many instances of formulation optimization point to ANNs as the surrogate method for optimization, primarily because of its effectiveness to use and learn from available experimental datasets [133]. Wu et al. [156] used a neural network framework to predict the release profile parameters of 1000 salbutanol sulfate osmotic pump tablet formulations using varying concentrations of excipients. Based on model outputs, the authors were able to further generate response surface, which was used to determine optimum parameters (concentration of excipients) for the desired release pattern. ANNs have been used to simultaneously optimize multiple parameters, including particle size distribution for micro and nanoscale formulations [157, 158]. For example, Dario et al. [157] developed a novel formulation for the treatment of the Chagas' disease using microparticles and optimizing process parameters such as encapsulation efficiency, size, and dissolution rate. ANNs have also been used to optimize formulation parameters to obtain zero-order release kinetics for press coated tablets [159].

ANNs have been implemented as hybrid methods, where the network is used as a method to evaluate critical data attributes while a sub-optimization technique performs optimum calculations. Takayama et al. [160, 161] developed a simultaneous optimization technique that incorporated both neural network methodology and conventional minimization techniques (i.e., direct search methods) to predict controlled release profiles of solid dose tablets. In their research, the authors proved that the newly developed surrogate method was superior to conventional statistical methods by means of more closely predicting formulation properties based on experimental datasets [161]. Turkoglu et al. [162] effectively optimized the formulation of direct compression products with maximized API concentration and minimized lubricant using a hybrid ANN-direct search optimization technique. Further formulation optimization methods using neural network have been previously discussed in sections of this chapter as hybrid methods [19, 20, 22, 23, 83].

RSM provides a mathematical expression that relates the process variables (e.g., compression force, API concentration) with a desired response (e.g., dissolution). It is one of the most referenced optimization methods for solid dose pharmaceuticals, given its development simplicity [11, 123, 124, 126, 163, 164]. The effects of lubricant concentration and compression force were studied by Rotthauser et al. [165] using response surface methodologies for the formulation of effervescent tablets. In the study, researchers modeled the effects of varying lubricant concentrations and compression force on tablet properties. Model inspection concluded that high lubricant concentrations and compression forces were detrimental for the dissolution of effervescent tablets. Behzadi et al. [166] used RSM to minimize the hydrolysis of sucrose in tablets containing citric acid. Stability testing was performed using tablets with varying ratios of sucrose to citric acid and formulation process parameters, such as compression force. Results were fitted to a response surface model, which was optimized using a gradient method on the model. Late and Banga [167] applied a response surface approach to investigate interaction effects between formulation parameters and optimize them to develop a novel fast disintegrating tablet formulation using β-cyclodextrin as a diluent.

Bose et al. [163] and Monajjemzadeh et al. [168] investigated different combinations of polymers and excipient fillers using RSM to formulate optimal sustained release tablet matrices. Bose et al. studied pre-compression and post-compression parameters used in selected formulations to find the optimal one based on response surface modeled dissolution behavior. Monajjemzadeh et al. optimized the formulation of the solid oral dosages using both tablet hardness and their release profiles as control variables for the model. The optimized tablet formulations were obtained by inspection of the RSM model and were validated using dissolution experimental studies. Zhang et al. [169] used a Box-Behnken experimental design to develop a correlation between process conditions and the formation of sodium alginate films. The goal of the research was to develop a lower dimension model (i.e., surrogate-model) from which to determine optimum process conditions for manufacturing film coated hard capsule shells.

Multivariate methods have also been used extensively in pharmaceutical processing for the design of chemometric methods for process analytical tools (PATs) [170]. More recently these methods have also been used to characterize and optimize formulations of solid dose products [144, 145, 171]. Tomba et al. [136–138] and Garcia-Munoz et al. [172–174] explain the use of latent variable methods to enable in silico design of new product formulations. The authors present and validate multivariate optimization methods for raw material selection by optimizing powder properties (i.e., bulk density, flow) of blends using combinations of API and excipient formulations. The resulting optimum blends were experimentally tested and the optimization results validated. This proposed methodology was further evaluated to accelerate product development while minimizing required experimentation and raw material consumption [144, 175].

Manufacturing Practices

Unit operation process modeling and optimization has also significantly benefited from the implementation of surrogate-based optimization techniques. Murtoniemi et al. [176–178] were among the first employers of ANN methods to model fluidized bed granulation processes. In their research, Murtoniemi et al. demonstrated ANNs were more accurate model predictors than existing statistical techniques and provided a framework for optimization of process parameters [177]. Watano et al. [179, 180] applied ANN to optimize and scale-up agitated fluidized bed granulators employing a similar methodology to Murtoniemi et al. These previous methods for optimizing fluid bed granulators were innovative for the industry, as they did not take into account design space concept later introduced in later FDA guidances. More recent publications on optimization of fluidized bed granulation processes use design space approaches to generate feasible optimal solutions that can be implemented in multiple unit processes. Djuris et al. [181] uses the design space approach to optimize the outcomes of both granulation and compression processes by identifying the critical formulation and process parameters. Dry granulation processes have also been modeled and optimized using surrogate-based methods [182–184]. Inghelbrecht et al. [182] implemented the ANN as a predictor method for granule friability based on roller compaction process input parameters. Rocksloh et al. [184] used ANN methodology as a characterization technique for determining the crushing strength and disintegration time of high-dose plant extract tablets. Turkoglu et al. [183] generated a framework for optimization of acetaminophen tablet manufacturing using roller compaction by combining the optimization capabilities of genetic algorithms with the predictive capabilities of the ANN.

Tablet manufacturing was optimized by Walsh et al. [185], Aksu et al. [154, 186], and Sovany et al. [187]. Walsh et al. maximized tablet production in multiple manufacturing environments using ANN and results from 100 pharmaceutical product batches. Each dataset was derived from an individual batch analysis and used for training the ANN. The dataset was evaluated and classified (weighted) based on their operating parameters and relative importance to the overall process outcome. The heavily weighted parameters were optimized using a genetic algorithm method, giving a maximized tablet yield from the manufacturing process. Aksu et al. [154] employed an ANN method to develop a multivariate approach yielding quality products in direct compression lines. This approach was based on cause-effect relationship understanding between formulation ingredients and final product attributes. In a proceeding publication, Aksu et al. expanded the optimization methodology for direct compression to wet granulation processes [186]. Sovany et al. [187] used ANN to capture and optimize post-compressional properties of scored tablets by integrating experimental knowledge of available datasets to an optimization metamodel. Other unit operations, such as milling agglomeration, have been optimized in the literature yet difficulties arise due to the inability of models to accurately predict critical quality attributes leading to false optimum results [188].

The RSM provided an efficient and low-cost way for API purification that can be used for laboratory and manufacturing scale pharmaceutical industrial applications [189]. Process optimization of pharmaceutical operations involving uniform coating of tablet [190], granulation processes [33, 129], and powder feeding [115] and mixing [191] has also been achieved using RSM methods. Jia et al. [115] compared response surface methodologies (RSM) against Kriging for the optimization of loss-in-weight feeders models used in continuous tablet manufacturing systems. The results pointed to the varying differences between the methods interpolation techniques, however, results were found to be within desired ranges of the study. A similar result was found by Boukouvala et al. [191], whom performed a similar study between RSM, High-Dimensional model representations (HDMR) and Kriging, with case studies involving development of design spaces for continuous convective mixers and feeders.

Dubey et al. [190] implemented the Kriging methodology to improve a tablet coating process, which presented intermittent problems. A set of designed experiments along with response surface modeling and Kriging methods was used to successfully arrive at an optimal set of operating conditions that would discontinue the presence of problems in the coating unit. Through a series of recent publications Boukouvala et al. [118, 119, 192] further described the different methods for implementing Kriging and other response surface methods as an optimization technique in the pharmaceutical industry. The proposed Kriging optimization approach in combination with response surface methodologies were used to describe two pharmaceutical case studies involving powder feeding and blending operations [192]. The result included successful implementation of the model along with the minimization of the variability of the final product properties. Kriging methods were also used in the analysis of drug release patterns and their optimization for drug-eluting stents in the arterial wall [193]. In this research, the authors optimized the release profile of drug, a process that was modeled using the finite element simulation-based model (FEM). Kriging was used as the FEM emulator method for the optimization of the drug lifetime in the arterial wall. The results pointed to the effective use of Kriging methodology to optimize controlled release profiles of drugs.

3 Challenges in Optimization

As seen in the vast applications of optimization methods in pharmaceutical development, there is a growing interest in using these methods for improving product quality while reducing research efforts. The use of optimization methods as part of the development process is an idea embedded in the Quality-by-Design framework, which can improve manufacturing of drug products significantly. However, it is important to note the inherent problems with the application of optimization methods involving their application in pharmaceutical systems.

The use of optimization tools in pharmaceutical problems is still rather limited compared to other engineering disciplines. This is attributed to the increasing mathematical complexity of predictive models used to simulate the pharmaceutical production units, both due to the nature of powder behaviors and process uncertainty. The challenges when dealing with pharmaceutical solid processes can be summarized as: (a) lack of powder properties characterization knowledge; (b) their relation to critical operating conditions; (c) high degree of uncertainty in measurements; and (d) variable-response effect between powder properties, process parameters and unit performance on the critical quality attributes of final products. Detailed pharmaceutical mathematical models, such as simulation-based elemental models (i.e., DEM, CFD and FEM), offer high reliability and process representation but are computationally expensive and do not allow for easy use of optimization methods. Lower dimensionality models, requiring assumptions and large datasets, can be used to reduce computational expense however they bring another degree of uncertainty to the problem. This uncertainty is due to the reduced order model assumptions and regression of data. The lack of reliable and validated process models, does not allow for easy application of optimization methodologies.

Although there are available guides for which optimization technique to implement [5, 11, 77, 112], there is no clear answer to which method is most suitable for a given problem. The selection of appropriate optimization tools is based on the following factors: (a) the reliability and validity of the model being used; (b) the number of functions being evaluated; (c) the presence of nonlinearities and/or discontinuities; and (d) the number of system constraints. Pharmaceutical processes, aside from being complex in nature, often require multiple equations in order to solve for an optimum solution. This high number of function calls can make optimization computationally expensive and yield results outside of real feasible regions. Constraints added to these equations can furthermore impart computational expense as bounds set for optimization algorithms lead to non-real solutions, which may lead to non-convergence of the optimization problem.

Nonlinearities and non-convexities are common in powder technology process models, due to the intrinsic nature of solids behaviors. Powders tendency to behave as both liquids and solids based on the environmental actions imparted, make the system highly nonlinear in nature with discontinuities present at critical transitions. These problems, often captured in complex simulations are difficult to evaluate by optimization algorithms and are assumed as continuums in lower dimensional space models.

4 Conclusions and Future Outlook

This chapter provided a short overview of current optimization methodologies encountered in design problems of pharmaceutical products and processes. Techniques were classified as direct search, gradient based and surrogate-based optimization methodologies. General concepts regarding implementation of these optimization methods were introduced along with examples of their application in manufacturing process design and formulation development. The relationship between optimization and the current pharmaceutical regulatory framework of Quality-by-Design (QbD) was also established along with specific literature cases where both were described in detail.

Challenges in the implementation of these methods were furthermore discussed and described with a future perspective of solving these some available challenges. Availability of computing power will increase application of optimization methods, however, not with further improvement of methods to simulate powder behavior. Powder modeling, specifically oriented towards pharmaceutical manufacturing, needs to be developed taking into account optimization goals. Lower dimensional models should be preferred during model development as well as systems with minimum number of constraints and discontinuities. Aims at improving the measurement of powder properties would greatly reduce the uncertainty around collected datasets.

Overall, it has been shown that optimization is a critical component in the design of effective pharmaceutical products. This, in accordance to the QbD framework, aligns with the need for developing necessary models and techniques that allow for their application across the industry.

Acknowledgments

The authors would like to thank the funding provided by the Engineering Research Center for Structure Organic Particulate Systems "ERC-SOPS" (NSF-0504497, NSF-ECC 0540855).

References

1. USFDA (2002) Pharmaceutical cGMPs for the 21st century: a risk-based approach. USDoHaH Services (ed). Rockville, MD: FDA
2. Administration UFaD (2004) Guidance for industry: PAT—A framework for innovative pharmaceutical development, manufacturing, and quality assurance. UDoHaH Services (ed). Rockville, MD
3. Castle BC, Forbes RA (2013) Impact of quality by design in process development on the analytical control strategy for a small-molecule drug substance. J Pharm Innov 8 (4):247–264
4. Sen M et al (2013) Multi-scale flowsheet simulation of an integrated continuous purification–downstream pharmaceutical manufacturing process. Int J Pharm 445 (1–2):29–38
5. Yu L (2008) Pharmaceutical quality by design: product and process development, understanding, and control. Pharm Res 25 (4):781–791
6. Belegundu A, Chandrupatla T (1999) Optimization concepts and applications in engineering. Prentice Hall, Upper Saddle River, NJ
7. Muzzio FJ (2006) Engineering approaches for pharmaceutical processes. In: Levin M (ed) Pharmaceutical process scale-up. Taylor & Francis, Boca Raton, FL, pp 57–69
8. Rogers AJ, Inamdar C, Ierapetritou MG (2013) An integrated approach to simulation of pharmaceutical processes for solid drug manufacture. Ind Eng Chem Res 53(13): 5128–5147
9. Montgomery D (2009) Design and analysis of experiments. Wiley, Hoboken, NJ
10. Singh B, Gupta RK, Ahuja N (2006) Computer-assisted optimization of pharmaceutical formulation and processes, in pharmaceutical product development. CBS, New Delhi
11. Lewis G (2006) Optimization methods, in encyclopedia of pharmaceutical technology. Informa Healthcare, England
12. Biegler L (2008) In: Green DW (ed) Perry's chemical engineering handbook. McGraw-Hill Companies, New York, NY
13. Doornbos DA (1981) Optimisation in pharmaceutical sciences. Pharm Weekbl 3(1): 549–577
14. Shek E, Ghani M, Jones RE (1980) Simplex search in optimization of capsule formulation. J Pharm Sci 69(10):1135–1142
15. De Saavedra MS, Cuadra IS (2001) Application of a mixed optimization strategy in the design of a pharmaceutical solid formulation at laboratory scale. Drug Dev Ind Pharm 27 (7):675
16. Worakul N, Wongpoowarak W, Boonme P (2002) Optimization in development of acetaminophen syrup formulation. Drug Dev Ind Pharm 28(3):345
17. Rajab M et al (2010) Optimization of a metformin effervescent floating tablet containing hydroxypropyl methylcellulose and stearic acid. Pharmazie 65:67
18. Patel DP, Patel NM, Pandya NN, Jogani PD (2007) Formulation and optimization of carbamazepine floating tablets. Ind J Pharm Sci 69:763
19. Plumb AP et al (2003) Effect of varying optimization parameters on optimization by guided evolutionary simulated annealing (GESA) using a tablet film coat as an example formulation. Eur J Pharm Sci 18(3-4): 259–266
20. Plumb AP et al (2005) Optimisation of the predictive ability of artificial neural network (ANN) models: a comparison of three ANN programs and four classes of training algorithm. Eur J Pharm Sci 25(4-5):395–405
21. Vaghani SS et al (2012) Design and optimization of a stomach-specific drug delivery system of repaglinide: application of simplex lattice design. Pharm Dev Technol 17(1): 55–65
22. Zhang JC, Chen YZ, Wu ZN, Liao WR (2012) Optimize the preparation process of Erigeron breviscapus sustained-release pellets based on artificial neural network and particle swarm optimization algorithm. J Chinese Med Mater 35:127–133
23. Abraham A, Grosan C, Ţigan Ş (2007) Ensemble of hybrid neural network learning approaches for designing pharmaceutical drugs. Neural Comput Appl 16(3):307–316
24. Nauman EB, Patel K, Karande P (2011) Design of optimized diffusion-controlled transdermal drug delivery systems. Drug Dev Ind Pharm 37(1):93–102
25. Bhattacharyya S et al (2007) Design, evaluation and statistical optimisation of a controlled release multiparticulate acyclovir delivery system. Lat Am J Pharm 26(6):852
26. Prajapatia SP, Patel L, Patel C (2011) Floating matrix tablets of domperidone formulation and optimization using simplex lattice design. Iran J Pharm Res 10:447

27. Varshosaz JM, Moazen E, Fathi M (2012) Preparation of carvedilol nanoparticles by emulsification method and optimization of drug release: surface response design versus genetic algorithm. J Dispers Sci Technol 33 (10):1480
28. Barmpalexis PK, Kachrimanis K, Georgarakis F (2011) Solid dispersions in the development of a nimodipine floating tablet formulation and optimization by artificial neural networks and genetic programming. Eur J Pharm Biopharm 77:122
29. Barmpalexis P et al (2011) Symbolic regression via genetic programming in the optimization of a controlled release pharmaceutical formulation. Chemom Intell Lab Syst 107(1): 75–82
30. Patel MS, Shah TJ, Amin AF, Shah NN (2009) Design, development and optimization of a novel time and pH-dependent colon targeted drug delivery system. Pharm Dev Technol 14(1):62–69
31. Qu N et al (2008) Radial basis function networks combined with genetic algorithm applied to nondestructive determination of compound erythromycin ethylsuccinate powder. Chemom Intell Lab Syst 90(2):145–152
32. Momenbeik F, Roosta M, Nikoukar AA (2010) Simultaneous microemulsion liquid chromatographic analysis of fat-soluble vitamins in pharmaceutical formulations: optimization using genetic algorithm. J Chromatogr A 1217(24):3770–3773
33. Chaudhury A et al (2014) Population balance model development, validation, and prediction of CQAs of a high-shear wet granulation process: towards QbD in drug product pharmaceutical manufacturing. J Pharm Innov 9 (1):53–64
34. Muteki K et al (2013) Mixture component prediction using iterative optimization technology (calibration-free/minimum approach). Ind Eng Chem Res 52(35): 12258–12268
35. Raaymakers WH, Hoogeveen J (2000) Scheduling multipurpose batch process industries with no-wait restrictions by simulated annealing. Eur J Oper Res 126:131
36. Graells MC, Canton J, Peschaud B, Puigjaner L (1998) General approach and tool for the scheduling of complex production systems. Comput Chem Eng 22:S395
37. Hamamoto SY, Yih Y, Salvendy G (1999) Development and validation of genetic algorithm-based facility layout – a case study in the pharmaceutical industry. Int J Prod Res 37(4):749–768
38. Andradóttir S (2006) Chapter 20: An overview of simulation optimization via random search. In: Shane GH, Barry LN (eds) Handbooks in operations research and management science. Elsevier, Amsterdam, pp 617–631
39. Begtra JB, Bengio Y (2012) Random search for hyper-parameter optimization. J Mach Learn Res 13:281
40. Calzolari D et al (2008) Search algorithms as a framework for the optimization of drug combinations. PLoS Comput Biol 4(12): e1000249
41. Maiser B, Dismer F, Hubbuch J (2014) Optimization of random PEGylation reactions by means of high throughput screening. Biotechnol Bioeng 111(1):104–114
42. Namasivayam V, Bajorath J (2012) Multiobjective particle swarm optimization: automated identification of structure-activity relationship-informative compounds with favorable physicochemical property distributions. J Chem Inf Model 52(11):2848–2855
43. Pandeya ST, Thakkar D (2005) Combinatorial chemistry: a novel method in drug discovery and its application. Indian J Chem 44B:335
44. Swisher J et al (2004) A survey of recent advances in discrete input parameter discrete-event simulation optimization. IIE Trans 36(6):591
45. Wang X, Tang L (2012) A discrete particle swarm optimization algorithm with self-adaptive diversity control for the permutation flowshop problem with blocking. Appl Soft Comput 12(2):652–662
46. Zhou Y-P et al (2006) Adaptive configuring of radial basis function network by hybrid particle swarm algorithm for QSAR studies of organic compounds. J Chem Inf Model 46:2494–2501
47. Bindschaedler C, Gurny R (1982) Optimization of pharmaceutical formulations by the simplex method using a TI 59 calculator. Pharm Acta Helv 57(9):251–255
48. Hayes SD, Dunne A, Smart T, Davis J (2003) Interpretation and optimization of the dissolution specifications for a modified release product with an in vivo–in vitro correlation (IVIVC). J Pharm Sci 93:57
49. Leonarski F et al (2013) Evolutionary algorithm in the optimization of a coarse-grained force field. J Chem Theory Comput 9(11): 4874–4889
50. Pulgarin JM, Molina AA, Pardo MA (2002) The use of modified simplex method to

optimize the room temperature phosphorescence variables in the determination of an antihypertensive drug. Talanta 57 (4):795–805
51. Nelder JA, Mead R (1965) A simplex method for function minimization. Comput J 7:308
52. Spendley WH, Hext GR, Himsworth FR (1962) Sequential application of simplex designs in optimization and evolutionary operation. Technometrics 4:441
53. Xu H et al (2000) Application of a modified simplex method to the multivariable optimization of a new FIA system for the determination of osmium. Fresenius J Anal Chem 368 (8):780
54. Coello C et al (1996) Multiobjective optimization using a micro-genetic algorithm. In: Evolutionary multi-criterion optimization, Lecture notes in computer science. Springer, Berlin, pp 126–140
55. Cruz-Monteagudo M et al (2008) Desirability-based methods of multiobjective optimization and ranking for global QSAR studies. Filtering safe and potent drug candidates from combinatorial libraries. J Comb Chem 10:897
56. da Silva GA, Augusto F, Poppi RJ (2007) Simultaneous optimization by neuro-genetic approach of a multiresidue method for determination of pesticides in Passiflora alata infuses using headspace solid phase microextraction and gas chromatography. J Chromatogr A 1138(1-2):251–261
57. Davis L (1987) Genetic algorithms and simulated annealing. Pitman, London
58. Leardi R (2001) Genetic algorithms in chemometrics and chemistry: a review. J Chemometr 15(7):559–569
59. Mandal A, Ranjan P, Wu CFJ (2009) G-SELC: optimization by sequential elimination of level combinations using genetic algorithms and Gaussian processes. Annals Appl Stat 3(1):398–421
60. Perez-Escobedo JL (2010) Multiobjective optimization of New Product Development in the pharmaceutical industry. In: MEGeP. Institut National Polytechnique de Toulouse, France
61. Smith BG, Gemperline P (2000) Wavelength selection and optimization of pattern recognition methods using the genetic algorithm. Anal Chim Acta 423:167–177
62. Veldhuizen DAV, Lamont GB (2000) Multiobjective evolutionary algorithms: analyzing the state-of-the-art. Evol Comput 8(2):125–147
63. Wang J et al (2006) Genetic algorithm-optimized QSPR models for bioavailability, protein binding, and urinary excretion. J Chem Inf Model 46(6):2674–2683
64. Zitzler E, Thiele L (1999) Multiobjective evolutionary algorithms: a comparative case study and the strength Pareto approach. IEEE Trans Evol Comput 3(4):257–271
65. Fonseca CM, Fleming PJ (1995) An overview of evolutionary algorithms in multiobjective optimization. Evol Comput 3(1):1–16
66. Brünger A et al (1990) Slow-cooling protocols for crystallographic refinement by simulated annealing. Acta Crystall A 46:585
67. Bulgak AASJL. Integrating a modified simulated annealing algorithm with the simulation of a manufacturing system to optimize buffer sizes in automatic assembly systems. Proceedings of the 1988 Winter Simulation Conference, San Diego, CA, 1988
68. Goffe WF, Ferrier GD, Rogers J (1994) Global optimization of statistical functions with simulated annealing. J Econ 60:65
69. Ingber L (1993) Simulated annealing: practice versus theory. Math Comp Model 18 (11):29
70. Kirkpatrick S, Gelatt C, Vecchi M (1983) Optimization by simulated annealing. Science 220:671
71. Kirkpatrick S (1984) Optimization by simulated annealing: quantitative studies. J Stat Phys 34(5-6):975–986
72. Laarhoven PM, Aarts EL (1987) Simulated annealing. In: Simulated annealing: theory and applications. Springer, Dordrecht, pp 7–15
73. Manz EH, Haddock J, Mittenthal, Optimization of an automated manufacturing systems simulation model using simulated annealing. Proceedings of the 1989 Winter Simulation Conference, Washington, DC, 1989
74. Ogata K, Umeyama H (2000) An automatic homology modeling method consisting of database searches and simulated annealing. J Mol Graph Model 18(3):258–272
75. Tekin E, Sabuncuoglu I (2004) Simulation optimization: a comprehensive review on theory and applications. IIE Trans 36(11): 1067–1081
76. Wales DJ, Scheraga HA (1999) Global optimization of clusters, crystals, and biomolecules. Science 285(5432):1368–1372
77. Venter G (2010) Review of optimization techniques. In: Blockley RS, Shyy W (eds) Encyclopedia of aerospace engineering. John Wiley & Sons, New York, NY
78. Nocedal J, Stephen W (1999) Numerical optimization. Springer, New York, NY

79. Ting N (2006) Dose finding in drug development. Springer, New York, NY
80. Masulli FT, Verkhivker G (2008) Computational intelligence methods for bioinformatics and biostatistics: 5th international meeting. In: CIBB international meeting. Springer, New York, NY
81. Dawoodbhai TS et al (1991) Optimization of tablet formulations containing talc. Drug Dev Ind Pharm 17(10):1343–1371
82. Li Y et al (2010) A computer algorithm for optimizing to extract effective diffusion coefficients of drug delivery from cylinders. Infom Technol J 9(8):647–1652
83. Ghaffari A et al (2006) Performance comparison of neural network training algorithms in modeling of bimodal drug delivery. Int J Pharm 327(1-2):126–138
84. Toumi A et al (2007) Efficient optimization of simulated moving bed processes. Chem Eng Process Process Intensif 46(11):1067–1084
85. Toulouse C et al (1996) Optimisation and scale-up of batch chemical reactors: impact of safety constraints. Chem Eng Sci 51(10): 2243
86. Makrydaki FG, Georgakis C, Saranteas K. Dynamic optimization of a batch pharmaceutical reaction using the design of dynamic experiments (DoDE): the case of an asymmetric catalytic hydrogenation reaction. In: Proceedings of the 9th International Symposium on Dynamics and Control of Process Systems, 2010, Leuven, Belgium
87. Sen M et al (2013) Flowsheet optimization of an integrated continuous purification-processing pharmaceutical manufacturing operation. Chem Eng Sci 102:56–66
88. Kim J et al (2010) A heuristic-embedded scheduling system for a pharmaceutical intermediates manufacturing plant. Ind Eng Chem Res 49:12646–12653
89. Chambers CE, Snir EM, Ata A (2009) The use of flexible manufacturing capacity in pharmaceutical product introductions. Decis Sci 40(2):243
90. Battiti R (1992) First- and second-order methods for learning: between steepest descent and Newton's method. Neural Comput 4(2):141–166
91. Dinçer S, Özdurmu S (1977) Mathematical model for enteric film coating of tablets. J Pharm Sci 66(8):1070–1073
92. Fletcher R, Powell MJD (1963) A rapidly convergent descent method for minimization. Comput J 6(2):163–168
93. Golfetto WA, Fernandes SS (2012) A review of gradient algorithms for numerical computation of optimal trajectories. J Aerosp Technol Manag 4(2):131–143
94. Graña Drummond LM, Svaiter BF (2005) A steepest descent method for vector optimization. J Comput Appl Math 175(2):395–414
95. Lemarechal C (2012) Cauchy and the gradient method. Doc Math Extra Volume: 251–254.
96. Byvatov E et al (2003) Comparison of support vector machine and artificial neural network systems for drug/nondrug classification. J Chem Inf Comput Sci 43(6): 1882–1889
97. Pulkkinen S (2008) A review of methods for unconstrained optimization: theory, implementation and testing, in department of mathematics and statistics. University of Helsinki, Finland
98. Andrews CW, Bennett L, Yu LX (2000) Pharm Res 17:639–644
99. Hou TW, Wang J, Zhang W, Xu X (2007) J Chem Inf Model 47:460–463
100. Ma CY et al (2008) Prediction models of human plasma protein binding rate and oral bioavailability derived by using GA-CG-SVM method. J Pharm Biomed Anal 47(4-5): 677–682
101. Peh K-K et al (2000) Use of artificial neural networks to predict drug dissolution profiles and evaluation of network performance using similarity factor. Pharm Res 17(11): 1384–1389
102. Sorich MJ et al (2003) Comparison of linear and nonlinear classification algorithms for the prediction of drug and chemical metabolism by human UDP-glucuronosyltransferase isoforms. J Chem Inf Comput Sci 43(6): 2019–2024
103. Trotter MH, Holden SB (2003) QSAR Comb Sci 22:533–548
104. Perry RH, Green DW (2005) Perry's chemical engineering handbook, 5th edn. McGraw-Hill, New York, NY
105. Fallgreen M (2006) On the robustness of conjugate-gradient methods and quasi-Newton methods. Department of Mathematics, Royal Institute of Technology (KTH), Sweden
106. Bissett B (2003) Practical pharmaceutical laboratory automation. CRC Press, Boca Raton, FL
107. Nguyen LB et al (2013) Combining genetic algorithm and Levenberg-Marquardt algorithm in training neural network for hypoglycemia detection using EEG signals. Conf Proc IEEE Eng Med Biol Soc 2013: 5386–5389

108. Sadegh Zadeh K (2011) A synergic simulation-optimization approach for analyzing biomolecular dynamics in living organisms. Comput Biol Med 41(1):24–36
109. Ueckert S, Nyberg J, Hooker A (2006) Comparison of different global optimal design approximations. Department of Pharmaceutical Biosciences, Uppsala University, Sweden
110. Xie D et al (2003) Principal component analysis combined with truncated-Newton minimization for dimensionality reduction of chemical databases. Math Program 95(1): 161–185
111. Yoshida K et al (2013) Estimation of feasible solution space using Cluster Newton Method: application to pharmacokinetic analysis of irinotecan with physiologically-based pharmacokinetic models. BMC Syst Biol 7 (Suppl 3):S3
112. Siddhaye S et al (2004) Pharmaceutical product design using combinatorial optimization. Comput Chem Eng 28(3):425–434
113. Han Z-H, Zhang K-S (2012) Surrogate-based optimization. In: Roeva O (ed) Real-world applications of genetic algorithms. InTech, Rijeka, Croatia
114. Lo Nigro G et al (2013) A user friendly real option based model to optimize pharmaceutical R&D portfolio. J Appl Oper Res 5(3): 83–95
115. Jia Z et al (2009) Predictive modeling for pharmaceutical processes using kriging and response surface. J Pharm Innov 4(4): 174–186
116. Forrester AIJ, Keane AJ (2009) Recent advances in surrogate-based optimization. Prog Aerosp Sci 45(1-3):50–79
117. Queipo NV et al (2005) Surrogate-based analysis and optimization. Prog Aerosp Sci 41(1):1–28
118. Boukouvala F (2013) Integrated simulation and optimization of continuous pharmaceutical manufacturing, in chemical and biochemical engineering. Rutgers, The State University of New Jersey, New Brunswick, NJ
119. Boukouvala F, Ierapetritou M (2013) Surrogate-based optimization of expensive flowsheet modeling for continuous pharmaceutical manufacturing. J Pharm Innov 8(2): 131–145
120. Cressie N (1985) Fitting variogram models by weighted least squares. J Int Assoc Math Geol 17(5):563–586
121. Kleijnen JPC (2009) Kriging metamodeling in simulation: a review. Eur J Oper Res 192 (3):707–716
122. Simpson TW et al (2001) Kriging models for global approximation in simulation-based multidisciplinary design optimization. AIAA J 39(12):2233–2241
123. Baş D, Boyacı İH (2007) Modeling and optimization I: usability of response surface methodology. J Food Eng 78(3):836–845
124. Bezerra MA et al (2008) Response surface methodology (RSM) as a tool for optimization in analytical chemistry. Talanta 76(5): 965–977
125. Khuri AI, Mukhopadhyay S (2010) Response surface methodology. Wiley Interdiscip Rev Comput Stat 2(2):128–149
126. Raymond HM, Montgomery DC (1995) Response surface methodology: process and product in optimization using designed experiments. Wiley, New York, NY
127. Li G, Wang S-W, Rabitzl H (2000) High dimensional model representations (HDMR): concepts and applications. In Proceedings of the Institute of Mathematics and Its Applications Workshop on Atmospheric Modeling, pp 15–19
128. Li G, Wang S-W, Rabitz H (2002) Practical approaches to construct RS-HDMR component functions. J Phys Chem A 106 (37):8721
129. Rogers A, Hashemi A, Ierapetritou M (2013) Modeling of particulate processes for the continuous manufacture of solid-based pharmaceutical dosage forms. Processes 1(2): 67–127
130. Fyfe C (2005) Artificial neural networks. In: Gabrys B, Leiviskä K, Strackeljan J (eds) Do smart adaptive systems exist? Springer, Berlin, pp 57–79
131. Abraham A (2005) Artificial neural networks. In: Handbook of measuring system design. John Wiley & Sons, New York, NY
132. Achanta AS, Kowalski JG, Rhodes CT (1995) Artificial neural networks: implications for pharmaceutical sciences. Drug Dev Ind Pharm 21(1):119–155
133. Agatonovic-Kustrin S, Beresford R (2000) Basic concepts of artificial neural network (ANN) modeling and its application in the pharmaceutical research. J Pharm Biomed Anal 22:717
134. Sutariya V et al (2013) Artificial neural network in drug delivery and pharmaceutical research. Open Bioinforma J 7:46–62
135. Cui Y et al (2012) Variable selection in multivariate modeling of drug product formula and manufacturing process. J Pharm Sci 101(12): 4597–4607

136. Tomba E, Barolo M, García-Muñoz S (2014) In-silico product formulation design through latent variable model inversion. Chem Eng Res Des 92(3):534–544
137. Tomba E et al (2013) General procedure to aid the development of continuous pharmaceutical processes using multivariate statistical modeling - an industrial case study. Int J Pharm 444(1-2):25–39
138. Tomba E et al (2013) Latent variable modeling to assist the implementation of Quality-by-Design paradigms in pharmaceutical development and manufacturing: a review. Int J Pharm 457(1):283–297
139. García Muñoz S, Padovani V, Mercado J (2014) A computer aided optimal inventory selection system for continuous quality improvement in drug product manufacture. Comput Chem Eng 60:396–402
140. Ende D et al (2007) API quality by design example from the torcetrapib manufacturing process. J Pharm Innov 2(3-4):71–86
141. Hayashi Y et al (2012) Reliability evaluation of nonlinear design space in pharmaceutical product development. J Pharm Sci 101(1): 333–341
142. Kikuchi S et al (2011) Latent structure analysis in pharmaceutical formulations using Kohonen's self-organizing map and a Bayesian network. J Pharm Sci 100 (3):964–975
143. Kourti T (2010) Pharmaceutical manufacturing: the role of multivariate analysis in design space, control strategy, process understanding, troubleshooting, and optimization. In: Ende DJa (ed) Chemical engineering in the pharmaceutical industry. John Wiley & Sons, New York, pp 853–878
144. Norioka T et al (2011) Optimization of the manufacturing process for oral formulations using multivariate statistical methods. J Pharm Innov 6(3):157–169
145. Onuki Y, Morishita M, Takayama K (2004) Formulation optimization of water-in oil water multiple emulsion for intestinal insulin delivery. J Control Release 97(1):91–99
146. Palabiyik M et al (2013) Multivariate optimization model in a partial least squares-1 method for simultaneous determination of dorzolamide hydrochloride and timolol maleate in eye drops. Curr Pharm Anal 9(4):404–412
147. Portillo P et al (2008) Quality by design methodology for development and scale-up of batch mixing processes. J Pharm Innov 3 (4):258–270
148. Takayama K et al (2004) Multivariate spline interpolation as a novel method to optimize pharmaceutical formulations. Pharmazie 59 (5):392–395
149. ten Berge J (1993) Least squares optimization in multivariate analysis. DSWO Press, Leiden
150. Varshosaz J, Eskandari S, Tabakhian M (2010) Production and optimization of valproic acid nanostructured lipid carriers by the Taguchi design. Pharm Dev Technol 15(1): 89–96
151. Vignaduzzo S et al (2011) Multivariate optimization and validation of a CZE method for the analysis of pridinol mesylate and meloxicam in tablets. Chromatographia 74(7-8): 609–617
152. Yasuda A et al (2010) Self-organizing map analysis using multivariate data from theophylline powders predicted by a thin-plate spline interpolation. J Pharm Sci 99(11): 4535–4542
153. Colbourn EA, Rowe RC (2006) Neural computing and formulation optimization. In: Encyclopedia of pharmaceutical technology. England, Informa Healthcare
154. Aksu B et al (2012) Quality by design approach: application of artificial intelligence techniques of tablets manufactured by direct compression. AAPS PharmSciTech 13(4): 1138–1146
155. Huang J, Goolcharran C, Ghosh K (2011) A quality by design approach to investigate tablet dissolution shift upon accelerated stability by multivariate methods. Eur J Pharm Biopharm 78(1):141–150
156. Wu TP, Pan W, Chen J, Shang R (2000) Formulation optimization technique based on artificial neural network in salbutamol sulfate osmotic pump tablets. Drug Dev Ind Pharm 26:211
157. Leonardi D et al (2009) Development of novel formulations for Chagas' disease: optimization of benznidazole chitosan microparticles based on artificial neural networks. Int J Pharm 367(1–2):140–147
158. Barmpalexis P et al (2010) Artificial neural networks in the optimization of a nimodipine controlled release tablet formulation. Eur J Pharm Biopharm 74(2):316–323
159. Ali AA, Ali AM (2013) Optimization of propranolol HCl release kinetics from press coated sustained release tablets. Pharm Dev Technol 18(5):1238–1246
160. Takayama KM, Morva A, Fujikawa M, Hattori Y, Obata Y, Nagai T (2000) Formula optimization of theophylline controlled release tablet based on artificial neural networks. J Control Release 68:175

161. Takayama K, Fujikawa M, Nagai T (1999) Artificial neural network as a novel method to optimize pharmaceutical formulations. Pharm Res 16(1):1–6
162. Turkoglu JO, Ozarslan R, Sakr A (1995) Artificial neural net- work analysis of a direct compression tableting study. Eur J Pharm Biopharm 41:315
163. Bose A, Wong TW, Singh N (2013) Formulation development and optimization of sustained release matrix tablet of Itopride HCl by response surface methodology and its evaluation of release kinetics. Saudi Pharm J 21 (2):201–213
164. Malakar J, Nayak AK, Goswami S (2012) Use of response surface methodology in the formulation and optimization of bisoprolol fumarate matrix tablets for sustained drug release. ISRN Pharm 2012:730624
165. Rotthauser B et al (1998) Optimization of an effervescent tablet formulation containing spray dried l-leucine and polyethylene glycol 6000 as lubricants using a central composite design. Eur J Pharm Biopharm 46:85
166. Behzadi SS et al (2006) Investigation of the stability of tablets prepared from sucrose and citric acid anhydride utilizing response surface methodology. Eur Food Res Technol 223 (2):238–245
167. Late SG, Banga AK (2010) Response surface methodology to optimize novel fast disintegrating tablets using beta cyclodextrin as diluent. AAPS PharmSciTech 11(4): 1627–1635
168. Monajjemzadeh F et al (2013) Design and optimization of sustained-release divalproex sodium tablets with response surface methodology. AAPS PharmSciTech 14(1):245–253
169. Zhang Q et al (2013) Optimized water vapor permeability of sodium alginate films using response surface methodology. Chinese J Oceanol Limnol 31(6):1196–1203
170. Gabrielsson J, Lindberg N-O, Lundstedt T (2002) Multivariate methods in pharmaceutical applications. J Chemometr 16(3):141–160
171. Zhang L, Henson MJ, Sekulic SS (2005) Multivariate data analysis for Raman imaging of a model pharmaceutical tablet. Anal Chim Acta 545(2):262–278
172. Polizzi MA, García-Muñoz S (2011) A framework for in-silico formulation design using multivariate latent variable regression methods. Int J Pharm 418(2):235–242
173. Garcia-Munoz S (2014) Two novel methods to analyze the combined effect of multiple raw-materials and processing conditions on the product's final attributes: JRPLS and TPLS. Chemom Intell Lab Syst 133:49–62
174. García-Muñoz S, Mercado J (2013) Optimal selection of raw materials for pharmaceutical drug product design and manufacture using mixed integer nonlinear programming and multivariate latent variable regression models. Ind Eng Chem Res 52 (17):5934–5942
175. Norioka T et al (2013) A novel approach to establishing the design space for the oral formulation manufacturing process. Chem Pharm Bull 61(1):39–49
176. Ghaffari A, Abdollahi H, Khoshayand MR, Bozchalooi IS, Dadgar A, Rafiee-Tehrani M (2006) Performance comparison of neural network training algorithms in modeling of bimodal drug delivery. Int J Pharm 327(1-2):126–38
177. Murtoniemi EY, Yliruusi Y, Kinnunen P, Merkku P, Leiviska K (1994) The advantages by the use of neural networks in modelling the fluidized bed granulation process. Int J Pharm 108:155
178. Murtoniemi EY, Merkku P, Kinnunen P, Leiviska K, Yilruusi J (1994) Effect of neural network topology and training end point in modelling the fludized bed granulation process. Int J Pharm 110:101
179. Watano ST, Takashima H, Miyanami K (1996) Scale-up of agitation fluidized bed granulation by neural networks. Chem Pharm Bull 45:1193
180. Watano SS, Sato Y, Miyanami K (1997) Application of a neural network to granulation scale-up. Powder Technol 90:153
181. Djuris J et al (2012) Design space approach in optimization of fluid bed granulation and tablets compression process. ScientificWorldJournal 2012:185085
182. Inghelbrecht SR, Remon J-P, Fernandes de Aguiar P, Walczak B, Massart DL, Van De Velde R, De Baets P, Vermeersch H, De Backer P (1997) Instrumentation of a roll compactor and the evaluation of the parameter setting by neural networks. Int J Pharm 148:103
183. Turkoglu MA, Aydin I, Murray M, Sakr A (1999) Modelling of a roller-compaction process using neural networks and genetic algorithms. Eur J Pharm Biopharm 48:239
184. Rocksloh K et al (1999) Optimization of crushing strength and disintegration time of a high-dose plant extract tablet by neural networks. Drug Dev Ind Pharm 25(9): 1015–1025

185. Walsh DE, Zaccari N (2001) Predictive statistical process controls—a neural network approach to maximizing tablet yield. Pharmaceut Tech Asia 13(9):18
186. Aksu B et al (2013) A quality by design approach using artificial intelligence techniques to control the critical quality attributes of ramipril tablets manufactured by wet granulation. Pharm Dev Technol 18 (1):236–245
187. Sovány T et al (2013) Application of physicochemical properties and process parameters in the development of a neural network model for prediction of tablet characteristics. AAPS PharmSciTech 14(2):511–516
188. Zain AM, Haron H, Sharif S (2010) Prediction of surface roughness in the end milling machining using artificial neural network. Expert Syst Appl 37(2):1755–1768
189. Zhang YB et al (2014) Ultrasound-assisted extraction and purification of schisandrin B from Schisandra chinensis (Turcz.) Baill seeds: optimization by response surface methodology. Ultrason Sonochem 21(2): 461–466
190. Dubey A et al (2012) Improvement of tablet coating uniformity using a quality by design approach. AAPS PharmSciTech 13(1): 231–246
191. Boukouvala F, Muzzio FJ, Ierapetritou MG (2010) Design space of pharmaceutical processes using data-driven-based methods. J Pharm Innov 5(3):119–137
192. Boukouvala F, Muzzio FJ, Ierapetritou MG (2011) Dynamic data-driven modeling of pharmaceutical processes. Ind Eng Chem Res 50(11):6743–6754
193. Li H et al (2013) Drug release analysis and optimization for drug-eluting stents. ScientificWorldJournal 2013:1–5

Chapter 10

An Overview of Pharmaceutical Manufacturing for Solid Dosage Forms

Stephan Sacher and Johannes G. Khinast

Abstract

In this chapter, the main processing steps and manufacturing aspects of solid dosage forms are described and the relevant literature is reviewed. Starting with powder feeding, powder blending, granulation (dry and wet), and drying the most important unit operations to make compactable granules are reviewed. As an alternative to granulation, hot-melt extrusion is introduced, together with the various downstream processing choices. Next, tableting and capsule filling—for making a final dosage form—are discussed, followed by a section on coating. In all sections scale-up methods are reviewed and an outlook for future developments is provided. The last two sections are devoted to process analytical technology (PAT) and continuous manufacturing.

Key words Pharmaceutical manufacturing, Process analytical technology, Continuous manufacturing, Particle processing

1 Introduction

Solid dosage forms, such as tablets and capsules, still constitute the most widespread form of drug products due to their ease of use and the associated high patient compliance. In this context, established manufacturing technology has been available for many decades. However, manufacturing of solid dosage forms poses several challenges and is far from understood, mainly due to the large number of different materials, the broad range of material properties, the many different therapeutic demands and delivery requirements, as well as the large variety on dosage forms. Therefore, even with established processes and unit operations, manufacturing of solid dosage forms remains challenging, although recent efforts to establish general rules for process selection and development have been reported [1].

Depending on the type of drug substances and excipients and their specific material properties different processing routes have to

Marianthi G. Ierapetritou and Rohit Ramachandran (eds.), *Process Simulation and Data Modeling in Solid Oral Drug Development and Manufacture*, Methods in Pharmacology and Toxicology,
DOI 10.1007/978-1-4939-2996-2_10, © Springer Science+Business Media New York 2016

be chosen in order ensure processability and to meet the requirements of administration. Those include direct compaction of tablets from powders, different routes of granulation followed by compaction, as well as filling of capsules with powder or pellets. All manufacturing lines have in common the goal to achieve a homogenous blend. Thus, feeding and blending are of utmost importance (*see* Sects. 2 and 3). To avoid segregation effects, i.e., in order to assure a constant amount of API in all dosage units, and to enhance flowability, agglomeration is often performed (*see* Sect. 4). In the case of wet granulation, the material has to be dried (Sect. 5). Alternatively, hot-melt extrusion can be used to make performance-enhanced materials (Sect. 6). Before tableting (Sect. 7), blending with lubricants (often MgSt) is necessary to avoid frictional overheating in the tableting machine. Powders and/or pellets can also be filled in capsules (Sect. 8). Finally, in case of tablets a layer of coating is applied in some cases to achieve specific release characteristics, to protect the drug from gastric juices, to add an active component, or to enhance the optical features (Sect. 9).

In the past pharmaceutical manufacturing processes were developed based on empirical knowledge. However, the Quality by Design (QbD) initiative provided a framework for risk-and science-based product and process development involving sound process understanding. The objective is to design quality into products instead of testing quality into them at the end of the manufacturing process. State-of-the-art technologies, such as advanced modeling and simulation, as well as Process Analytical Technology (PAT, *see* Sect. 10), enable the generation of profound process understanding. Therefore, the creation of robust manufacturing processes is becoming reality. In-line process monitoring and control concepts allow to react to variation of raw material quality and other process conditions and to minimize time consuming and costly end of the pipe, off-line, quality testing.

Although most solid drug products are still manufactured in batch mode, continuous manufacturing is gaining importance (*see* Sect. 12). A wide range of manufacturing equipment is able to be run in continuous mode, such as tablet presses. Even unit operations such as dosing, blending and agglomeration, which historically have been operated in batch, are nowadays available as state of the art technology for continuous mode. Combined with smart process design and control real time release can be realized and efficiency can be maximized. Personalized manufacturing is another important trend, allowing tailoring medications for specific patients or groups. Modern manufacturing lines will allow this production of patient centric dosage forms. Miniaturized processes will even increase the capability to follow customer specific needs or to produce small series for clinical studies in an economically acceptable way.

In the following sections all main processed for pharmaceutical solid dosage form manufacturing are described in more detail to provide the reader with an overview of the current state of the art.

2 Powder Feeding

Accurate powder feeders are critical elements in every pharmaceutical process setup, ranging from feeding of a tablet press to the powder delivery to hot-melt extruders. Their purpose is to constantly deliver a certain quantity and quality of matter. Hence, the necessary attributes of feeders are:

- to provide an accurate and constant solid mass flow over extended times, even for small powder flow rates down to a few hundred grams per hour.
- to minimize changes in density, porosity, particle size distribution, composition by avoiding milling, agglomeration, or attrition.
- to eliminate dust development.
- to minimize segregation as much as possible.
- to provide flexibility with regard to different materials, e.g., cohesive and non-cohesive or sticky powders.

A variety of different devices and methods for dosing or feeding exist, working discontinuously or continuously, with or without measurement of mass flow rate. The specific material properties and the process demands determine which devices are suitable for a specific feeding task [2, 3]. A broad overview is given in Fig. 1, and in the following parts the main categories of feeding systems are defined.

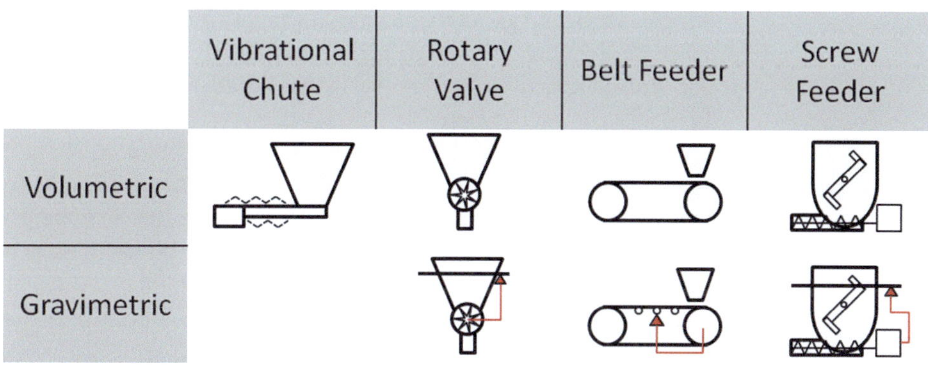

Fig. 1 An overview of important feeder types. The main distinction is between volumetric and gravimetric based feeders

2.1 Volumetric Feeding

Volumetric feeding is based on the feeding of confined volumes and not mass. The measurement of the fed mass is not performed. Thus, typical problems are density changes due to powder compaction and a reproducible fill level of the volume units. For volumetric feeding, a calibration relating the mass flux to the feeder settings has to be developed. Often a linear range around a set-point can be defined by tests. The range, however, cannot be expanded during operations.

Typically, volumetric feeding units are screw feeders, with one or two screws, where a constant delivery rate "of volume" is provided, defined by the rotation rate and the screw geometry. Screw geometries can vary widely and are chosen to suit the product. A coarse classification distinguishes worms, paddle screws, full-blade screws, and double-concave screws. Usually supporting intake elements such as shovels are used in combination. Screws are suitable for cohesive to free flowing materials and a wide mass flow range, which can be controlled by geometry and rotational speed. Screw feeders exhibit a pulsating behavior, which is further influenced by the outlet geometry [4–6].

In contrast to volumetric screw feeders, vibrating chutes and pipes are pure conveying elements without accurate dosing. Further complexities arise for the transport of small particles because of limited transfer of momentum between the device and the particles [7]. Pneumatic transport of fluidized materials can be used to transfer solids from one unit operation to another, when for example the processing vessels are located at the same floor level, or if simultaneous drying is desired.

Belt feeders are suitable for goods sensitive to attrition. Unfortunately, due to adherence to the belt, dust settling and variations in layer thickness belt feeders tend to be inaccurate [7]. Rotary valves are suitable for free flowing granular systems. The critical step here is the correct and reproducible filling of the individual chambers and the complete emptying. Often this has to be supported by additional devices [7]. For hopper discharging, simple devices, as vertical-gates, shaker feeders, and revolving plate feeders are preferred. Other frequently encountered devices are annular grove feeders for very small volumes, and others [7–10].

In batch processing, volumetric feeding systems can be used for filling a container or hopper (e.g., for tableting). To ensure the correct mass the container is then weighed. As the drop height can influence the measured weight, often the last bit of mass is added in a more gentle way (the so-called fine-flow filling). Alternatively, the material container can be weighed before and after a discharging step.

2.2 Gravimetric Feeding

A continuous gravimetric feeding system consists of three main elements [11]:

1. Devices to create a mass flow, as described previously.
2. Devices to measure the mass flow.
3. A control system based on comparison of actual and target mass flow.

Differential balances have become the dominant measurement devices for mass flow. During outflow of a feeder hopper the decrease in mass (dM/dt) is measured and used to control the feeding device, such as the screw rotation speed. As the bulk density changes, the revolution speed has to be adjusted during discharging. When a minimum fill level is reached, refilling takes place. During hopper refilling, volumetric dosing is performed. To avoid long times of uncontrolled mass flow, refilling takes place rapidly or regularly smaller portions are refilled [12, 13].

Apart from balances, indirect measurement methods can be exploited to determine the mass flow. The exerted force of a mass flow can be measured by a paddle-wheel. Radiance absorption can be exploited to measure the passing material volume and density. With optical, acoustical, electrostatic, or capacitance sensors and a proper calibration, the mass flow can be determined [14]. However, these systems are still in a state of development.

2.3 Accuracy of Feeding

The deviations of the dosing rate in kg/h have to be evaluated in comparison to the actual mass flow. Deviations are a result of pulsation (e.g., particle agglomerates falling from the screw outlet instead of individual grains), errors and dynamic behavior of the balance, dosing, discharging and preload, as well as dust settling or vibrations. Deviations in the mass flow rate can be classified according to

- Periodic changes due to pulsation, where target and actual mean are the same.
- Stochastic changes, where target and actual mean are close to each other.
- Target and actual mean differences persist over time.

An example for the mass flow rate of a feeder is given in Fig. 2. Steady state operation after a start-up phase of 100 s is shown.

For characterization of feeders, the empirical mean and standard deviation of a number of samples of powder mass with the same dosing interval are used. Hence, a confidence interval for the mean mass flow can be constructed [4, 10].

Typically, low-dose feeding in the range of less than a kg/h poses problems, especially for cohesive materials. Difficulties arise due to the predominance of electrostatic and cohesive effects,

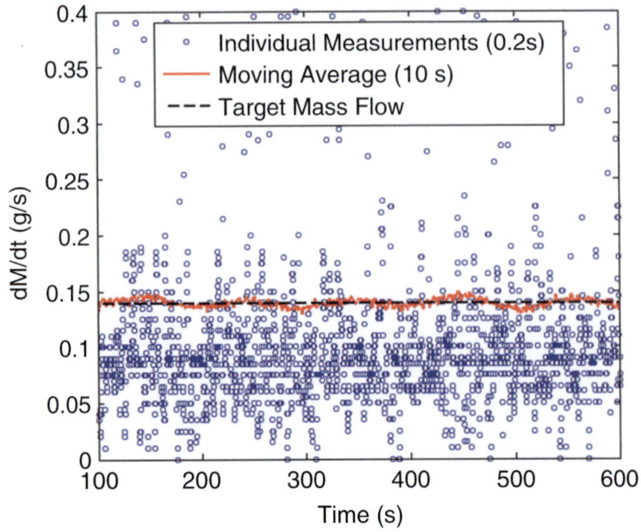

Fig. 2 Mass flow measurements at the outlet of a gravimetric feeder (K-TRON K-SFS-24). Individual measurements, taken at intervals of 0.2 s show strong scattering caused (a) by discretization errors of the balance, (b) powder clumps at the outlet of the feeder

humidity, and variations in particle packing density. These properties and the interaction of powders with the environment complicate the dosing and measurement of powder flow. Vibrational shakers (e.g., Microdose by MG2 based on vibrational dosing and dose measurement via a capacitance system or Capsugel's precision powder micro-dosing systems Xcelodose using the pepper-pot shaker principle) have been successfully developed, yet applications in day-to-day manufacturing are still rare. Moreover, promising results for very small doses have been achieved by the application of acoustic and ultrasonic vibration methods, as well as by electrostatic manipulation, which is used in printing applications already [2].

2.4 Effect on Product Quality

Content uniformity is a critical quality attribute of pharmaceutical products. Accurate feeding is thus highly important, and in fact—according to our experience—most content uniformity problems in a continuous manufacturing environment are caused by feeder deviations and/or problems [15]. Similar observations hold for processes that employ split feeding of the API and excipients. While typically high-frequency variations are dampened by subsequent processing steps, low-frequency fluctuations result in out-of-spec products. Often several feeders are used to dose a number of different components at the same time. Typically, a "master–slave" control scheme is used, where all slave feeders are controlled by a single master feeder. Hence, when changing concentrations of feed rates, all feeders must have the same change characteristics to avoid over- or under-feeding of a component during a certain time interval. This is hard to achieve in practice.

2.5 Outlook

In the past feeding systems often have been used without extensive considerations of accuracy and impact on product quality. Due to an increasing demand for process understanding the awareness of choosing appropriate feeders is also increasing. Especially with the shift to continuous processes, the importance of accurate feeding gains importance. In particular providing constant mass flows at a very low level is challenging. Here, more work and innovative solutions are necessary, especially as there is no reliable method available for directly measuring powder flow rates in a system.

3 Powder Blending

Powder mixing (or blending) is one of the most critical pharmaceutical unit operations and aims at the homogenization of a product, the blending of active ingredients and excipients (fillers, disintegrants, lubricants, etc.) or the distribution of liquids in a powder [16]. Blending is strongly affected by the particle properties (size distribution, density, cohesivity, shape) as well as moisture and electrostatics [17]. Necessarily this leads to different strategies for blending. This is especially true since there are three main effects that compete at every time during the mixing process. These are (1) the mixing of particles by moving particles with respect to each others, (2) segregation due to a wide variety of mechanisms and (3) cohesive effects, including van der Waals, electrostatic and capillary forces, but also mechanical interlocking or stickiness in partially melted systems. These three main effects determine the kinetics of the blending process and understanding these effects is the basis for successful design of blending operations. Moreover, mechanical properties of the materials such as high attrition tendency may limit the use of certain mixers, for example high-shear blenders.

3.1 Mixing Mechanisms

Three fundamental mechanisms are taking place during blending. Their extent is determined by particle mobility and mixer type [18]:

- Diffuse mixing is the random motion of individual particles in the powder bed. In contrast to fluids, where individual molecules diffuse also at rest, diffusion in powders occurs only when the particle assembly flows. High shear rates increase diffusion. Particle transport occurs perpendicular to the flow.
- Convective mixing is the transport of bulk material in a certain direction within the blender. Here, particle transport is in the direction of the flow. Convection alone does not induce mixing. Only when particles of type A are conveyed in a region with type B particles, mixing occurs.
- Shear mixing takes place due to the different velocities of different layers of the flow. Here, particles in close proximity are

removed from each other, leading to mixing. Shear is also critical in breaking apart agglomerates. Through the frictional forces that are occurring due to the different velocities of adjacent layers, random particle motion (i.e., diffusion) is induced, typically measured by the granular temperature.

Segregation causes the formation of spatial concentration differences of different particle types. Segregation may be caused by a variety of different properties including difference of size, density, shape or surface characteristics. Typically, in blending operations particle size differences and cohesivity have the largest impact. For example, for free-flowing (non-cohesive) powders differences in diameter of more than 30 % lead to strong segregation. In contrast, cohesion may reduce the susceptibility of poly-disperse systems to size-induced segregation. However, the addition of even very small quantities of liquid (to enhance cohesion) may drastically reduce segregation. In powder mixing, sieving segregation is the most important mechanism, whereby larger particles move to the surface and smaller particles move to the bottom. The mechanism is complex and many effects are occurring. In hopper filling (but also die filling), sifting segregation (i.e., the concentration of smaller particles in the center and large particles at the outer region) or fluidization segregation for smaller particles (i.e., particles that are entrained in the replaced air) occur. Dusting is the formation of fine dusts during hopper filling.

Thus, choosing the right type of mixer suitable for the task at hand (tendency to segregate and cohesion) can be quite challenging, as demonstrated by ample literature in this field, for example in refs. [19, 20].

3.2 Mixture Characterization

In a stochastic mixture the probability of finding a particle in a sample is proportional to its concentration. This is the best achievable target level for mixing [21]. Although a variety of different indices exist for classifying the homogeneity of a mixture, most are based on the variance of component concentration in between drawn samples of the mixture.

For a number of N drawn samples, with the measured concentration c_i respectively, the empirical variance s^2 is either given by

$$s^2 = \frac{1}{N}\sum_{i=1}^{N}(c_i - \mu)^2,$$

when the target concentration μ is known, or by

$$s^2 = \frac{1}{N-1}\sum_{i=1}^{N}(c_i - \overline{c})^2; \quad \overline{c} = \frac{1}{N}\sum_{i=1}^{N}c_i$$

when the target concentration has to be estimated by the empirical mean $\bar{c}$. Often, the relative standard deviation (RSD), also called Coefficient of Variation (CoV), is given as indicator of mixture quality:

$$\mathrm{RSD} = \mathrm{CoV} = \frac{s}{\bar{c}}$$

Another common index is the Lacey index M_L. The Lacey index has the advantage to be normalized between 0 and 1 and is calculated by

$$M_L = \frac{\sigma_0^2 - s^2}{\sigma_0^2 - \sigma_r^2},$$

with σ_0^2 representing the theoretical variance of the completely separated (i.e., umixed) blend, and σ_r^2 the theoretical variance of the stochastic mixture (hence dependent on particle and sample size) [22].

Furthermore, the obtained empirical variance is the sum of [21]:

- Variance of the mixture (i.e., the mixture's heterogeneity).
- Variance introduced by sampling.
- Variance of the analytical method.

An example for the assessment of mixture quality during blending is shown in Fig. 3.

Practical guidelines on how to correctly sample a mixture should be followed closely to avoid bias [23, 24]. In the experience of the authors, the difficulties associated with correct sampling cannot be overstated, as most sampling methods induce significant bias [25]. Another important issue is the size of the samples. Optimally, the sample size is equal to the "depth of scrutiny." This is the amount of mixture, at which homogeneity has to be ensured. In pharmaceutical applications this is often the size of a single dosage form. For stochastic mixtures a minimum theoretical variance exists for a certain sample size. Consequentially, a large number of samples must be taken to reduce uncertainty, if those are small [26–28].

3.3 Mixer Types

Depending on purpose and materials a suitable blender has to be chosen. An overview of different mixer types is given in Fig. 4. Those can coarsely be classified [16] as:

- Tumble blenders, where the entire outer shell is tumbling or turning, leading to avalanching of the enclosed material. Typical tumble blenders include V-, double cone or IBC-blenders. For further dispersion internal baffles or counter-rotating impellers

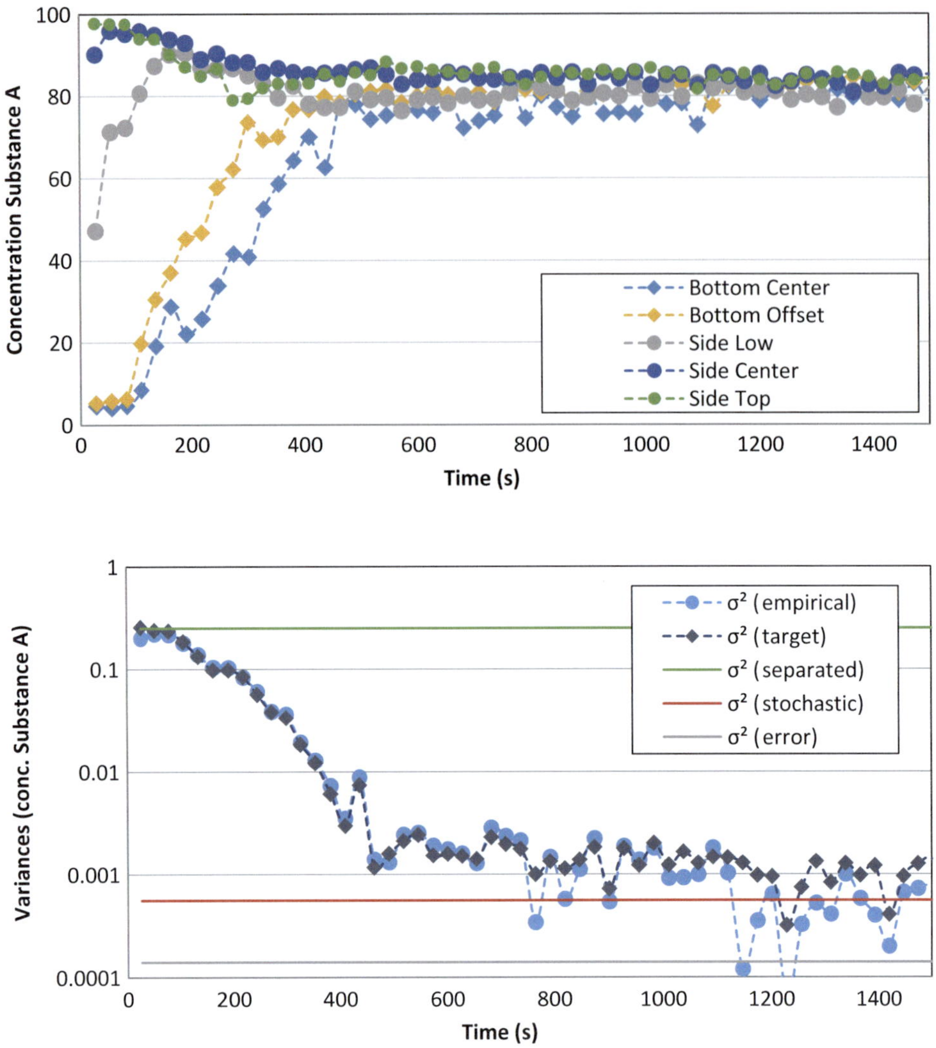

Fig. 3 Batch blending process in an agitated vessel of two substances with the mass ratio 80/20. Samples were taken with NIR-spectroscopy in regular intervals at different positions and the estimated concentration of substance A is plotted above. In the lower graph the calculated variance of all positions compared to the empirical (σ^2 empirical) and target value (σ^2 target) is depicted. Furthermore, the level of the completely separated mixture (σ^2 separated), the stochastic mixture at sample size (σ^2 stochastic), and the level of measurement error (σ^2 error) are shown. It can be seen that the initially segregated mixture is becoming more homogenous over time, until a steady state is reached at roughly 800 s and the end point is reached

can be included [29]. Their effect is limited to breaking agglomerates. For mildly cohesive powders and free-flowing powders with minimal segregation tendency these blenders are ideally suited.

- Agitated mixers use mechanical force, which is applied by impellers, paddles, plows, or ribbons to move the powder bed and induce convective mixing and shear mixing. Depending on the

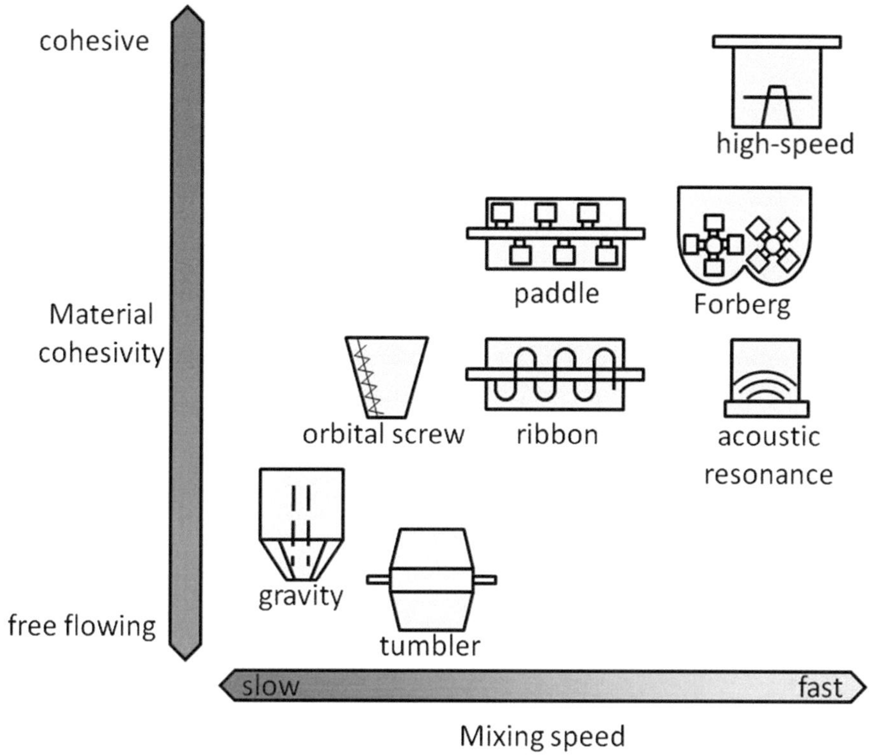

Fig. 4 Different blender types, roughly classified according to their mixing speed and material cohesivity that can be handled. Choosing the correct mixer type is a task requiring quite some experience

tip speed the material is sheared or even fluidized. A large number of different agitator shapes and systems exist, leading to known paddle, plow, Forberg, ribbon, screw, sigma, and Z-bladed mixers. A characterization of those mixers is often done according to the Froude number, which describes the ratio of centrifugal and gravitational force [16]. This, however, is not a scientifically based criterion for blender scale-up. For segregating and strongly cohesive systems, these blenders are suited as high levels of shear can be imparted on the material.

Further types are gravitational blenders, which can blend huge amounts of materials by controlled discharging of a hopper, and pneumatic blenders, which are based on aeration and mixing of the powder bed by air streams.

High intensity and high shear mixers stress the material by different means, thus leading to grinding and destruction of agglomerates [7, 16].

Resonant acoustic mixers apply a low-frequency high-intensity acoustic field, leading to fluidization and dispersion of the material in the blender, causing mixing at the micro-scale throughout the volume, without the need for any moving elements [30].

3.4 Batch vs. Continuous Blending

Another distinction of mixers can be done with respect to batch wise or continuous operation mode. For batch mixers the geometry and size of the blending vessel, and the operating conditions and mixing time are affecting the mixing quality. Nearly any type of powder can be processed with an adequate mixer; however, low quantities might be deposited on the mixer walls and be lost. Moreover, segregation is unavoidable during discharge [16].

In continuous mixers the goods are mixed in a single pass through the system (typically agitated by a screw or plows). Mixing is achieved in axial and radial direction to a different extent, while axial mixing is usually less effective. Hence, controlled and precise feeding is much more important for good mixing quality. Continuous mixers can be very compact and have a small hold-up time, enabling plant designers to place them close to subsequent unit operations, thus minimizing segregation. However, the complete system of feeders and mixers cannot be readily used for other materials [11, 12]. A continuous blender can be characterized by its residence time distribution, describing the average residence time and variance of particles regarding uniformity in the mixer. Axial mixing will dampen the effect of feed fluctuations. Time periods at the input larger than the residence time in the mixer cannot be dampened. A larger residence time might be undesirable as well, because of attrition or drying of particles inside the blender [31, 32].

3.5 Scaling

Blenders are notoriously hard to scale-up. In tumble blenders, typically only a layer of material is flowing, whose thickness is determined by the particle properties. Upon scale-up the thickness does not change, and thus for larger systems the ratio of flowing to stagnant material decreases. For agitated mixers the spatial distribution of shear in the small- and large-scale system is very different (as it is in fluid stirred tanks). Additionally, the Froude number scaling ($Fr = n^2 \cdot d/g$) leads to different shear level, as the maximum velocity at the agitator tip scales with $\sqrt{d}$ and thus the shear rate with $1/\sqrt{d}$. Thus, large blenders at the same Froude number impart less shear than the small-scale system. In summary, scale-up of powder blenders is far from trivial (and to a large extent empirical) and most large-scale systems must be thoroughly tested before deployment [33]. Recent progress in the simulation of blenders via the Discrete Element Method has provided design engineers with a tool for understanding the impact of scale on blending performance [34].

3.6 Outlook

Simulation is a valuable tool for understanding the behavior and flow patterns of material during blending [34–36]. Recent trends suggest that simulation will continue to provide clearer and more accurate representations of the mixing process. This is further enhanced by the implementation of different PAT-tools, providing experimental insight into mixing and segregation mechanisms [26, 37, 38]. Nevertheless, significant progress in powder blending, especially with respect to low-throughput continuous mixing would be desirable from a design engineers point of view.

4 Granulation

Agglomeration is a size enlargement process to form solid and dry agglomerates (granules) with controlled properties, typically ranging between 0.1 and 1 mm. Agglomeration is widely used in different industries, including the consumer-product, chemical and food industries. This chapter focuses on the pharmaceutical application. Agglomeration in the pharmaceutical industry is widely used and only few processes are based on direct compression, i.e., direct tableting of API and excipient without an agglomeration step.

The objectives of agglomeration are manifold: First, the main goal is to ensure content uniformity (CU) of the product, containing, in most of the cases, small amounts of active pharmaceutical ingredient (API) and a larger amount of excipients. Agglomeration inhibits further segregation of the mixture, thus ensuring a uniform concentration in the final product. Second, agglomeration improves the flowability of material, thus enhancing dosage-accuracy. Third, the compactibility of the material can be enhanced. Other objectives include the minimization of dust, the increase in density or the reduction of problems associated with moisture uptake and caking. Granules should be non-friable and should exhibit a suitable mechanical strength for further processing. Moreover, the porous structure of granules increases their wettability and disintegration. The agglomerated material can be used as pharmaceutical end-product, such as multi particulate dosage forms in a single dose (sachets and vials) or multi-dose preparations, or as pharmaceutical intermediate-product for further processing, e.g., compaction into tablets or filling in capsules.

In pharmaceutical manufacturing, two main methods are used for granulation, i.e., wet and dry granulation, where the first method is dominating manufacturing.

4.1 Wet Granulation and System Types

Wet granulation is the process of agglomerating fine powders by spreading a liquid binder solution into the bed of particles, using manufacturing devices such as shear granulators (with wet-massing and granulator device), high-shear granulators, fluid-bed granulators, extruders, tumbling drums or other, etc.. Today, the most commonly used manufacturing processes for batch-wise and semi-continuous granulation are high-shear and fluid-bed granulation (Fig. 5a, b). Nevertheless, twin-screw extruders have gained

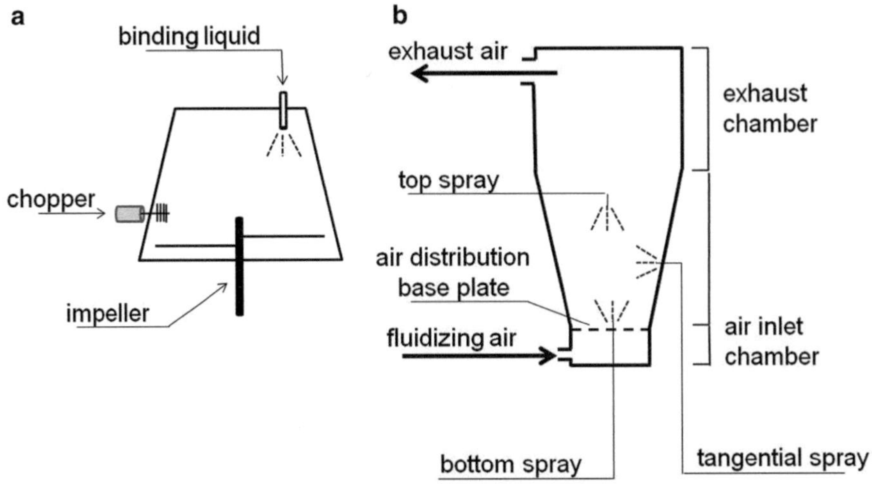

Fig. 5 Schematic of the (**a**) high shear granulator, (**b**) fluid bed granulator

interest over the last years for the ability to be integrated in a continuous manufacturing setup.

High-shear granulation: A high-shear granulator consists of a mixing container, in which the unmixed feed material is placed and mixed using rotating impeller. The liquid binder can be added to the feed material either by spraying or by pouring, and the wet mass is agitated vigorously by impeller. Impellers rotate at high speeds (up to 1500 rpm) on either a vertical or horizontal axis to create the agitation required for the granulation and compaction of the wet granules. Typically, a secondary smaller impeller, called chopper, is used with higher speeds to fracture larger agglomerates, which rise to the surface of the powder bed. After granulation, the product is discharged, passed through a mesh and dried. Granules produces with high-shear granulators are rather round high-density products with low porosity and a wide granule size distribution, compared to fluid bed granulation. Batch sizes range from small-scale lab granulators in the liter range to large-scale systems with a few thousand liters of processing volume [39–42].

Fluid-bed granulation: Here, the feed material is placed into a processing chamber and fluidized by a controlled airflow of conditioned preprocessed air. As can be seen in Fig. 5b, the main parts are: air inlet chamber, air distribution base plate, process chamber, spraying system, and exhaust chamber including the filters. In fluid-bed (FB) technology, processing methods are mainly characterized by the spraying direction and the design principle of fluidization air distributed into the processing chamber. Different processing methods are: top spray, bottom spray, Wurster and rotor systems (Fig. 5b). Wurster systems contain an internal tube which is designed to control the circulation of granules in the system. Technical solutions on the market have been developed to carry out different processes, such as granulation, pelletization, drying, coating, and layering [43]. Air flow rates, process temperature, fill level of the container, spray rate, and droplet size of atomized binder are critical parameters, affecting the granulation process and the final granule properties.

Fluid-bed granulation offers the advantages of a good control of process temperature. Moreover, the process combines mixing, granulating and drying within a single piece of equipment, shortening the manufacturing time and reducing cost. Nevertheless, additional costs associated with air processing and handling and filter change and cleaning arise. Moreover, scale-up of FB granulators is not trivial.

Granules produced via FB technology are less dense than the high-shear counter-parts and granules are not as rounded due to the lack of intense collision with agitators. Granule size distribution may be narrower if operated correctly. Batch sizes range from a few grams to a few tons of material, depending on the scale [39, 41, 43, 44]

Twin-screw granulation: Due to the inherent benefits of continuous manufacturing (see below) extrusion by using single or twin-screw extruders has become of increasing interest over the last years. Extruders are used to produce granules without a die-plate, i.e., granules drop from the exit of the screw section into a container and are dried in a subsequent drying step, typically in a (sectored semi-continuous) fluid-bed. The critical process and material parameters and their impact on the twin-screw granulation product quality have been studied by several groups [45–48]. Tu et al. also developed a regime map for continuous twin-screw granulation [48].

4.2 Mechanism of Granule Growth in Wet Granulation

Wet granulation is based on three kinetic processes: (1) wetting and nucleation, (2) granule growth and compaction and (3) breakage and attrition. All processes occur simultaneous, with varying degrees. Wetting and nucleation are defined as formation of the initial granule nuclei when liquid binder is added and spread into the powder bed. The binder consists of a solvent (typically water, or sometimes ethanol and isopropanol) and a polymeric component that upon drying creates solid bridges and that increases the viscosity of the binder. Typical binders are PVP, starch, gum, or modified celluloses, but also sugar solutions are used. The nuclei formation kinetics are quantified by liquid penetration time (the time required for a single drop to penetrate into the powder surface), which depend on both wetting thermodynamics and wetting kinetics. Wetting thermodynamics are affected by the fluid surface tension and the powder/liquid contact angle. The liquid viscosity and effective pore size of the powder bed have strong impact on the wetting kinetics. After nucleation, consolidation and growth occur as the nuclei are agitated and collide with each other. These collisions densify the granules, and the granulation liquid is increasingly filling the void spaces between primary particles, thus increasing granule mechanical strength.

Two main forms of granule growth are defined as *steady growth*, in which the rate of growth is approximately constant, and, *induction phase*, in which the granules consolidate for a long period of time without any granule growth. In the latter case a rapid growth can be observed after sufficient consolidation. Depending on the material properties and the granulation process parameters, other granulation types can be observed as well, for instance *nucleation only*, where nuclei form during adding the binder but no further growth occurs, *crumb behavior*, where the formulation is too weak to form permanent granules, and *overwetting*, by excess binder and an oversaturated powder bed. Breakage and attrition is the step when granules grow too large or dry, resulting in a weak and brittle product [49–54].

4.3 Dry Granulation

Dry granulation is performed either by a process where large tablets (slugs) are produced in a heavy-duty tablet press or by a roller compactor between two counter-rotating rolls to produce a continuous sheet, the ribbon. In both cases materials are compressed to form binding surfaces (by plastic deformation and/or brittle fracture).

Roller compaction is more common. Here, the ribbon is milled and sieved afterwards to produce the granules of a desired size spectrum. The method offers several advantages. There is no need of wetting (solvent free) and drying steps, making the process suitable for materials sensitive to moisture. Moreover, roll compaction is applicable for a wide range of materials and is easily scaled-up. Moreover, the method can be easily included in a continuous manufacturing environment. Nevertheless, granules produced via dry granulation typically have a worse compactibility than there wet-granulated counterparts.

Roller compactors consist of some basic elements with differences in the configuration (Fig. 6). The rolls may be mounted in a horizontal, vertical, or inclined position and the material may be moved either by gravity or by screw feeder, depending on the feeding system. Variations in the screw feeder's feeding rate may occur and can be responsible for axial density variations of the ribbon. The rolls are available in different geometries such as smooth, fluted or pocket design. The mechanism of the compaction is based on (a) solid bridges, (b) intermolecular forces such as van der Waals forces, electrostatic forces, and hydrogen bonding, and (c) interlocking forces, depending on the compressibility, deformation, and fragmentation behavior of used material [55–57].

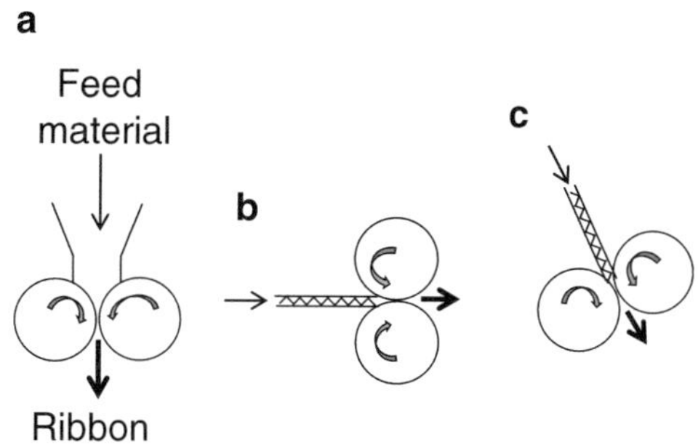

Fig. 6 Schematic of roll compactors (**a**) horizontal, (**b**) vertical, (**c**) inclined position of rolls

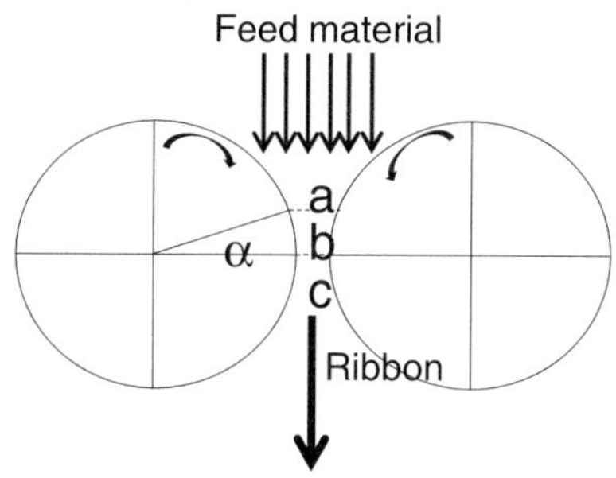

Fig. 7 Three regions of roller compaction process: (**a**) slip region, (**b**) nip region, (**c**) release region

The compaction process consists of three regions: the slip region, the nip region, and the release region (Fig. 7). In the slip region, due to faster velocity of the roll surface than the material, the powder slips at the roll surface and flows into the roll gap. Particle rearrangement and de-aeration occur in this region and a small degree of densification is observed. The behavior of particles in this region depends on the wall friction and inter-particle friction of the material. The nip region starts at the nip angle α, which defines the angular location of the onset of the nip region (powder and roll-surface velocity match) and specifies the transition from slip to no-slip wall boundary conditions. The powder is gripped by the rollers and undergoes compaction due to the decreasing roll gap and the friction along the surfaces of the rolls. The maximum compression pressure and nip angle are two important parameters and define the extent of powder densification. The nip region corresponds to a small portion of the roll, which is often less than 10°, depending on the material characteristics and process parameters [58, 59]. The release is started when the roll gap starts to increase again. The size of the release region depends on the elastic behavior of the ribbon and the roll speed. After release the ribbon may increases in size due to the elastic recovery. Generally, the maximum pressure applied in the nip increases with increasing nip angle, decreasing roll gap and decreasing roll speed.

The major challenges of roller compaction are the large amount of fine particle fraction after processing and the loss in compactibility, which is possibly due to a work-hardening phenomenon. Work hardening occurs predominantly for plastically deforming materials but it is also observed in processing of brittle and fragmenting materials including inorganic materials [55–57, 60–63]. A mechanistic understanding of the material behavior, especially at the exit of the screw and in the nip region is necessary to optimize the ribbon strength and homogeneity and to minimize the fine particle fraction.

4.4 Scaling

Due to the fact that roller compaction is inherently a continuous manufacturing, the scale-up is not considered a highly critical step. The science-based scale-up of wet granulation is, however, more challenging, both for high-shear and FB granulation. One of the earliest approaches for the scale-up of high-shear granulation is the establishment of a dimensionless relationship between the power consumption and the wet mass consistency. Having this, the relationship becomes general for geometrically similar high-shear mixers. The power consumption of the granulation end-point is the value that corresponds to the optimal wet mass consistency. For scale-up, the dimensionless relationship can be applied again to determine the target power consumption at the new scale [64]. Another approach is the development of nucleation regime maps based on the dimensionless spray flux and drop penetration time for given formulation. Scale-up can then be undertaken by defining the optimal design space as a function of dimensionless numbers and keeping these numbers constant [65]. Scale-up of fluid-bed granulators is mostly based on mass and heat balances and on the bed moisture content and droplet size measurements [66–68]. Recent progress in the simulation of wet granulation via CFD-Discrete Element Method (DEM) has, however, provided a powerful tool for understanding the impact of scale on the granulation process [9, 69, 70].

4.5 Outlook

Increasing the understanding of the granulation process and the granule growth mechanism is essential for ensuring the product quality during the manufacturing process. Clearly, also equipment-scale processes, such as fluidization, spray formation, mixing, and agitation have to be included in the analysis.

For understanding and controlling the wet granulation, different regime maps have been developed, starting with Iveson and Litster and Hapgood et al. [49–51]. These regime maps, together with modeling and simulation, have been further improved for understanding and prediction of granulation behavior and have been adapted to other granule manufacturing processes [2, 9, 14, 70–79]. Further developments will include a broader range of materials and processing conditions. Several models have been also developed for the prediction of material behavior during the dry compaction. The most important models are Johnson's model as well as finite element methods. Artificial neural networks and genetic algorithms have been used as well [80–85].

Current efforts include the use of combined DEM and FEM models to understand ribbon formation in roller compaction and CFD-DEM in fluid bed modeling. In addition, different analytical methods have been developed for monitoring the process and the end-point of wet granulation. End point monitoring is of special importance for manufacturing the product with predefined quality. In the case of high-shear granulation, monitoring of the impeller amperage, power consumption, torque, or indirect torque is common [64]. Novel application of spectroscopic techniques (particularly near infrared and Raman spectroscopy) can improve the end point prediction. The popularity of these methods is based on the versatility, high measurement speeds and their non-destructive nature. NIR displays a cross-sensitivity to chemical sample properties which is quantitatively and qualitatively interpretable [44, 86]. However, the interpretation of spectra is complicated, and detailed signal pretreatment is often necessary to remove irrelevant spectra. Raman spectroscopy has the drawback of the occurrence of fluorescence [87, 88]. Real time particle size measurements can also be applied, e.g., by focused beam reflectance measurements (FBRM) or spatial filter velocimetry (SFV) (using Parsum probe). Both methods are suitable for at-line monitoring. However, window fouling can cause problems during the in-line measurements [44, 89, 90]. Thus, in the field of on-line particle size distribution monitoring progress would be desirable.

Another powerful tool is acoustic emission (AE), which can be used for gathering information from the vibrations due to the particle–particle or particle–wall collisions and frictions during wet granulation. These vibrations are measured by applying AE sensors to the manufacturing equipment. The resulting acoustic spectra contain information about the granulation process. For prediction of this information, multivariate principal component analysis (PCA) and partial least squares (PLS) models are required. However, the method is promising because it is noninvasive and cheap [44, 91, 92]. The described methods are widely used for the monitoring of fluid-bed granulation. However, their application for high shear granulation requires more development.

5 Drying

Reduction of product moisture or solvent content below a certain limit is a crucial process step during pharmaceutical manufacturing. Reasons are [93]:

- To obtain free-flowing material to facilitate transport, dosing, or packaging.
- To enable subsequent manufacturing steps, e.g., direct compaction.
- To remove toxic solvents and to ensure minimal impurity levels in the final drug product.
- To create a certain morphological structure, e.g., an amorphous material.
- To guarantee long-term stability by prevention of hydrolysis or other reactions.
- To extend shelf-life by prevention of mildew and bacterial growth.

Drying usually follows steps such as crystallization/filtration/washing, wet granulation, or wet extrusion, but is also used to create particles of a desired morphological structure (spray drying). Protein solutions are dried as well, typically via freeze-drying.

5.1 Drying Fundamentals

Drying of material is a complex process, its intricacies often being underestimated. For example, during drying of a porous material, several processes are occurring at the same time. Consider a porous pellet: First heat and mass-transfer occur at the external gas-solid boundary layer of the particle. Second, in the internal pore space liquid and gas phases exists that exhibit complex transport phenomena of the gaseous components, liquid phase and heat. Moreover, a local equilibrium between the liquid phase and the gas phase exists, which in the case of free liquid water is described by the vapor pressure, but at low moisture contents becomes an adsorption isotherm. Thus, the temperature and humidity of the drying air has a large impact on the process.

All these effects can under certain conditions become rate-limiting, leading to "drying regimes." The most common regimes include the "constant-rate period" at high liquid content, where the drying rate of a porous pellet stays constant due to limitation by the external heat transfer. In this case the material is rather cool, maintaining the so-called "wet-bulb temperature." Once the transfer of liquid to the surface of the pellet becomes rate limiting, the "falling-rate period" is observed at low moisture contents, with a constant reduction of the drying rate. Inside the pellet strong temperature and concentration gradients develop until the material is dry. Note, however, that in addition to heat and mass transport also a redistribution of the API in the pellet or swelling/shrinking behavior of the porous substance may occur. Drying of non-porous substances involves only a constant-rate period. During drying also precipitation and/or crystallization of dissolved materials may occur, e.g., during spray drying. In the latter process, removal of solvent is typically fast, such that amorphous materials are obtained.

In general, drying, implying heating and evaporation (or vacuum generation), is highly energy consuming. Moreover, wet powder needs to be agitated in some form, to increase heat and mass transfer. This, however, may lead to agglomeration and/or attrition. Therefore, choosing the most efficient drying system in accordance to product requirements and integration into previous and subsequent manufacturing steps is an important task.

5.2 Dryer Types

For pharmaceutical production a large number of dryers has been developed, both batch and continuously operated. Reasons for the variety in drying equipment are [93]:

- Large variations in drying kinetics: kinetics strongly depend on system properties, such as type of moisture-material bonds (e.g., free, adsorbed vs. inclusion in the crystal lattice), particle size and shape, porosity, and pore size [94].
- Large variations in product requirements: specific dryer types or modes are required to achieve desired product structures, forms or shapes, or must accommodate sensitive materials or sticky powders.
- Large variety of materials: As suggested in ref. [94] pharmaceutical products can be roughly classified into moist, free-flowing granular materials and powders, paste-like materials and liquids (solutions and suspensions). Thus, materials range from filter and centrifuge cakes to granules, single moist particle assemblies, pastes, solutions, or suspensions.

A classification of dryer types can be made according to the underlying mechanism of heat transfer. Accordingly, three major classes exist, including *conductive*, *convective*, and *radiative* drying. However, in practice always combinations of several heat transfer mechanisms within one drying equipment are common [93]. Also widespread are filter-bed dryers, allowing for initial filtration and washing steps prior to thermal drying. That is, mechanical dewatering and thermal drying can be realized in one single device [95, 96].

5.2.1 Convective Dryers

In convective dryers heat is supplied via a gaseous phase. The gas also serves as carrier for the evaporated liquid. If inert drying gases are used or valuable solvents need to be recovered, closed cycle systems for the gas circulation are employed [94]. Most common dryers of this class are spray dryers, (spin-)flash dryers, fluidized bed dryers, or belt dryers.

- *Spray dryers*: In spray drying, liquids or suspension are dispersed into small droplets by means of an atomizer. This is the main challenge of spray drying, i.e., to generate a homogenous spray with a narrow droplet-size distribution even for suspensions with higher solids content. The droplets are subsequently dried in a hot air stream while falling down a spray tower. The large surface area of the formed droplets enables rapid solvent evaporation (30–160 kg/h m^2 of particle surface area). In the co-current drying mode, where hot air is injected close to the atomizer, overheating of the drug product is avoided, as the heat of the drying gas is consumed by evaporation [94]. Spray drying enables a strong influence on the final physical morphology of the drug product, e.g., by creation of amorphous phases, or certain particle sizes and shapes (spherical, hollow, etc.). That is, drug stability and release characteristics can be designed by choosing specific process parameters, such as drying gas flow rate, temperature, or droplet size [97–99]. A schematic is shown in Fig. 8. Thus, spray drying is more often used as a tool for particle engineering rather than a drying technology.

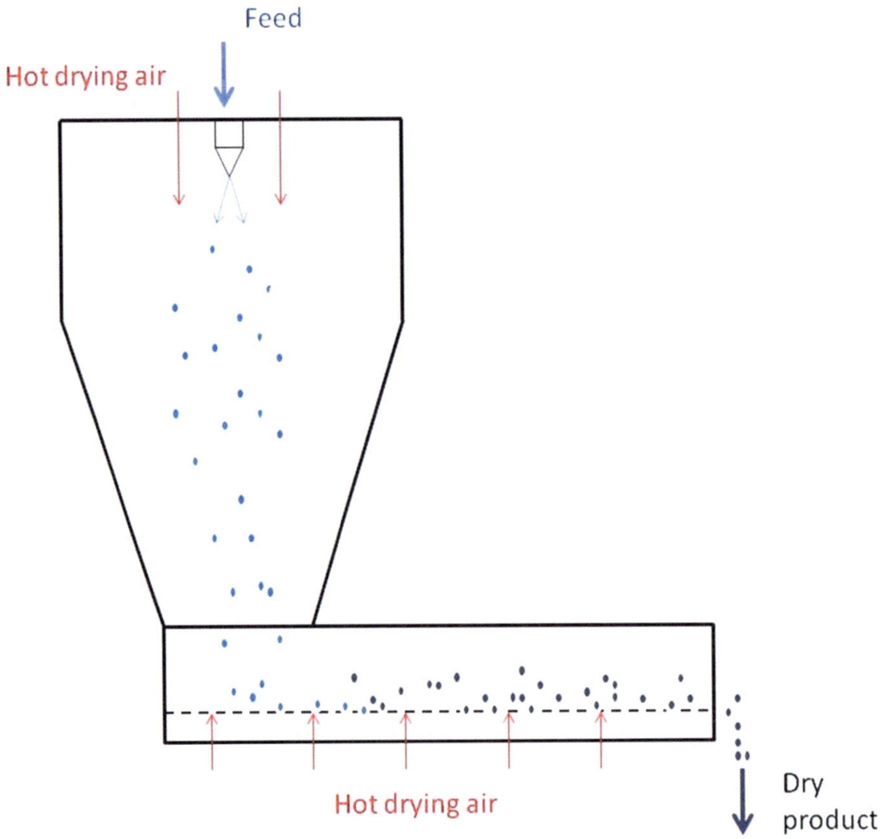

Fig. 8 Schematic of a spray dryer with integrated fluid bed

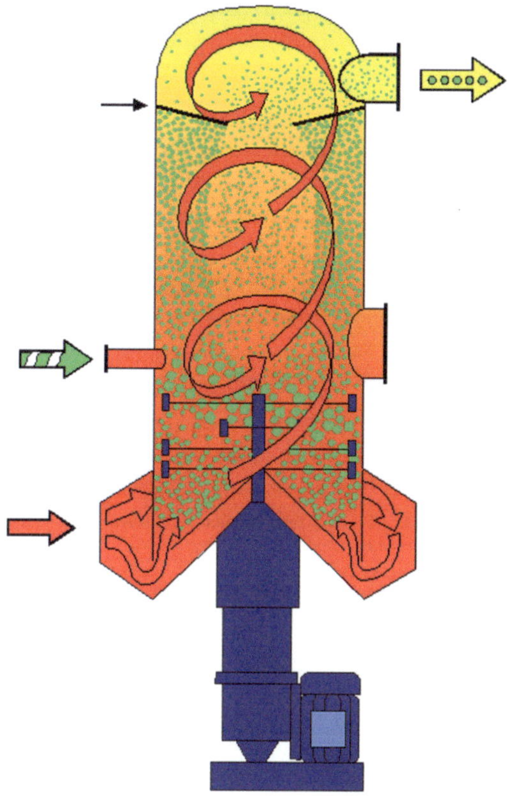

Fig. 9 Schematic of a spin flash dryer (courtesy of Anhydro SPX Flow Technology, Denmark)

- *Flash dryers*: In flash drying easy to disperse, free-flowing material is introduced into an upward-flowing drying air stream. The drying air stream also acts as transport medium for the product. This co-current drying mode and very high evaporation rates (200–1200 kg/h m^2) prevent the materials from overheating [94]. A variation of this dryer type is a spin-flash dryer that also allowing pasty material to be dried by providing intense agitation by blades at the bottom of the dryer (*see* Fig. 9). Note, however, that particles must be removed from the drying air stream, typically via cyclones. Thus, some particles, mostly smaller ones, are lost in the process. Flash dryers can be operated continuously.
- *Fluidized-bed dryers*: In fluidized-bed dryers materials are dispersed and dried in an air flow creating a fluidized particle bed with liquid-like properties. In contrast to flash dryers the particle bed is much denser and the system is operated below the particle-entrainment gas velocity, i.e., particles remain in the bed. Heat and mass transfer, as well as mixing characteristics, are very good. This dryer type can process fluidize-able granular

materials with mean sizes ranging roughly from 10 μm to 1 cm [100], with particles ranging from 25 to 1000 μm being best suited for smooth fluidization [100] based on Geldart's classification for A and B powders [101, 102]. A variety of this process is fluidized bed granulation and fluidized bed spray coating, where in addition to drying coating and granulation are performed. Fluidized-bed drying is a batch process. However, segmented fluid beds have been developed where particles move (or are moved) through segments, such as the GEA and Glatt systems, emulating a continuous process.

- *Belt dryers*: In belt dryers the product is as evenly spread as possible on a permeable belt conveyer and continuously moved through a drying chamber where hot air dries the product on the belt [103]. Radiation drying elements (IR) can be implemented to enhance the drying effect. Problems are associated with agglomeration due to the lack of shear forces acting on the agglomerates. In addition, cleaning and validation are complex. Belt dryers are continuous systems. A schematic of a belt dryer with multiple conveyer belts for large throughput and a long residence time is shown in Fig. 10.

5.2.2 Conductive Dryers

In these dryers, heat is supplied via contact drying. An additional gas stream may serve as carrier for the evaporated phases. Examples for conductive dryers are drum dryers, vacuum dryers, freeze dryers, which are typically operated in batch mode:

- *Drum dryers*: In rotating drum driers usually liquids and pastes are dried, forming a layer on the dryer walls. Blades are used to scrape off the dried product. Residence time distribution, as well as energy consumption are kept low [104]. This dryer type is

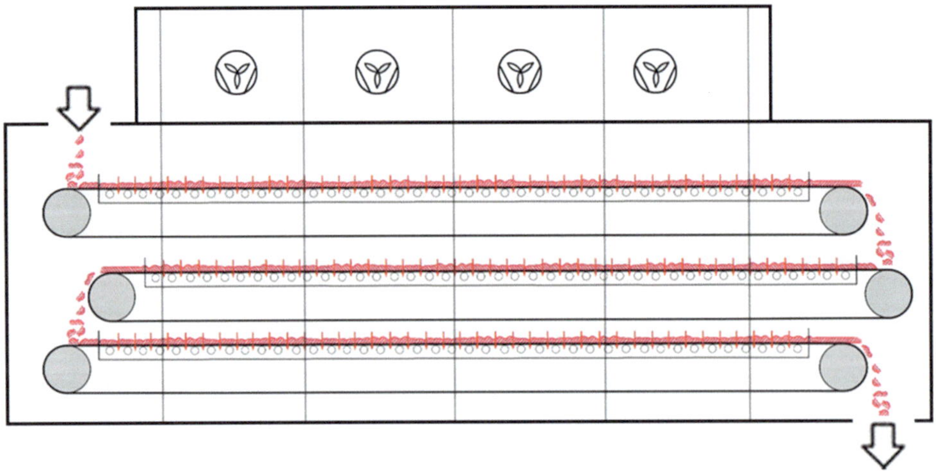

Fig. 10 Schematic of a belt dryer with multiple conveyor belts for large throughputs and long residence times

often used for sticky materials that would be difficult to dry using other process technologies [94].

- *Filter(-bed) dryers*: In filter dryers the slurry (typically washed crystals) is charged and then, filtration takes place, followed by contact drying. Thus, the following steps are included: slurry charging, pressurized filtration, controlled heating and cooling, cake smoothing and drying, discharging and sampling, Clean-In-Place (CIP). A filter dryer consists of a main insulated jacketed vessel and filtration base which can be lowered for inspection and replacement of the filter media.
- *Vacuum dryers*: These driers allow solvent removal at relatively low temperatures due to decreased ambient pressure. They are widely used to dry thermolabile or easily oxidizing products [94].
- *Freeze dryers*: Freeze drying (lyophilization) is a multistage process. In a first phase low process temperatures lead to crystallization of the solvent [105] below its triple point, the lowest temperature at which the solid and liquid phases of the material can coexist. This ensures that sublimation rather than melting will occur in the following steps. Then, during the primary drying phase, the pressure is lowered (to the range of a few millibars) and enough heat is supplied for the solvent to sublime [106]. Due to low drying temperatures and the absence of air, product damage caused by oxidation and chemical modification is avoided. A possible secondary drying phase aims to remove adsorbed water molecules. This part of the freeze-drying process is governed by the material's adsorption isotherms. In this phase, the temperature is raised higher than in the primary drying phase, and can even be above 0 °C. The original properties of pharmaceutical and biopharmaceutical products are preserved. Furthermore, freeze-dried products often exhibit extended long-term storage stability [107–110].

5.2.3 Radiative Drying

Energy is transported directly to the material via radiation. Heat is generated simultaneously at the surface, as well as inside the material. The following systems are commonly used:

- *Microwave dryers*: Applicability of microwave driers strongly depends on the materials dielectric properties [111]. The most important factors affecting the drying rate are material properties and the wetness of the material, i.e., wet areas experience greater heating. This yields almost uniform moisture levels within the final drug products [112]. However, due to volumetric heating temperature control in the product is difficult, in contrast to hot-air drying where the drying temperatures cannot exceed the drying air temperature. Rapid local heat and mass transport can lead to deterioration of the final product quality. To overcome nonuniform temperature distributions and due to

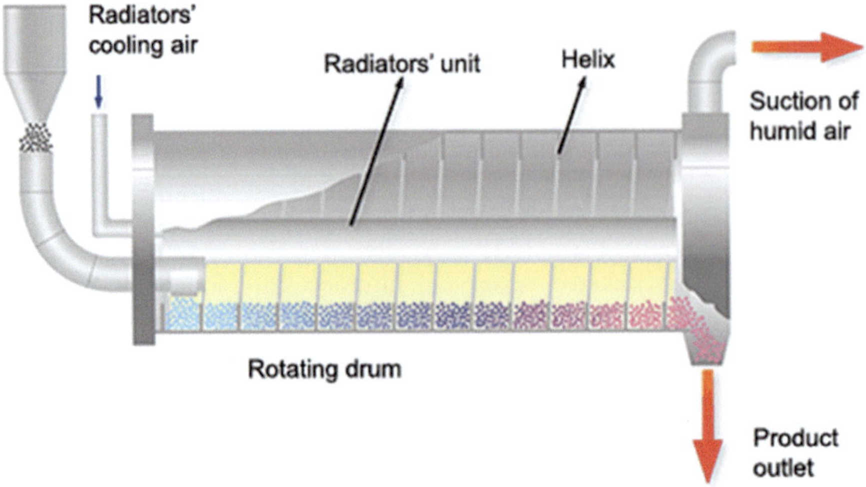

Fig. 11 Schematic of an Infrared rotary drum process (courtesy of Sunwell Global LTD, China)

a limited penetration depth microwave drying is often combined with other conventional dryer types, i.e., microwave assisted air drying, microwave assisted freeze drying etc. [113].

- *Infrared dryers*: Infrared dryers have proven to be very energy efficient. However, radiation ranges should be adapted to the material's maximum absorption ranges [114]. External agitation of the material is often needed due to the limited penetration depth of infrared radiation. Yet, the simplicity of infrared equipment allows easy integration in convective, conductive or other radiation based systems. The high versatility of the equipment, as well as fast transient responses, make infrared dryers interesting for pharmaceutical production [115]. A schematic is shown in Fig. 11.

5.3 Scaling

During the early stages of drug development material batches, e.g., for clinical trials are very small, ranging from several hundred milligrams to a few kilos. During later stages, however, production volumes of a batch may be up to a 1000 times larger [116]. Scale-up considerations are thus highly important when transferring a process developed on a lab- or pilot-plant-scale to the final large commercial scale production line. Equipment selection, process parameters, process conditions and process control strategies have to be carefully planned in order to avoid changes in product quality at varying scales. To avoid cost intensive experiments on full-scale equipment scientific-based scaling principles need to be applied [117]. Essentially, dimensionless heat and mass transfer rates have to be similar together with the mixing properties. Models range from global heat and mass balances to detailed incremental methods, tracking local conditions of solids and gases throughout the drying process. Mechanistic, first-principle based models allow exact predictions over a wide range of process conditions. However, detailed rigorous models often demand a large number of parameters that difficult to obtain. Therefore, lumped parameter models and simple correlations are often used as an alternative [118]. To identify relevant correlations dimensional analysis can be a useful tool. Keeping the process relevant dimensionless numbers constant largely simplifies the scale-up process ensuring geometric, kinematic, dynamic and thermal similarity [117].

5.4 Outlook

With continuous manufacturing increasingly used in pharmaceutical production, continuous dryers are of significant interest. However, few systems exist that can be used for low-throughput materials in the order to 1–50 kg/h. Moreover, dryers typically lead to agglomeration and/or attrition of materials which is typically undesirable [119, 120]. Moreover, for smaller particles (<30 μm) fluid bed drying is not an option. Such materials must thus be dried using other systems, such as belt driers, which may lead to caking. Thus, the implementation of novel and continuous low-throughout drying technology would be of significant interest to pharmaceutical manufacturing.

For automated continuous production not only new equipment and extensive process understanding are necessary, but also process analytical tools (PAT) for drying end-point control. With residual solvent or moisture contents being a crucial quality attribute in downstream manufacturing on-line tools for measuring the solvent content are very important. As an example of recent developments in this field electrical capacitance tomography used to survey moisture distributions inside fluidized bed reactors can be mentioned [121, 122] as well as spectroscopic methods (NIR).

6 Hot Melt Extrusion and Injection Molding

Established in the polymer and food industry for decades, hot melt extrusion (HME) attracted increasing interest in the pharmaceutical industry during recent years due to its capability for the production of performance-enhanced materials, such as solid solutions and dispersions. These materials offer the potential to enhance the bioavailability of poorly soluble drugs, a growing challenge in formulation development. Moreover, complexly structured materials with tailored release characteristics can be produced. HME is a solvent-free process. Beyond that, possible objectives of HME are the manufacture of controlled release systems ([123]) or the incorporation of nanoparticles in a solid matrix [124]. Moreover, HME can be easily integrated in a continuous manufacturing environment to achieve high productivity and a constant product quality in an efficient way. For more details about goals and applications of HME we refer to the literature (e.g., [125–129]).

6.1 Extrusion Basics

Extrusion is defined as forcing a material through a die. In HME, first the granular feed material has to be molten and mixed prior to the actual extrusion. To achieve the required melting (often called plastification) of the granular material, screw machines have been established in HME to exert high shear forces and the corresponding energy input to the processed material. Barrel heating provides only a small fraction of the energy required for melting. Polymer melts and other matrix materials used for HME formulations are typically highly viscous (in the range of 100–10,000 Pas). Thus, mixing occurs in the laminar flow regime, which requires more complex mixing equipment than in the turbulent flow regime, where mixing down to the molecular scale is intensively supported by turbulent eddies. Screw machines provide the required mixing capabilities for highly viscous materials, and, depending on the actual screw geometry, also the possibilities of kneading and pressure generation.

6.2 Screw Configurations

Different types of screw extruders exist and can be classified based on the number of screws into single-screw, twin-screw, and multi-screw extruders. The single screw is the simplest possible design of a screw extruder, which typically provides sufficient conditions for melting, however, yields a poor mixing performance compared to the more complex types. Thus, single screws are used as melting devices for injection molding rather than for HME processes, where appropriate mixing capabilities are typically required. Twin screws can be co-rotating or counter-rotating, as well as intermeshing or non-intermeshing (intermeshing means that the screws almost touch each other). For pharmaceutical HME typically the co-rotating intermeshing twin-screw design (Fig. 12) is preferred due to its self-cleaning screw profile and its excellent mixing capabilities. The more complex multi-screw designs are rarely used. Only for specific applications with extreme devolatilization requirements multi-screws can achieve a higher specific surface than single and twin-screws. For more details about the basic extruder types refer the literature (e.g., [126, 130–132]).

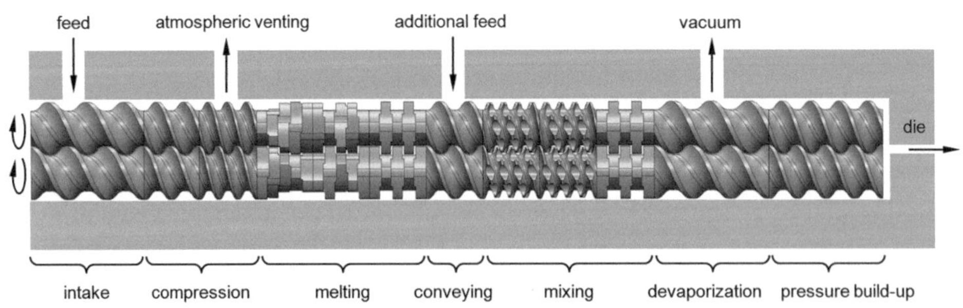

Fig. 12 Schematic of the co-rotating, intermeshing twin-screw extruder and an example arrangement of different process zones

The HME process comprises a number of subsequent process steps, performed in a single device. Depending on the actual application, the following process steps occur:

- Intake.
- Compression.
- Atmospheric venting.
- Melting (plastification).
- Conveying.
- Feeding additional excipients (solid or liquid).
- Dispersive and distributive mixing.
- Devaporization under vacuum (devolatilization).
- Pressure build-up followed by final extrusion through the die hole(s).
- Downstream processing of the molten strands.

The intake zone conveys the powder, received from the feeding device, sufficiently fast to avoid clogging. Typically, extruders are starved fed (i.e., the feed rate is constant and below the maximum intake capability). Thus, most parts of the screw sections in the extruder are not fully filled. A decreasing screw pitch is used to compress the powder, i.e., to remove entrapped air. This can be supported by atmospheric venting. Plastification happens due to heat generated by friction and shear forces exerted by the rotating screws to the compressed material and—to a lesser extent—by the barrel heating system. Dispersive mixing causes a size reduction of dispersed particles or droplets by shear forces, while distributive mixing means homogenization by distributing elements among each other. The devaporization zone applies vacuum to remove volatile components, e.g., entrapped air or water steam. In order to create a vacuum, fully filled screw sections (by back-feeding elements) have to be created to insulate the vacuum zone. The pressure build-up zone generates the required die pressure for extrusion. Finally, the molten strands are processed by calandering, hot-die-face cutting, or cold-strand cutting.

In reality, the process steps cannot be strictly isolated from each other, e.g., mixing effects occur practically in every step (not only in the designated mixing zones). Pressure is only generated in all completely filled sections (not only at the die) and conveying occurs everywhere, except in non-conveying elements (e.g., kneading blocks with 90° offset angle, *see* Fig. 14). To provide appropriate temperature conditions for each zone, the barrel is typically divided into a number of temperature-controlled zones along the screws. The material temperature is usually significantly higher than the surrounding barrel temperature (up to 30 K and more, mainly depending on the viscosity and the screw speed), since the viscous

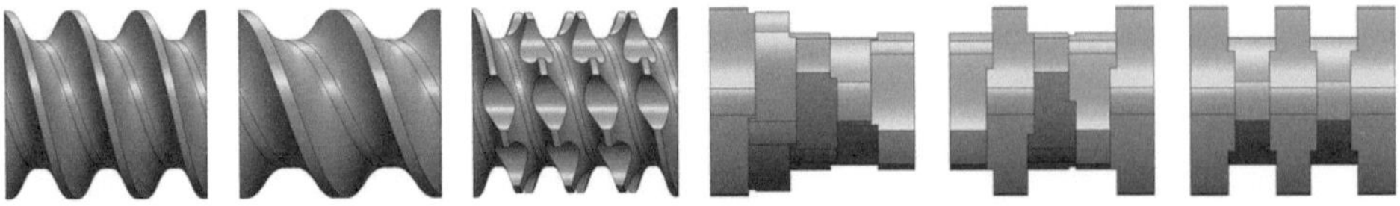

Fig. 13 Different screw elements (**a** and **b**: conveying elements with different pitch, **c**: mixing element, **d–f**: kneading elements with offset angles of 30°, 60°, and 90°)

dissipation heat is significant in case of the typical highly viscous materials. To achieve the required flow conditions for each process step, co-rotating twin-screws usually consist of a modular screw design, i.e., the screws can be arranged of different types of individual screw elements. The most important types are conveying elements, kneading elements, and mixing elements (Fig. 13).

6.3 Process Parameters in HME

In addition to the geometry of the individual screw elements and the used screw configuration, many other parameters impact the HME process. The clearance distances (between both screws as well as between screws and barrel in the order of 100–250 μm) are crucial for the amount of viscous dissipation and, thus, have a considerable impact on the generated heat and the achieved melt temperature. The die geometry (e.g., number of die channels, diameter and length) determines the required die pressure and the length of the pressure build-up zone. It is also a major factor for the occurrence of flow instabilities, such as shark-skinning and pulsating flows. Also the operation parameters (screw speed, throughput, barrel zone temperatures) and the material properties (e.g., density, rheology, melting or glass transition point, melting enthalpy, specific heat capacity, thermal conductivity, miscibility of different components, phase transitions, etc.) have essential influence on the process.

A variety of parameters can be used to characterize the HME process, e.g., melt temperature (typically spatially distributed due to the viscous dissipation), screw filling ratio (distributed along the screw, depending on the used screw elements), pressure (spatially distributed, ambient in partially filled sections and increased in completely filled sections), power consumption and torque, specific mechanical energy consumption, degree of mixedness (evolution along the screw), residence time distribution. In Fig. 14 computed profiles of temperature, pressure and filling ratio along the screw are shown.

However, only a small amount of these parameters is experimentally accessible, e.g., pressure, power consumption or residence time distribution. The majority of relevant parameters cannot be measured easily, e.g., the screw filling ratio, the local melt temperature, the local degree of mixedness. Nevertheless, knowledge about the local mixing phenomena caused by different types of screw

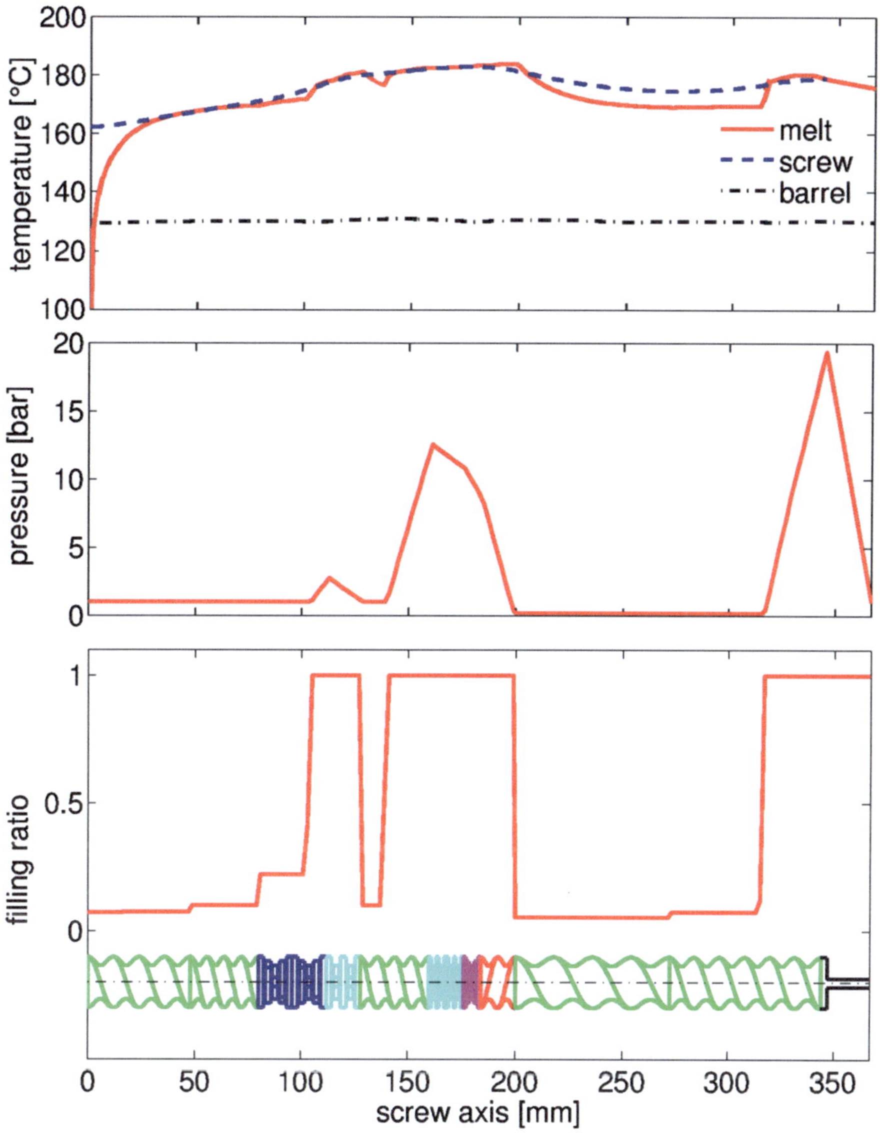

Fig. 14 Computed profiles of temperature, pressure, and filling ratio along the screw axis

elements, and the impact of the material properties is crucial for a proper process design with respect to the given materials and the specified critical quality attributes of the product (e.g., morphology or limits for impurities, degradation and moisture). Today mostly empirical knowledge and experience are available to overcome that challenge, and no systematic design approaches exist.

6.4 Co-Hot-Melt-Extrusion

In co-extrusion two or more materials are extruded through a single die but using a separate extruder for each material, arranged in a way that the extrudates merge and form a laminar-layered structure [129]. The die design is critical in co-extrusion, and various shapes can be designed based on the actual requirements, e.g., flat dies, split dies, annular dies, side fed and spiral mandrel dies [133, 134]. An example for the application of co-extrusion is the production of an intravaginal contraceptive thermoplastic ring (NuvaRing®) [135]. In this application, the core material contains the API, which is covered by a layer of another material that forms a diffusion barrier and controls the API release.

6.5 In-Line Monitoring of HME

The product quality in pharmaceutical applications is usually highly critical and continuous monitoring systems are required to avoid large amounts of non-conforming product. In the literature it was shown that existing monitoring tools (NIRS, Raman spectroscopy, optical particle size analysis, etc.) can be utilized to acquire critical quality attributes, such as API content and pellet size [86, 136–139].

6.6 Downstream Processing

At the die the extruder supplies a molten strand which has to be processed further by suitable subsequent downstream units. Different options exist to make either intermediates (by hot-die cutting, cold-strand cutting, milling, film drawing) or final products (shaping calanders, injection molding). The selection of the downstream process is based on the target dosage form, the material's rheology, the product's purity, and the production costs. Often, intermediates (pellets, powders, flakes) are desired which are then transferred to standard pharmaceutical manufacturing processes (capsule filling, milling, compaction). For more information we refer to ref. [140].

6.7 Injection Molding

Similar to HME, injection molding (IM) is an very widely used process in the polymer industry and has significant potential for solid-dosage-form manufacturing of pharmaceuticals due to its capability for producing well defined shapes and sizes in a single step. Due to the similar process conditions, the potential to produce solid dispersions and to enhance the bioavailability of poorly soluble drugs is comparable to hot melt extrusion (HME). A recent review about pharmaceutical IM was given by Zema et al. [141].

The IM process exhibits similarities to HME in some aspects. Firstly, the granular feed is molten. Mixing is usually not conducted in IM, rather in a HME unit prior to IM. Instead of the continuous extrusion in HME, in IM the melt is injected semi-continuously into a shape-giving mold (tooling) with high pressure. The melting device of the IM process is typically a single screw, which works similar to a simplified HME process, comprising the process steps intake, compression, melting and pressure generation. However, to achieve the semi-continuous injection capability, the screw is moveable in axial direction and acts like a piston during the injection to

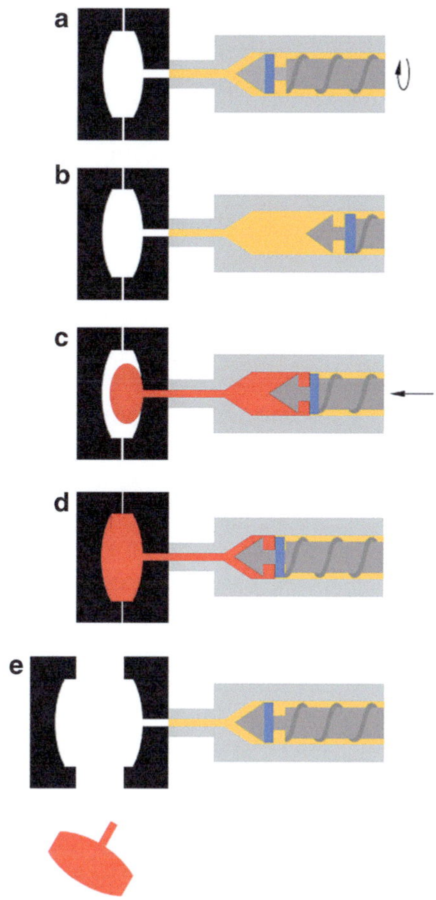

Fig. 15 Schematic of the injection molding cycle (**a**: plasticizing I, **b**: plasticizing II, **c**: injection, **d**: packing and cooling, **e**: ejection). The mold is illustrated in *black color*, the barrel and the die in *grey*. The *blue part* symbolizes the non-return valve

achieve the required pressure and flow rate. Thus, the IM cycle consists of the following parts (Fig. 15) [140]:

- Plasticizing I: the screw rotates in the forward position and melts the material due to friction and viscous dissipation, supported by heat transfer from the barrel.
- Plasticizing II: the rotating screw moves backward and the melt accumulates in the emerging clearance in front of the screw.
- Injection: the screw moves forward like a piston and pushes the melt into the mold cavity. A special valve prevents backflow into the screw.
- Packing and cooling: the pressure is retained during cooling to compensate the volume shrinkage with fresh material.
- Ejection: the mold opens and ejects the product.

The operating pressure in IM can reach up to several 1000 bar (which is not suitable for all APIs), and depends on the shape of the cavity and the rheological properties of the melt. The number of pieces per cycle can be easily adapted by the geometry of the mold cavity. Depending on the shape of the product, it is possible to achieve a quantity of 100 pieces per cycle and even more. The cycle time depends on the material properties of the formulation and is typically in the order of seconds. Thus, a production rate up to 100,000 pieces per hour is reasonable. Note, that IM is suitable to produce personalized drug products due to capability of efficiently making small batches in a single step.

6.8 Scaling of HME and IM

Important fundamentals for the scale-up of screw machines, based on dimensional analysis, were reported by Pawlowski [142] for single screw extruders, extensively confirmed by experimental data. Specifically, dimensionless parameters for throughput, pressure drop and power consumption were presented, which are independent of the actual length scale, and thus, relevant for scale-up. Kohlgrüber [131] extended these considerations to co-rotating twin-screw extruders, and pointed out, that various criteria have to be fulfilled in order to achieve comparable conditions at different length scales, as geometrical similarity, energetic similarity, similar shear conditions and constant discharge pressure.

IM machines are designed as a modular concept. It consists of three modules, the mold, the clamping unit and plasticizing unit. The mold design constitutes the dimensioning of the entire machine. The requested quantity per hour in combination with the cycle time determines the number of cavities in the mold. Then, the mold geometry is generated and further key figures, e.g., shot volume and clamping force, are derived. The shot volume describes the required amount of material to completely fill the molding. The clamping force acts on the mold halves and must be higher as the pressures at the parting plane during the injection cycle to prevent material leakage. Thus, subsequently the size of clamping and plasticizing unit can be chosen. For deeper understanding we refer readers to standard IM textbooks from the plastics industry [143].

The mold is usually a single-piece production and accordingly causes high development and manufacturing costs. Thus, CFD approaches are widely applied to investigate the flow within the molds to avoid production problems [144]. The simulations require several material properties, which are typically not available for pharmaceutical applications as every formulation is unique and have to be determined case by case.

6.9 Outlook

HME and IM are complex processes where the product performance (i.e., the critical quality attributes) depend on many process parameters, the screw configuration, melt rheology and materials behavior and the thermodynamic properties of the API/excipient mixture. Today, only simple process models and empirical knowledge are available.

Modeling and simulation methods can help to increase the understanding of HME. However, this is an extremely challenging task due to the complexity of the geometry, the corresponding scale challenges (the clearance distances are four orders of magnitude below the length of an entire screw), and the complexity of the underlying physical phenomena (e.g., phase transition from solid to molten, complex melt rheology, partially filled screw sections). Simplified model approaches are still of interest, since they can yield approximate insights into the entire HME process. For example, a one-dimensional (1D) flow model can be applied to different types of screw elements with a typical polymer used in pharmaceutical HME [145, 146]. However, more advanced approaches are needed to capture the full complexity of HME and IM. A novel approach based on advanced fluid dynamical simulations via Smoothed Particle Hydrodynamics (SPH) is presented in refs. [147, 148]. Figure 16 shows a snapshot of an SPH simulation of a typical screw elements used in co-rotating twin-screw extruders, involving completely filled and partially filled regions. This yields the detailed flow field and can be used to study mixing, e.g., via tracer particles (represented by white spots in Fig. 16).

In addition, advanced PAT tools are critical prerequisites for advanced process control. However, the major challenges exist, e.g., to monitor the melt and product concentration (and even temperature) in this constrained geometry. Due to the limited penetration depth of spectroscopic tools (<1 mm), only the concentration in a thin layer close to the barrel wall can be monitored.

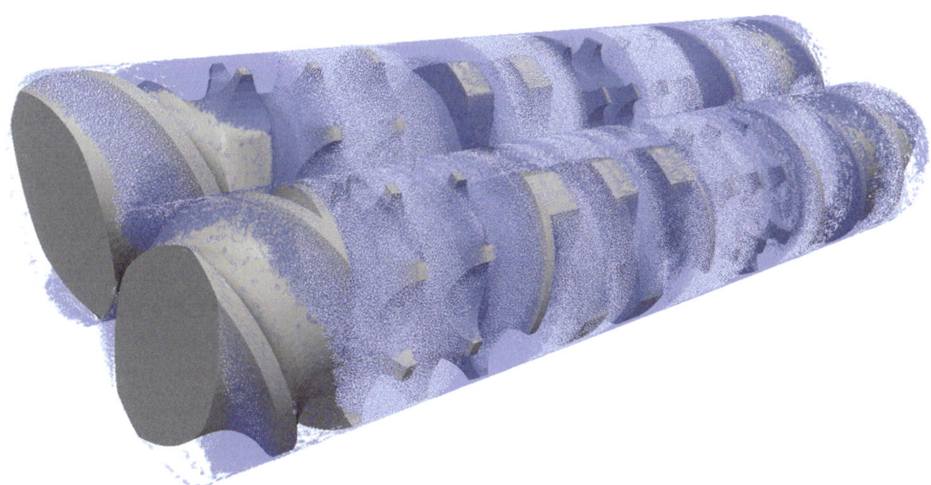

Fig. 16 Snapshot of an SPH simulation of typical screw elements used in co-rotating twin-screw extruders

This is also true for the melt temperature, which is known to significantly vary across the channel and where only a mixture between barrel and melt temperature may be measured. In summary, new developments for advanced PAT tools are required.

7 Tableting

Tableting is the compaction of powders into a single unit and it is one of the most important unit operations in the manufacturing of solid pharmaceutical dosage forms, as the overwhelming majority of solid dosage forms are tables. The quality of pharmaceutical tablets is assessed in terms of assay, content uniformity (CU), disintegration, dissolution as well as appearance. Moreover, mechanical properties are critical quality attributes, such as the crushing strength, friability and the tendency to develop structural defects, such as chipping (of corners), lamination (disintegration in layers) or capping (a top layer separates from the tablet body). These mechanical tablet properties determine the post-processing ability fluidized beds, coaters or during packaging.

The tablet quality attributes are the result of formulation parameters (i.e., addition of lubricants) and tableting process parameters (i.e., disk speed and dwell time). Therefore, early knowledge during process/product development is critical to obtain a product with a desired performance (i.e., dissolution) and to avoid problems during transfer to the manufacturing site. Thus, scale-up and process-transfer studies involving compaction simulators are highly advisable.

7.1 Compaction Physics

For an accurate discussion of the generation of a pharmaceutical tablet, it is important to differentiate between compressibility, compactibility, and tabletability.

Compressibility is the ability of a powder to decrease in volume under compaction pressure. The decrease in volume under compaction pressure (compressibility) has been extensively modeled with different success, most notably by countless empirical equations, such as the Heckel or Kawakita models [149]. To a certain extent, parameters can be estimated and mixing rules can be developed. More complex is the prediction of *compactibility*, which is the ability of granular material to form a tablet with a specific strength under compaction pressure. It is characterized by plotting tensile strength versus tablet porosity. *Tabletability* shares the definition of compactibility; however, it is evaluated by plotting tablet tensile strength versus compaction pressure. Leuenberger [150] studied models to predict the ability of granular material to form tablets with a specific strength, i.e., compactibility and tabletability models. This issue is more challenging than the modeling of

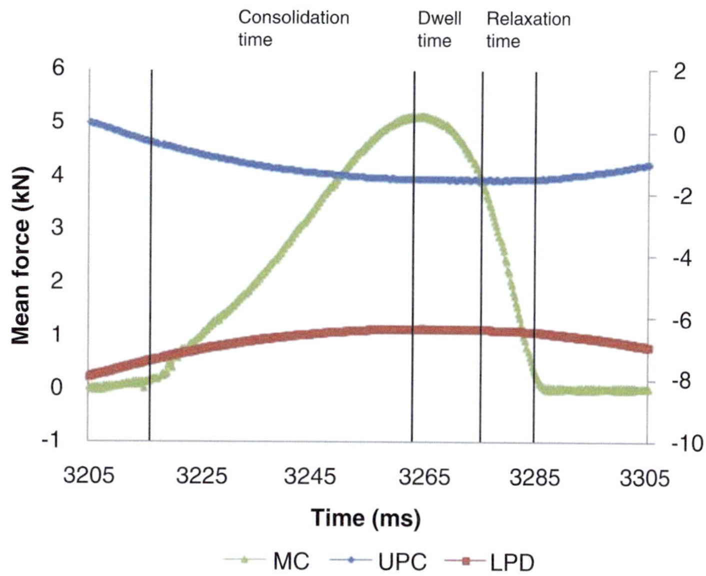

Fig. 17 Compression cycle of a single tablet indicating the consolidation, dwell, and relaxation time. MC: mean compression force; UPD (upper punch displacement); LPD (lower punch displacement)

compressibility because the bonding forces between particles that are the cause of tablet strength depend also on interparticle bonding area and chemical nature of particles. Sun [151] describes these contributions and proposes a system to predict tablet strength.

Compactibility and tabletability are affected by a number of process parameters and material attributes. The major parameters in the tableting process are speed, force and also the force profile of the main compaction event. The force profile experience by the material is presented in Fig. 17, and consists of three consecutive stages in the order of several tens of milliseconds: consolidation, dwell and relaxation (decompression). During the consolidation, the punches change their vertical position and compact the powder. During the dwell time, the punches are not changing their vertical position, and during the relaxation times, the punches increasing the distance between upper and lower punches. The compression time is the addition of consolidation and dwell time.

Compression of granular material takes place through a sequence that involves particle rearrangement, deformation, compaction and relaxation. Rearrangement involves the motion of particles to fill the voids among them. Particle deformation takes place when particles cannot rearrange any longer and start to deform elastically. Compaction is highly dependent on material properties. There are two main mechanism for compaction: (1) Plastic deformation: there is an increase in the contact area between particles under larger compaction forces that contributes to tablet strength (typical example: MCC), (2) Brittle fracture: particles fragment under larger compaction forces, which also leads to an increase in inter-particle contact area (typical example: lactose). Most of the pharmaceutical materials compact via a combination of these two mechanisms. Finally, relaxation takes place when the distance between punches increases and also after ejection from the die. If the elastic forces during relaxation exceed the tensile strength of the tablet, its physical integrity will be compromised.

7.2 Tableting Equipment

7.2.1 Single-Station Presses: Eccentric Tablet Press and the Compaction Simulator

There are two types of single station presses that are typically used for development of a tableting process with a limited amount of formulation: Eccentric presses and compaction simulators. The process is developed in single station presses and transferred to rotary presses.

Eccentric presses are presses where the upper punch penetrates the die while the position of the lower punch does not change (except ejection) during the tablet formation. However, in rotary presses (described in more detail below), the lower and upper punches both move into the die. In addition, these presses differ in terms of punch speed (low velocities for eccentric presses and the higher velocities for rotary presses), precompression (only applied in rotary presses) and dwell time. The differences in punch speed and a lack of precompression may lead to differences in the lamination and capping of tablets produced by these presses. These operation differences also impact the evaluation of powder compactibility and Heckel parameters. In fact, the differences in Heckel parameters for materials are larger when evaluated in the eccentric machine. However, the same ranking of materials was encountered in both the eccentric and rotary presses [152].

Compaction simulators are an advanced type of single-station presses because they can reproduce upper and lower punches displacement profiles of any rotary tablet press with variable speed. The compaction simulators are classified in linear simulators and simulators with rotating cams. Precompression and ejection steps can also be included in the simulation. Compaction simulators are the best tool for process development and scale-up although they do not currently address the issues of feeding and die fill at high speeds, or speed-related temperature fluctuations.

7.2.2 Rotary Presses

The manufacturing of tablets in industrial scale is performed in rotary presses that can produce up to 1–2 million tablets per hour. A tableting event involves several stages and each features process parameters that affect the quality attributes of tablets [153]. The following steps that take place in the turret of a rotary press:

1. Feed frame, where the powder is introduced into the die.
2. The several sets of upper punch–die–lower punch that move under pre-compression and main compression rollers. As a result, the powder is compressed into a tablet.
3. Ejection cam, where the tablet is ejected from the die.

The presses available on the market (typical suppliers being Fette, Korsch, IMA, Romaco-Kilian, GEA-Courtoy, and many others) differ in the construction details and mechanisms for each stage. For example, the filling of dies in the feed frames is typically driven by gravitational force. However, there are presses where die filling is driven by the centrifugal force of the turret [154]. The ejection mechanism presents also some differences: although most presses eject at the top of the die, IMA Comprima ejects at the bottom of the die and uses gravity to eject the tablet from the press. Tableting in single punch machine and especially in rotary presses can be successfully performed thanks to lubrication of the formulation. Lubrication minimizes friction with die walls, and therefore, ejection forces from the tablet die, enhancing tablet appearance and minimizing defects (i.e., capping). Lubrication also minimizes problems such as picking (surface material from the tablet becoming stuck and removed by the punch) and sticking (tablet material adhering to the die wall). Lubrication enhances powder flowability, which is a key parameter for the consistent and uniform die filling, which leads to minimizing tablet weight variability. However, the feed frame parameters of operation and concentration of hydrophobic lubricant (i.e., magnesium stearate) are critical because they negatively affect the powder wettability, tablet hardness and tablet dissolution. Thus, systems have been developed to spray a lubricant solution directly on the die surface, eliminating the need to add the lubricant to the formulation (e.g., by Fette). Many more advances have reached the market, e.g., specialized housings for effervescent tablets or highly actives and systems that allow easy cleaning and assembly of the machines. Moreover, all tableting machines are highly instrumented to measure all forces acting on the punches, including ejection forces. Even radial die forces can be measured on some systems.

7.3 PAT Tools

Tablet quality testing (dissolution, assay and hardness) is typically performed after manufacturing a batch via destructive methods. However, PAT applications should enable process controls and product/process optimization via on-line or in-line measurements. For example, PAT tools have been used to monitor blend properties in the feed frame of rotary presses and to predict tablet quality attributes and their deviations from target. For example, NIRS has been applied to monitor the concentration of API and excipients in the feed frame and successfully predict API content in tablets [155]. Moreover, NIRS has been successfully used to test the properties of finished products, for example, tablet hardness [156]. Testing hardness of finished product may be used to predict quality attributes of tablets and to control the compression parameters of a tablet press. Hardness, porosity and content have been successfully characterized with Raman and NIR spectroscopy for extended release tablets [157]. Chemical imaging is also useful to predict content in tablets [158] and it potentially can assess all tablets in a production batch, i.e., 100 % quality control may be achieved. Terahertz (THz) is a technique that has received increased attention recently for the assessment of polymorphism and for crystallinity characterization via PAT. Moreover, a methodology for implementing THz as PAT tool has been recently presented for pharmaceutical tablets [159, 160] and subsequently used to predict API and excipient content of tablets [159].

7.4 Scale-Up

In order to scale-up a process, three similarity principles must be matched: geometric, kinematic and dynamic ratios of characteristic variables. In contrast to most other processes, the length scale of the process remains the same, as tablets do not change size upon scale-up. Thus, similar punches need to be used in the large scale system. The most important scale-up consideration is usually tableting speed. Thus, matching consolidation time and dwell time, as well as the corresponding pressures, between R&D-scale and industrial scale will ensure the same or a very similar tablet quality [154]. Note, however, that reduced consolidation and dwell times upon scale-up may have detrimental effects, even at the same compression forces. In addition, tablet ejection, die filling or frictional heating may impact the scale-up process.

Compaction simulators are very useful in investigating the effect of punch velocity and force on the compaction properties of powders. Thus, scale-up (and process transfer) of a tableting process can be ideally performed using compaction simulators. They can simulate the punch movement of rotary presses at high speeds, while making only one a few tablets for research and analysis.

7.5 Outlook

Tablets are the main drug delivery system today. Solid dosage forms are still the preferred product form, even for an increasing number of APIs with dissolution and bioavailability issues. In order to overcome these issues, amorphous dispersions are an advantageous formulation approach [161, 162]. The processes for solid dispersions include, among others, hot-melt extrusion [124] and spray drying [97]. Compaction of solid dispersions may affect the structural and physical stability of solid dispersions. For example, compaction may lead to demixing or nucleation of API [163]. As a result, homogeneity and, most importantly, the dissolution benefit of solid dispersions are negatively affected. However, compaction may sometimes have a positive effect on the stability of the product due to increasing the API/polymer interactions [164]. Thus, more research in this field is needed.

Moreover, efforts are underway to add new functionalities to tablets, e.g., to enhance and monitor patient compliance (i.e., the integration of edible sensors) [165]. In addition, further developments in the field of PAT for tableting are expected. Techniques with fast evaluation capabilities, offering the possibility to monitor every tablet, such as chemical imaging, will receive considerable attention [158].

8 Filling of Hard Capsules

In the pharmaceutical industry, a wide range of capsule filling systems, which employ different technologies and principles are in use today. Capsule filling is sometimes also referred to as "encapsulation." Hard-shell capsules, which are typically made using gelatin (non-animal sources do exist as well) mainly contain dry, powdered ingredients or small pellets. But also mini-tablets and pastes or various combinations of different materials can be filled into capsules [166] for blinding purpose in clinical studies, to separate incompatible products, or to achieve specific goals in terms stability and modified release, only to name a few options. Clearly, the filling material must be compatible with the shell, and therefore, no deliquescent or hygroscopic materials can be used. Hard-shell capsules consist of two halves: a lower-diameter body that is filled and then sealed using a higher-diameter cap [167]. Filling of hard-shell capsules is the focus of this section. Soft-gel capsules, which often contain liquids, are not discussed here.

Depending on the powder density, different fill weights can be obtained. Capsule sizes range from 5 (smallest) to 000 (largest) and maximum fill volume ranges from 0.1 to 1.37 ml. Thus, up to 1.5 g of powder or pellets can be filled in capsules. However, an important current trend is to manufacture small doses (<50 mg) of pure potent active pharmaceutical ingredient (API) for early research clinical trials using the so-called "drug-in-capsule" approach and for inhalation purpose (dry-powder inhalers, DPI) for the treatment of respiratory disease or if the API is not readily absorbed orally [168].

Formulations relevant for capsule filling typically involve a combination of materials to achieve accurate dosing, good bioavailability, ease of filling and production, product stability and also elegance for the patient. They must have good flowability, be non-adhesive but compressible enough to form plugs if required. To enhance the powder flow properties lubricants or glidants are added. Disintegrants are used to assist the break up and disintegration of the capsule content in the stomach. For highly potent APIs, which are administered in very low doses, the bulk volume has to be increased to allow accurate filling. For this, fillers or diluents are included in the formulation. Coloring and taste masking is performed to enhance safety and the patient compliance [169].

Hard capsules can be filled in several ways from manual preparation in the lab or in the pharmacy to fully automated industrial

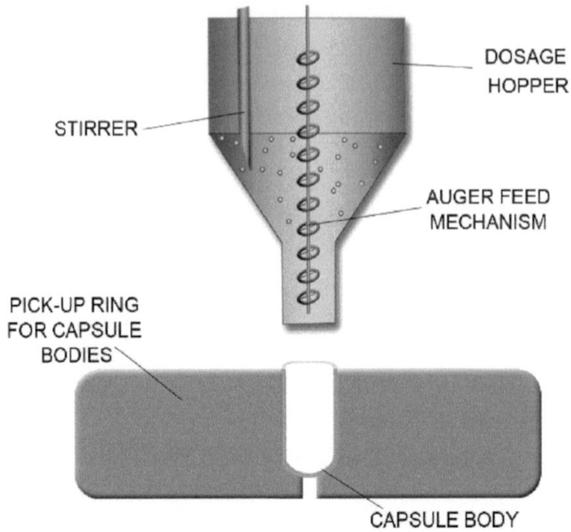

Fig. 18 Auger filling mechanism

production. Although capsule-filling machines may vary widely in their engineering design the major difference between them is the dosing technique. The most common classification is: (1) direct and (2) indirect filing methods. Principles that dose directly into the capsule are direct methods and machines that implement dosing techniques outside the capsule before filling are considered indirect methods. The major challenge for indirect systems is that although doses need to be specified by weight these filling systems work on a volumetric basis [170].

8.1 Filling Principles

8.1.1 Direct Filling

Auger filling: This principle is based on the semi-automatic and automatic equipment, where the powder in a hopper is filled into capsules continuously by a rotating auger in conjunction with a stirrer. This principle is shown in Fig. 18. The empty capsule bodies are placed below the auger into a rotating turntable. The dosed weight is dependent on auger speed, the twist angle of the auger and the time the capsule body spends under the hopper outlet. Fill weight is also dependent on the powder density, which evolves from initial bulk density in the auger until reaching steady state [170]. Thus, fill weight may vary over the course of the filling process. For example, Mettler Toledo is producing the Quantos MicroDosing System™, which uses the above described Auger filling principle.

Vibration assisted filling: In this gravimetric dosing principle, the capsule body is filled directly through a mesh, which is connected to a vibration plate. This vibration assists powder flow and therefore dosing ("pepper shaker principle"). In addition, the equipment includes a microbalance, a load cell or a capacitance system to control fill weight, even for very low doses. Current systems on the markets include for example MG2's Microdose (1–40 mg) or

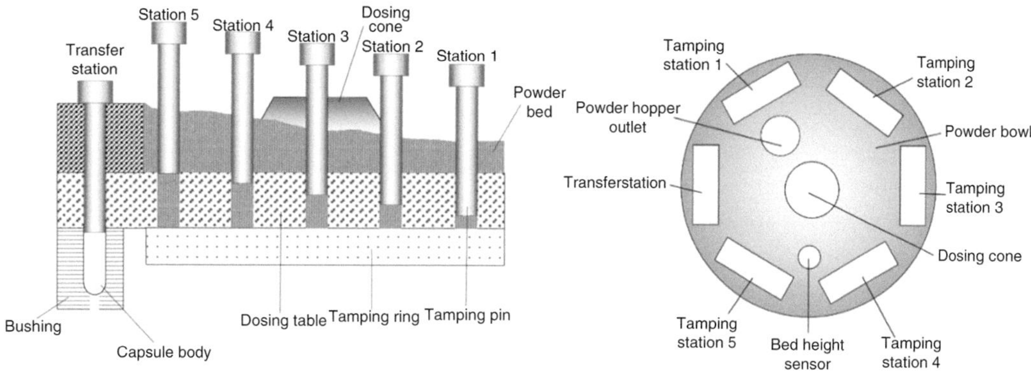

Fig. 19 Schematic diagram of a dosing-disk and tamp-filling system

Capsugel's Xcelodose®S (0.1–100 mg). These machines are of special interest for research purposes and clinical trials batches and allow the filling of several hundred capsules per hour with doses in the range between 0.1 mg up to about 100 mg.

8.1.2 Indirect Filling

Tamp Filling: In dosing-disks or tamp-filling machines, the powder is in a cylindrical powder bowl that contains a removal dosing disk with six dosing holes (Fig. 19). The powder bowl rotates 360° stopping at six stations with matching dosing holes. The material is fed from a hopper, to a dosing cone, which helps to distribute the powder horizontally into the powder bowl. As the dosing disk rotates, the first hole is partially filled with powder and then is tapped by a pin or tamping fingers. This process of partially filling and tamping is repeated until the last hole is reached. After excess powder is scraped off, the dosating disk positions the plug of powder over a capsule body and injects it into the capsule. The fill weight can be controlled by the thickness of the dosing-disk, the powder bed depth and the tamping pressure. The tamping pins are spring loaded in lab and medium scale or have a cushion of compressed air at industrial scale to minimize the tamping force to keep the plug density low [170–172]. Tamping machines such as the Bosch GKF 2500 (Fig. 20) adjusted with up to 18 tamping fingers (industrial scale) can produce up to 150.000 capsules per hour. Other manufacturers of industrial-scale tamping machines are IMA and Romaco (Italy) and Harro Höfliger (Germany).

Dosator-nozzle filling: In dosator-nozzle machines (MG2, Zanasi, IMA, Matic, Marcofar), the dosator moves into the powder bed and collects the desired volume of powder from the powder layer. During dosing, compaction is applied to form a stable plug. For inhalation products, the capsules are filled with a controlled degree of compaction or even without compaction, to ensure that the plug is turned back into a powder for efficient drug delivery (plugs cannot be inhaled). The cylindrical volume (dosing chamber)

Fig. 20 Bosch Packaging Technology—GKF 2500

is determined by the dosing chamber length (determined by a movable piston) and dosator diameter (Fig. 21). After collecting the powder the dosator nozzle is lifted from the powder bed and moves towards the empty capsule body into which the dose is ejected. The rotary product container and the dosators have different axes of rotation to enable the dosators first to sink into the product layer and then to discharge the product into the capsule bodies. Due to the dosing principle, the powder has to be retained in the dosing chamber, while this section is in motion. Thus, for the retention of the powder in the nozzle during transfer, the powder must be able to form an arch. Hence, the requirements for powders and granules to be used in dosator-nozzle machines vary significantly from those used in tamping pin machines. Fill weight of capsules is controlled by adjusting the dosing chamber length, as well as varying the powder bed height in the bowl. Compared to tamp-filling, the dosator-nozzle system allows a wide range of fill weights by simple adjustment of the piston position for the choice of nozzle [170]. The instrumented continuous production machine G250 from MG2 can dose a variety of dosage forms accurately into capsules and can reach a maximum output of 200,000 capsules per hour.

A further development in dosator-nozzle design is the vacuum-operated system that implements a static piston with a porous plate

Fig. 21 Dosator filling mechanism (MG2, Italy)

at its product-touching end. The powder is sucked into the nozzle by vacuum and ejected into the capsule by reversal of the airflow. In that way, no compaction is performed, the nozzle does not contain any moving parts, resulting in less demand for lubrication and less densified powder plugs, and therefore, very small doses can be filled. Romaco produces the Macofar series and Harro Höfliger offers vacuum-assisted dosator nozzles, which are able to dose 10–600 mg with a maximum output of 4500 capsules per hour [170].

8.2 Low-Dose Filling

As mentioned above, low-dose filling (<50 mg) of pure APIs is becoming of significant interest for various reasons. However, low-dose filling leads to challenges during manufacturing. Modern capsule-filling technologies [173–175] take into account these challenges and apply special adjustment to the machines for accurate dosing. For example, high-end continuous dosator machines like the Planeta 100 (Fig. 22) with two dosing units and 16 dosators mounted (MG2) offers accurate capsule filling at an industrial output of up to 100,000 capsules per hour even for very low doses for inhalation purpose. Other systems exist as well, e.g., the GKF2500 (Bosch packaging technology) production machine with a microdosing-wheel adjusted can reach a maximum output of 150,000 capsules per hour. The Modu C (Harro Höfliger) uses a drum dosing system with vacuum and compaction free (Fig. 23). It can reach an output up to 200,000 capsules per hour and achieves doses down to 1 mg depending on the formulation.

Fig. 22 Industrial dosator capsule filler—Planeta 100 (MG2, Italy)

8.3 Scale-Up

The scale-up of a capsule filing process must consider the design and operating principle of the filling technology, like different powder handling and plug formation mechanism as well as the formulation requirements. Similar to tablets, capsule size does not change upon scale-up. However, the filling speed increases potentially, leading to different effects during the powder sampling. Most filling principles form plugs via pistons, compression or tamping fingers, equal to tableting, and then eject the plug into the capsule body. Capsule plugs are considerably different from compressed tablets. The plug height to diameter ratio is bigger and the compression forces are much lower than for tablets. Another difference to tableting is that the increased output of capsules is achieved by increasing the number of dosing units, whereas in tableting

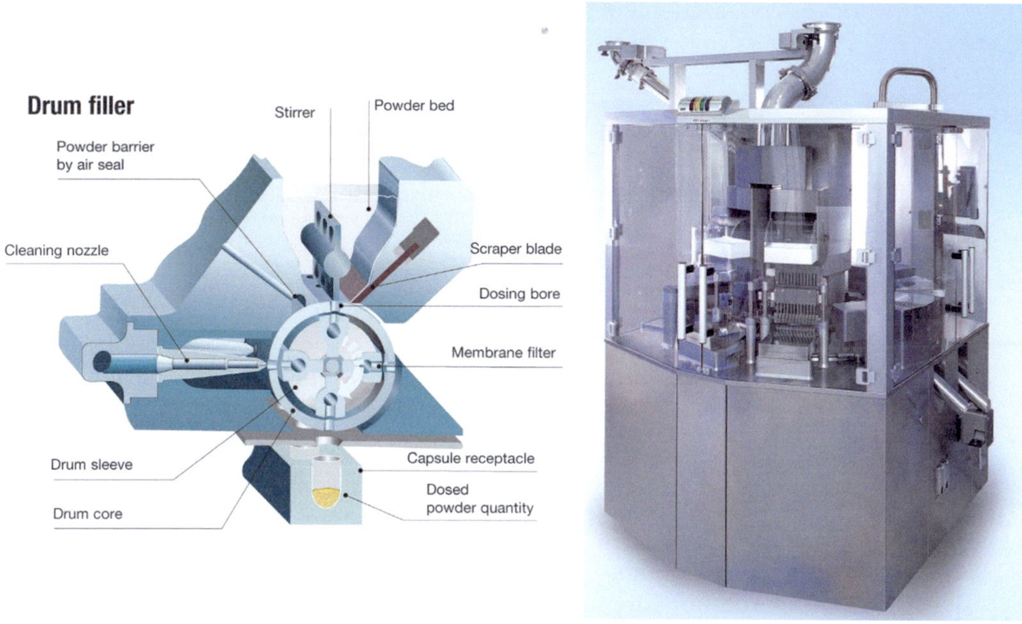

Fig. 23 Drum filling principle and the Modu C high-speed machine (Harro Höfliger, Germany) with exchangeable dosing systems (tamping pin, dosator, and vacuum drum)

compaction tends to be faster and dwell and contact time tend to be shorter [167]. Thus, scale-up of capsule filling is often a (more or less) straightforward process.

8.4 Outlook

An important challenge concerning capsule filling is the scale-up of low dose filling process of 1 mg or less. For accurate long-term and high-output production, machine improvements, an associated PAT strategy, as well as improved particle engineering for low-fill weight applications are critical. Improved instrumentation of capsule filling machines would be desirable, as well as the development of entirely new filling principles, based on weight rather than volume. Modeling and simulation can help to gain more insight in the complex process, yet are under-developed as of yet.

9 Coating

The coating of tablets (or pellets) in general is a common unit operation in the pharmaceutical industry. The immediate goal of the coating process is to enclose the core in one or more layers. This is done to eliminate disadvantages connected with the pure core and/or to introduce new features. Functions of the coating include [176].

- Masking of an unpleasant taste or odor of substances in the core (most likely the API).
- Protection from environmental influences, such as light and oxygen, to improve stability.
- Increasing the elegance, for example by giving a glossy finish or by applying a color. This also helps to distinguish different types of tablets.
- Reduce friction for a faster packing process.
- Modify the drug release profile, for example in enteric coating (a coating that is resistant to gastric juice, but dissolves in the lower pH values of the small intestine) or generally any form of sustained or delayed release coating.
- Separate incompatible APIs or other substances by adding one or more of them with the coating (rather than having it in the core). This is termed "active coating."

Common components of most modern film coating solutions are:

- A solvent, which can be aqueous or organic.
- A polymer, which often contains cellulose, methacrylic acid, methacrylate esters, polyvinyl acetate phthalate, or shellac.
- A plasticizer, to reduce the glass transition temperature of the polymer making it more elastic and deformable (flexible).

Optional ingredients may be:

- Anti- tack agents to reduce the sticking during/after the coating process.
- Anti-foam agents to prevent foam building during the spray process.
- Colorant.
- Flavors for taste masking.
- Surfactant(s) to lower the surface tension of the solution/ dispersion.

9.1 Coating Methods

Pharmaceutical coatings can be roughly divided into three main groups [176]: sugar coating, film coating in drums, and film coating in fluid beds, the latter two using aqueous or organic solutions or dispersions.

Sugar coating is the oldest form, and was already well-known in the confectionery industry before its application in the pharmaceutical industry. Nowadays, mainly due to the high dependence on operator skill and the thick multi-layer coating involving many coating steps, it is rarely applied.

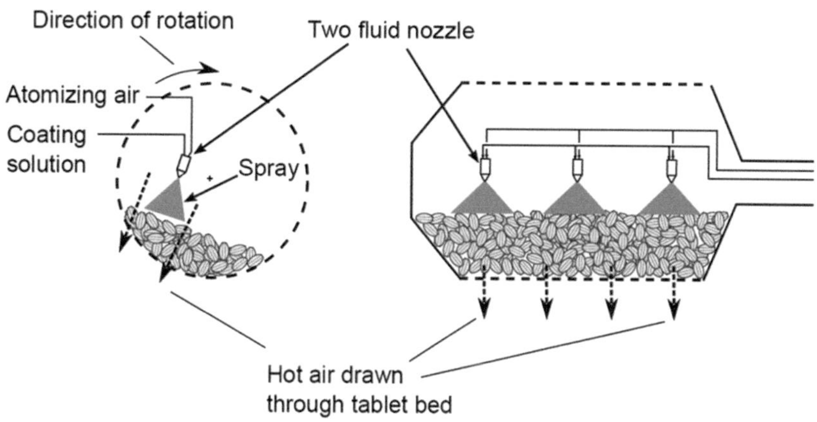

Fig. 24 Schematic of drum coating. A perforated drum is shown, as nowadays most drums apply this principle

Film coating in drums is the most common choice for the coating of tablets. The cores are placed in a rotating drum, and are sprayed from the top (*see* Fig. 24). To enhance drying, hot air is introduced. In older systems this is done in large pans that open at the front (and rear), with the drying air moving inside the coating drum. In modern coaters, a significant quality increase is achieved by using perforated pans, where the air stream passes through the tablet bed. Traditionally, coating is a batch process, but many manufacturers also offer solutions for continuous or quasi-continuous operation.

Fluid bed coating is applied mostly for granules and pellets (more or less spherical particles of smaller size) and rarely for tablets, as fluidization of the comparably larger tablets is not possible and only spouted bed regimes (with the associated high mechanical stress) can be obtained. The fluidization and drying air is introduced from below to fluidize the particles and to move them through the spray zone(s). Four main subtypes are defined, depending on the position of the spray nozzles: top spray, bottom spray, "Wurster" tube, side (tangential) spray as shown in Fig. 25. The most important process parameters that affect product quality for drum and fluid-bed coating are summarized in Table 1.

For both, drum coating and fluid bed coating, the properties of the coating liquid (such as type of solvent, surface tension, viscosity, density, drop size of the spray) influence the film formation. Nowadays, mainly aqueous coating solutions are used.

9.2 Single-Tablet View of Coating

From a single-tablet view, the coating process means a repeated exposure of tablet cores to a film-building liquid containing solute and aqueous or organic solvent. This liquid (or suspension) is typically introduced via spray nozzles where different suppliers offer competing products. In industrial coaters multiple spray

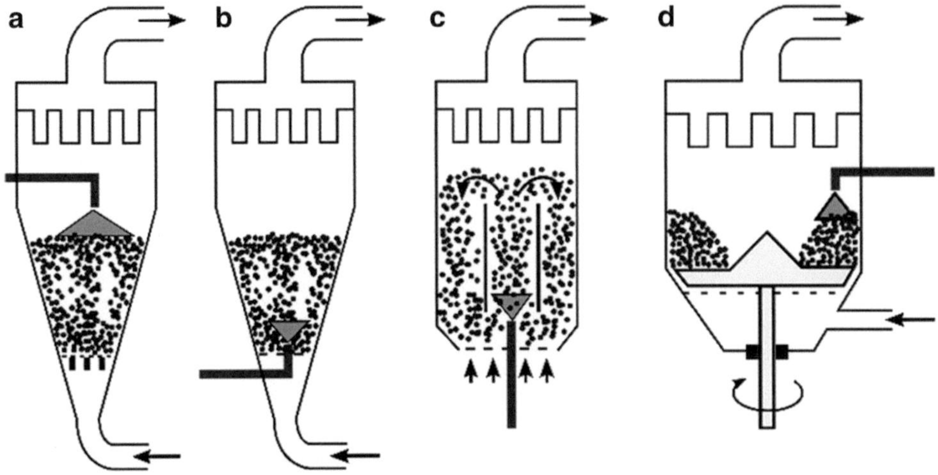

Fig. 25 Different types of fluid bed coating: (**a**) top spray or granulator, (**b**) bottom spray, (**c**) "Wurster" type, (**d**) side spray with rotating disk. From ref. [177]

Table 1
List of process parameters for drum coating and fluid bed coating. (The "x" denotes which parameters apply for which process; most apply for both types.)

Process parameter	Drum	Fluid bed
Drum rotation speed	x	
Baffle setup	x	
General spray position (top, bottom, . . .)		x
Presence of Wurster tube		x
Fill level	x	x
Amount of inlet air	x	x
Temperature and humidity of inlet air	x	x
Temperature and humidity of outlet air	x	x
Coating liquid flow rate	x	x
Spray nozzles:		
Number and placement of spray nozzles	x	x
Distance nozzle–bed	x	x
Atomizing air pressure	x	x
Pattern air pressure (if available)	x	x

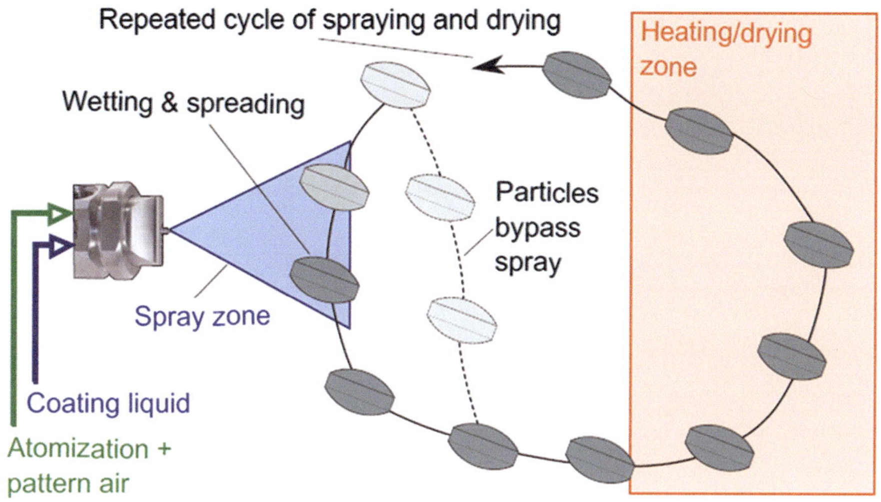

Fig. 26 General principle underlying almost all types of particle coating processes, from ref. [177]

nozzles (guns) are used to uniformly spray the bed. When a particle (tablet) travels through the spray region or spray zone, a partial coating is applied (depending on the local conditions in the spray zone). Immediately after its application, the liquid coating should spread on the core and—to some extent—penetrate the core. There may also be (undesirable) transfer to other particles. After this visit to the spray zone, the tablet moves into the drying region, where the solvent is evaporated by heated drying air, and the partial coating is thus solidified. This process is complex, as the film formation depends on many parameters (viscosity of wet film, temperature, etc.). Nevertheless, this step is critical as it determines the morphology and structure of the film and thus, its quality and performance. This repeated cycle, i.e., coating in the spray zone, drying, and re-entering the spray, is the underlying process of all coating processes, and is the key to the coating mass end-point and coating uniformity of the process [178]. An overview is given in Fig. 26.

9.3 Coating Quality and Variability

A coating that does not comply with its intended performance may have serious impact on patient health and safety. Therefore, the coated tablets have to consistently fulfill the intended critical quality attributes. To achieve this, a detailed understanding of the process is needed [179]. To this end, a large number of experimental works have been done (e.g., [179–184]).

In many applications, the most important quality parameters are the average coating thickness and the coating thickness variation, i.e., variability. Specifically, one distinguishes between intra- and inter-tablet coating variability. Intra-tablet coating variability describes the variation of the coating film over the single tablet.

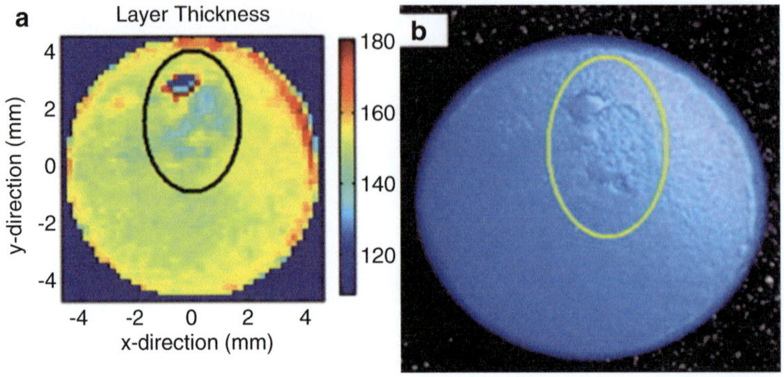

Fig. 27 Intra tablet uniformity. (**a**): false color 2D map of the coating layer gathered by terahertz pulsed imaging, (**b**): photo of the tablet, from ref. [185]

It can manifest in diverse forms, and is often related to the film building process, being an interplay of spraying/wetting and drying. One example of high intra-tablet variability is shown in Fig. 27. Inter-tablet coating variability means the variation of coating mass between the tablets in a batch. The inter-tablet variability is influenced by the mixing performance in the drum, which in turn is influenced by parameters such as the geometry of the drum and baffles, and the geometry of the spray zone(s).

In addition to the geometry, the process parameters influence the uniformity [177]. In the following several findings from our (and other) research are summarized:

- The rotational velocity has little influence on the uniformity given the same amount of revolutions.
- Increasing the number of nozzles has a large potential for improve the inter-tablet variability, especially in drums with slow axial mixing.
- The nozzle geometry has an influence on the drop size distribution and spray uniformity, which influences the intra tablet variability.
- For constant total process time, a higher fill level leads to an increase in the inter-tablet variability, yet corresponds to a better process equipment utilization and thus reduced cost.
- The spray rate itself (without considering a change in total process time) normally has reduced influence on the inter-tablet variability, but may affect the intra-tablet uniformity.
- A longer total process time for the same total coating mass leads to better uniformity, while increasing the process cost.

Different methods for the determination of the coating variability exist. A simple assessment of coating mass can be done by weighing the coated tablets and subtracting the mean weight of the tablet core. Similarly, the thickness can be determined by measuring the tablet radius. Accuracy (but also experimental effort) can be increased by labeling individual tablets and measuring the same tablet before and after coating. A source of error is that a weight/size increase of the core (e.g., due to solvent) is interpreted as an increase in coating mass/thickness.

For direct, yet destructive measurements, microscopic techniques can be used, most commonly scanning electron microscopy (SEM). Tablets are cut or broken in half, and the cut plane is examined. For non-destructive measurements, typical methods are Terahertz imaging, X-ray computed tomography (CT), and magnetic resonance imaging (MRI). A newer method that is non-destructive but significantly faster is Optical Coherence Tomography (OCT). OCT is based on low coherence interferometry; the image contrast is due to inhomogeneity in the refractive index. It can gather two- or three-dimensional cross-sectional image data in situ and in real time with good axial (depth) resolution, *see* Fig. 28. Most measurement techniques yield information on both intra-tablet and inter-tablet variability. However, a large enough number of tablets has to be investigated for meaningful results, and the measurement speed can be critical for routine application.

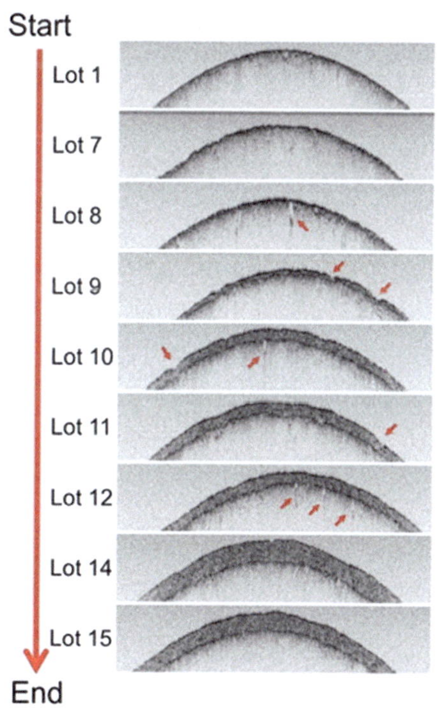

Fig. 28 OCT images of tablets from different stages of the coating process. The image size is 4.3×0.36 mm^2 (in air) with a resolution of 4.3 μm and <4 μm in lateral and axial direction, respectively. From ref. [186]

9.4 Modeling of the Coating Process

In recent years, computer simulations have been increasingly used to study coating [177, 187]. While simulations have their downsides (e.g., the quality of the simulation depends on the input parameters), a valuable advantage of simulations is that they can provide a large amount of high-fidelity data, and enable to extract quantities which are hard or impossible to measure. Different simulation approaches exist and were applied for coating processes depending on the focus of the work, including studied by [188–194]. Moreover, simulations allow an understanding of different mechanisms and effects on the product quality. Thus, even semi-accurate simulation tools can offer a significant benefit by learning about the process.

9.5 Scale-Up

As for most processes in the pharmaceutical industry, tablet coating is done on different scales. A process is typically designed on the lab scale (in the order of a few kilograms or of 10,000 tablets). It is then scaled up to the pilot /technical scale (In the order of 50 kg or 250,000 tablets) and further to industrial / full scale (in the order of 300 kg or more, or 1 million tablets). Most manufacturers offer their coating system in all scales, using geometrically similar drums and baffle setups for reliable scale-up.

As a first guideline, normally the relative fill level is kept constant across scales, and the rotation speed n (in rpm) is calculated either by keeping the dimensionless Froude number constant, or by keeping the peripheral velocity constant. The Froude number Fr is given as:

$$\mathrm{Fr} = \frac{r\omega^2}{g} = \frac{r\tilde{n}(2\pi n)^2}{g} = \text{const.}$$

with r the radius of the drum, g the gravitational acceleration, and ω and n the rotation speed in units of rad/s and rotation per second, respectively. This may lead to too high tablet velocities in the bigger drums. Alternatively, the peripheral velocity v_u can be kept constant:

$$v_u = r\omega = 2\pi rn = \text{const.}$$

In this case, there are methods to predict the velocity on the top of the tablet bed [195]. Keeping the peripheral velocity constant leads to a lower tablet velocity and to less tablet damage during the scale-up process.

9.6 Outlook

On one hand, the coating has a long history, and is considered a well-understood process. On the other hand, a consistently high level of quality is demanded from the product; and often it is realized too late that significant problems can arise, leading to the destruction of whole batches or forcing companies to pull a product off the market, especially in connection with extended-release or active coating products. Application of computer simulation has shown to be a valuable tool and will continue to give new insights. Furthermore, new measurement techniques will be applied both on-line and off-line; a promising candidate is Optical coherence tomography (OCT), which has been developed by us into a on-line monitoring method [196]. The ultimate goal is to go beyond taking samples of limited size towards a 100 % control of all tablets, especially for continuous processes.

10 Process Analytical Technology

Starting in 2004, the FDA released a series of guidelines, with the aim to promote "a risk-based approach" [197] to pharmaceutical development and manufacturing. As part of this approach, the Process Analytical Technology (PAT) initiative deals with real-time process analytics, as opposed to traditional end-of-line testing. The goal of PAT is to enhance understanding and control of the manufacturing process [198]. This should ultimately lead to real-time release (RTR) of manufactured drugs. Therefore, the PAT framework is described as a system for designing, analyzing and controlling manufacturing through timely measurements (i.e., during processing) of critical quality and performance attributes of raw and in-process materials and processes, with the goal of ensuring final product quality [198]. This includes monitoring of the process state, as well as raw and processed materials. Thereby, a sound scientific knowledge of the processes and their impact on the products needs to be gained. Only by then, an engineering approach of designing pharmaceutical processes such that the final products meet the critical quality attributes (CQAs), in contrast to a trial-and-error approach, will be possible. This is condensed in the Quality by Design (QbD) paradigm: "quality cannot be tested into products, it should be built-in or should be by design" [198]. With ICH Q8R [199] the (potential) ability to freely operate processes within a design space was introduced, which is important for process control. Thus, by showing profound scientific understanding of the process, optimizations became possible. In order to fully take advantage of this freedom, control strategies based on in-line measurements are needed, as stated by ICH Q10 [200]. In summary, a process is well understood, when: (1) all critical sources of variability are identified and explained, (2) the encountered variability is managed by the process according to control strategies and (3) the product quality attributes can be accurately and reliably predicted [198].

Along with the PAT initiative continuous manufacturing became of significant interest to industry. As opposed to batch manufacturing, in continuous production usually only a small product volume is processed at once. Thus, even short fluctuations might impact the final product quality [201]. As a consequence, PAT tools are necessary to monitor the process in real-time and control strategies have to be in place to keep the process within a desired range of operation. In contrast to continuous manufacturing, the aim of a control strategy for batch is quite different. Here the process is actively manipulated to follow a predefined trajectory, until the PAT-determined end-point is reached. The end-point is defined as the achievement of desired material attributes (e.g., particle size for granulation or water content for drying).

A key driver for PAT implementation in the pharmaceutical industry is the availability of advanced sensing devices. A wide range of sensors is available. The sensors can be divided in several groups, according to their integration in the process (in-line, on-line, at-line) and the number of output variables (univariate and multivariate). More important for controlling a process is to know what information can be obtained by a sensor. Therefore, the overview shown in Fig. 30 is based on the properties of interest and associated techniques. This overview is by no means exhaustive, but covers many important aspects of pharmaceutical manufacturing, such as chemical composition (spectroscopic techniques such as infrared [202], near infrared (NIR) [86, 137, 203, 204], Raman [86, 204–207], UV–Vis [208], ultrasound [209, 210], laser induced fluorescence [211, 212] and time of flight [213]), particle size and shape (laser diffraction [214–216], image analysis [138], focused beam reflectance measurement [217], spatial filter velocimetry [90, 218]), crystallinity and polymorphism (small- and wide-angle X-ray scattering [123, 219, 220], NIR [221], Raman [221–223]) as well as physical properties (Terahertz [224, 225], optical coherence tomography [224, 226], focus variation [227], ultrasound [228]).

PAT tools often deliver multivariate (highly correlated) data which need to be processed to extract relevant information about the process and/or product state as schematically illustrated in Fig. 29. In general, data processing is performed either by a calibration model which relates the raw data to one or several product quality parameters or by an automatic evaluation algorithm. More complex and advanced data processing might combine both cases.

Calibration models are typically data-driven models (e.g., multiple linear regression (MLR), projection to latent structures (PLS), support vector machine (SVM)), which needs to be developed a priori. Such models are mostly applied for spectroscopic data. A typical example in the pharmaceutical industry is the use of NIR spectroscopy. Figure 30 depicts NIR spectra of a molten extrudate, which is processed by hot melt extrusion (HME) [137]. Spectra might contain variations in baseline offset and multiplicative effects,

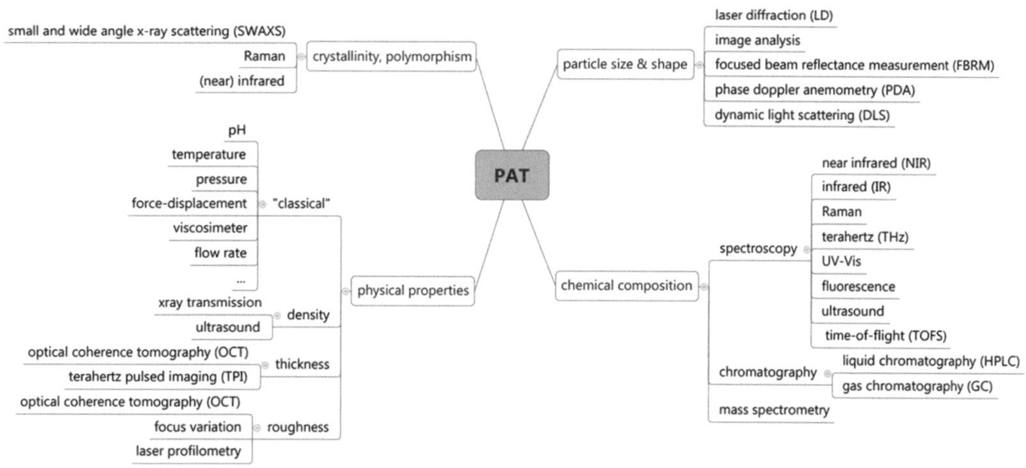

Fig. 29 Overview of important process PAT tools to measure important material properties

that are usually irrelevant for determining the property of interest. In case of HME these effects include the impact of material temperature and its opacity on the NIR spectra, among others. Therefore, preprocessing methods (e.g., derivatives, standard normal variate) are used to reduce or eliminate such undesirable effects in the data, which are not correlated to the property of interest and might influence the model performance.

The other processing step is associated with algorithms, which extract the informative data without the need of an a priori developed data-driven model. Such processing is specifically the case for imaging techniques, e.g., particle size measurement or optical coherence tomography (OCT). OCT provides cross-sectional images of samples (*see* Fig. 30 for OCT images of film-coated pellets) to reveal the internal structure in a non-destructive manner. This was used to determine the film thickness of coated pellets [196]. However, such a method requires an algorithm for the segmentation of the coating layer and an automated calculation of the thickness, enabling real-time process monitoring.

With the extended use of in-line analytics the amount of information about (approved) production processes will clearly increase. This might raise the fear within pharmaceutical companies to detect unwanted deviations, which might result in punishment from regulatory authorities. In case of an experimental PAT tool used to evaluate the suitability of an analyzer, the FDA stated that the collected data should be considered as research data. The FDA does not intend to inspect such research data collected on an existing product [198]. Within such a suitability study the experimental PAT tool should provide information to enhance process and product understanding. When finally pursuing RTR, validated process analyzers need to be used and the monitored data has to withstand FDA inspections. Therefore, the high level of gained

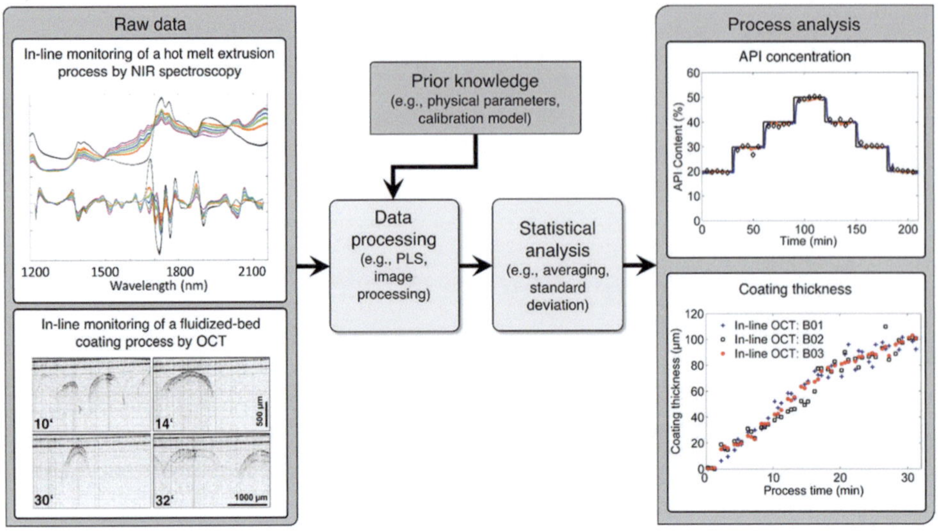

Fig. 30 Schematic overview of typical sequence from multivariate raw data (e.g., spectra, images) to product quality parameters (e.g., API concentration, coating thickness of pellets or tablets)

process understanding must be transformed into robust, tightly control processes. In this setting PAT can help to constantly produce highest quality pharmaceuticals.

10.1 Outlook

In the field of PAT several new developments are highly needed. These include:

- Use of entirely new measurement principles (e.g., for powder flow rate).
- Making current sensors calibration-free to (at least) simplify calibration and model development.
- To reduce costs of spectroscopic tools and other sensors, allowing routine application.
- To improve robustness and ease of use.
- To reduce sensor fouling problems.
- To allow easy (plug-and-play) integration of sensors and PAT tools in process control environments.

In summary, significant challenges still exist, providing ample opportunities for companies and academia.

11 Continuous Manufacturing

Traditional batch manufacturing follows a sequential approach. The materials are charged before the start of processing in a specific unit operation, transformed in a processed intermediate product and discharged at the end of processing. After each production step the intermediate products are collected, if needed analyzed and transferred manually to the next process step in various containers (IBC). Usually, the final product is tested extensively off-line in the quality-assurance laboratory. Depending on the number and nature of the unit operations (typically 5–15) the batch manufacturing process takes several days up to weeks.

In contrast, a continuous manufacturing (CM) process, while consisting of the same unit operations as the batch process, takes in the order of hours to make the final product. Material is simultaneously charged and discharged from the process and is transferred automatically, monitored and controlled in-line along the manufacturing path. Based on the implemented control strategy the process can be adjusted by means of in-process measurements. The quality is assured (QA) in real-time and—in theory—real-time release should be possible. Furthermore, many pharmaceutical unit operations, such as roller compaction [229, 230], tablet compression, extrusion, and capsule filling are inherently continuous process steps, that are operated in a batch mode, not utilizing their full manufacturing potential. In contrast, design efforts have to be made to convert blending, granulation, drying, and coating into continuous unit operations.

11.1 Advantages of CM

In addition to the reduced process time and possibility for real-time QA, continuous manufacturing has several more advantages that justify its application. First one has to distinguish between primary and secondary manufacturing. In the first case, i.e., the manufacturing of the API advantages are not as clear. Typically, the translation of batch synthesis chemistry into continuous flow chemistry does not yield advantages. Rather the opposite is true. Thus, in order to reap the benefits of CM, special chemistries have to be developed that gain from the characteristics of a flow-through system. Examples include fast, highly exothermic reactions that can be carried out in the small volume of a flow-through reactor (which would not be possible in batch) or using much more active (and thus selective) catalysts.

In contrast, the advantages of secondary manufacturing are quite obvious. For example, the possibility to increase the production volume without time-consuming scale-up is a major benefit. Thus, scale-up can be done by increasing processing time [231]. Scale-down, which is rarely addressed in classical pharma-development, is possible too. Thus, production of small batches is within reach, allowing efficient and flexible production of individualized medicines. This is referred to as "agility," allowing faster product development and a shorter time-to-market of new drugs [232]. Furthermore, continuous manufacturing provides increased "flexibility" in production, as the process parameter can be optimized in real-time to achieve high quality. In batch manufacturing strict production and qualification guidelines, that permit only minor adjustments to the process parameters, lead to so-called "frozen processes." As a consequence, variations in final product quality caused by varying raw material quality cannot be prevented. Thus, continuous manufacturing may provide increased product quality and decreased quality associated costs [232, 233]. Lastly, investment cost, floor space and labor costs may be reduced significantly, and product transfer steps are eliminated or minimized, improving safety and environmental impact.

11.2 Challenges of CM

Several challenges exist and CM may not be right for a specific product. The first challenge is the integration of traditional batch-processes (i.e., high-shear granulation, drying, or coating) into an automated and monitored continuous production line. However, novel approaches (e.g., granulation in twin-screw extruders) do exist. Furthermore, the material has to be transferred between processes without interruptions, avoiding segregation and including on-line QA via NIR, Raman, or other on-line analyzers. Many interactions between consecutive process steps have to be described and understood, leading to complex process-development experiments. The proper definition of the design space for an integrated process chain is thus more challenging and cost intensive [234]. Here, a promising approach to reduce experiments is the system-based approach. Process understanding is generated by mechanistic models and process monitoring is performed by integrated in-line and on-line PAT tools. Control is performed via modern model-predictive control (MPC) methods, including exceptional events management and strategies for start-up and shut down. The challenge of this approach is that it is relatively new, not widely applied in pharmaceutical industry and reliable process models are still missing or under development. Moreover, regulatory aspects need to be addressed, including batch definition, recall possibility, proper documentation and the possibility to investigate discrepancies in the product quality.

11.3 Existing Systems

One approach to the integration of multiple continuous unit operations in a continuous downstream line is the Consigma conti-line of GEA. Consigma is an integrated tableting line including continuous wet granulation via twin-screw extrusion, semi-continuous drying in a segmented fluid bed, and tableting equipped with state of the art on-line monitoring systems [235]. Most recently Glatt introduced the "MODCOS" system which is a continuous rotary chamber insert that can be used to convert Glatt's GPCG drying batch system into a continuous fluidized bed drying system. That in combination with various associated continuous process equipment from other companies like feeder, PAT and continuous granulation system make an integrated continuous wet granulation line possible. Moreover, additional industrial systems are in development, e.g., the conti-line by Bohle.

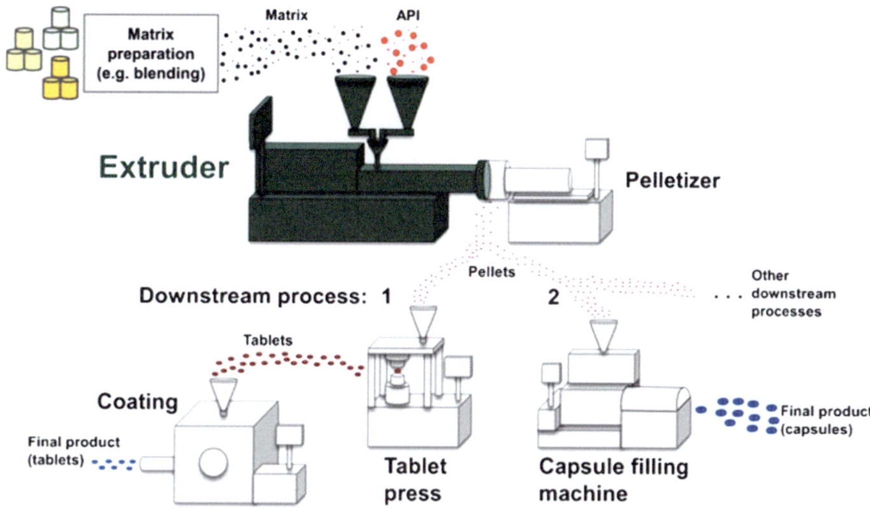

Fig. 31 Schematic illustration of a continuous plant, including the extruder and various

A group at University of Eastern Finland (Jarkko Ketolainen) replaced wet granulation by a dry granulation system. Rutgers University has developed a continuous manufacturing plant based on blending and direct compaction [85]. Hot-melt extrusion (HME) is a granulation technique with a wide range of possible application. Especially, the wide range of downstream possibilities to shape the extrudates is a significant opportunity for pharmaceutical industry. RCPE and the Graz University of Technology developed a line integrating continuous hot-melt extrusion (*see* Fig. 31).

The Novartis-MIT center for continuous manufacturing included also the primary manufacturing into the continuous production line [234]. CMAC at Strathclyde University (UK) is focusing on similar problems, ranging from synthesis to crystallization. All these promising concepts are up to date already realized on lab-scale but there is still no use case established on production scale.

11.4 Regulatory Environment

In general, regulators in the US and EU are supporting CM: One of the focus areas of the FDA has been paving the way for CM in pharmaceutical industry by the release of various quality-related guidance and initiatives. This started in 2003 with the release of FDA's initiative "Pharmaceutical CGMPs for the 21st century – a risk based approach" [197, 236] followed by other relevant documents like the PAT Guidance 2004 [198]. Also the ICH has contributed via the ICH Q8 and Q9 in 2006, ICH Q8 R1 in 2009 or ICH Q11 from 2013 to name a few. The FDA-EMA pilot program on QbD also supports these development, i.e., the shift of focus from experimental based towards mechanistic process understanding provides manufacturers with more freedom and enables continuous manufacturing approaches including model based control strategies [229].

However, a key factor for the successful implementation of continuous manufacturing in production remains the design of an effective plant-wide control strategy, which has to optimize the performance of the plant as a whole instead of isolated unit operations [234, 237]. ICH Q10 defines a control strategy as “a planned set of controls derived from current product and process understanding that assures process performance and product quality” [238, 239]. The main objective here is an excellent and constant product quality and the reduction of quality costs caused by off-line end product testing. To develop a fully automated and controlled process concept, Real Time Release Testing (RTRT) has to be realized. ICH Q8 (R2) defines RTRT as “the ability to evaluate and ensure the quality of in-process and/or final product based on process data, which typically includes a valid combination of measured material attributes and process controls.” Only the combination of mechanistic process understanding, PAT-tools that enable RTRT and a proper control strategy enables an automated CM processes. Only then the process can react to variability in raw material attributes and changing environmental conditions, for example when transferring a process to another site without deviations in final product quality [85, 239].

11.5 Outlook

Benefits of CM are obvious. However, there are still some hurdles to overcome. Advanced control strategies and real-time release testing are challenging to implement. Furthermore, the definition of a batch and the tracking of the material through the system by a fully understood residence time distribution is a task to be solved by developers and equipment suppliers. Here modeling and simulation must play an important role.

From regulatory perspective no specific guidance for continuous manufacturing exists up to day. Here new developments are expected. To bring continuous manufacturing to the production scale not only the regulatory uncertainties have to be eliminated but also economic benefit over traditional batch processing has to be proven [240].

In R&D continuous manufacturing enables a more rapidly development of novel dosage forms. Experiments can be performed automated so that the influence of process parameters on product quality can be screened easily over many conditions. Furthermore product development can be carried out on production scale what leads to a decreased time to market of new drug substances.

Lastly, CM must also encompass the continuous production of individualized, patient-centric drug products. Printing of drugs directly for the individual patient may be an option in this regard, as shown in Fig. 32. In this field, however, significant development work needs to be done.

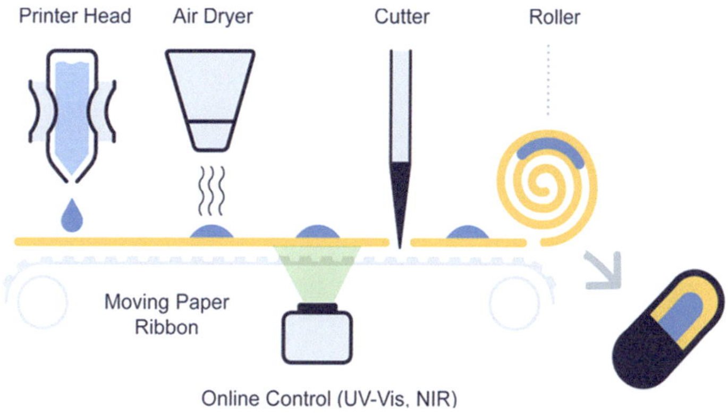

Fig. 32 API solution/suspension printing on excipient substrate [241]

12 Summary

In the last sections an overview of modern manufacturing approaches for solid dosage forms is given, with a focus on current developments. As can be seen from the above, the pharmaceutical industry is in a transition. Patient-centric, individualized drug products are becoming increasingly important, together with continuous manufacturing and advanced processing concepts, such as injection molding or low-dose capsule filling. Modern process sensors allow obtaining a very detailed picture of the quality of products directly in real time within the process. The regulatory environment demands better engineering science and process understanding. These are drivers for change, and thus, modeling and simulation is increasingly important.

Acknowledgement

The authors acknowledge the significant contributions of Diana Dujmovic, Andreas Eitzlmayr, Eva Faulhammer, Johannes Gursch, Dr. Gerold Koscher, Dr. Marcos Llusa, Daniel Markl, Dr. Sharareh Salar-Behzadi, Dr. Otto Scheibelhofer, Daniel Treffer, Dr. Gregor Toschkoff, and Dr. Patrick Wahl. Much of the work presented in this chapter was funded through the Austrian COMET Program by the Austrian Federal Ministry of Transport, Innovation and Technology (BMVIT), the Austrian Federal Ministry of Economy, Family and Youth (BMWFJ) and by the State of Styria (Styrian Funding Agency SFG).

References

1. Leane M, Pitt K, Reynolds G (2014) A proposal for a drug product Manufacturing Classification System (MCS) for oral solid dosage forms. Pharm Dev Technol 27:1–10
2. Roberts AW (2006) An historical overview and current developments. Bulk Solids Handl 26(6):392–411
3. Yang S, Evans JRG (2007) Metering and dispensing of powder; the quest for new solid freeforming techniques. Powder Technol 178(1):56–72
4. Kehlenbeck V (2006) Continuous dynamic mixing of cohesive powders. Dissertation, Technical University Munich
5. Uchida K, Okamoto K (2008) Measurement technique on the diffusion coefficient of powder flow in a screw feeder by X-ray visualization. Powder Technol 187(2):138–145
6. Fernandez JW, Cleary PW, McBride W (2011) Effect of screw design on hopper drawdown of spherical particles in a horizontal screw feeder. Chem Eng Sci 66(22): 5585–5601
7. Schulze D (2007) Powders and bulk solids: behavior, characterization, storage and flow. Springer Berlin Heidelberg New York, Berlin Heidelberg
8. Francis TM, Gump CJ, Weimer AW (2006) Spinning wheel powder feeding device — fundamentals and applications. Powder Technol 170(1):36–44
9. Mracek M, Wallaschek J (2005) A system for powder transport based on piezoelectrically excited ultrasonic progressive waves. Mater Chem Phys 90(2–3):378–380
10. Engisch W, Muzzio F (2012) Method for characterization of loss-in-weight feeder equipment. Powder Technol 228:395–403
11. Pernenkil L, Cooney CL (2006) A review on the continuous blending of powders. Chem Eng Sci 61(2):720–742
12. Berthiaux H, Marikh K, Gatumel C (2008) Continuous mixing of powder mixtures with pharmaceutical process constraints. Chem Eng Process Process Intensif 47(12): 2315–2322
13. Sato T (2010) Design of a GPC-based PID controller for controlling a weigh feeder. Control Eng Pract 18(2):105–113
14. Pahl MH (2000) Mischen, Rühren, Kneten und Dosieren. Chemie Ing Tech 12(72): 1451–1459
15. Clark JP (2009) Case studies in food engineering. In: Case studies in food engineering. Springer, London
16. Manjunath K, Dhodapkar S, Jacob K (2004) Mixing of particulate solids in the process industries. In: Paul EL, Atiome-Obeng VA, Kresta SM (eds) Handbook of Industrial Mixing. Wiley, Hoboken, NJ, USA, pp 924–986
17. Pu YU, Mazumder M, Cooney C (2009) Effects of electrostatic charging on pharmaceutical powder blending homogeneity. J Pharm Sci 98(7):2412–2421
18. Hogg R (2009) Mixing and segregation in powders: evaluation, mechanisms and processes. Powder Technol 27(27):3–17
19. Mueller M (2007) Feststoffmischen. Chemie Ing Tech 79(7):1015–1023
20. Poux M, Fayolle P, Bertrand J, Bridoux D, Bousquet J (1991) Powder mixing: some practical rules applied to agitated systems. Powder Technol 68(3):213–234
21. Fitzpatrick JJ (2009) Particulate and powder mixing. In: Cullen PJ (ed.), Food mixing. Blackwell Publishing Ltd., Oxford, p 269–287
22. Lacy P (1954) Developments in the theory of particle mixing. J Appl Chem 4(5):257
23. Deveswaran R, Bharath S, Basavaraj BV, Abraham S, Furtado S, Madhavan V (2009) Concepts and techniques of pharmaceutical powder mixing process: a current update. Res J Pharm Technol 2(2):245–249
24. Esbensen KH, Mortensen P (2010) Process sampling (Theory of Sampling, TOS)-the missing link in process analytical technology. In: Baakev KA (ed) Process analytical technology, 2nd edn. Wiley, Hoboken, pp 37–80
25. Muzzio FJ, Robinson P, Wightman C (1997) Sampling practices in powder blending. Int J Pharm 155(2):153–178
26. Huang A-N, Kuo H (2014) Developments in the tools for the investigation of mixing in particulate systems–a review. Adv Powder Technol 25(1):163–173
27. Scheibelhofer O, Balak N, Koller DM, Khinast JG (2013) Spatially resolved monitoring of powder mixing processes via multiple NIR-probes. Powder Technol 243: 161–170
28. Rosas JG, Blanco M (2012) A criterion for assessing homogeneity distribution in hyperspectral images. Part 1: homogeneity index bases and blending processes. J Pharm Biomed Anal 70:680–690
29. Mehrotra A, Muzzio FFJ (2009) Comparing mixing performance of uniaxial and biaxial bin blenders. Powder Technol 196(1):1–7
30. (2014) http://www.resodynmixer.com

31. Martínez L, Peinado A, Liesum L, Betz G (2013) Use of near-infrared spectroscopy to quantify drug content on a continuous blending process: influence of mass flow and rotation speed variations. Eur J Pharm Biopharm 84(3):606–615
32. Portillo PM, Ierapetritou MG, Muzzio FJ (2008) Characterization of continuous convective powder mixing processes. Powder Technol 182(3):368–378
33. Muzzio FJ, Ierapetritou MG (2013) Scale-up strategy for continuous powder blending process. Powder Technol 235:55–69
34. Radeke CA, Glasser BJ, Khinast JG (2010) Large-scale powder mixer simulations using massively parallel GPUarchitectures. Chem Eng Sci 65(24):6435–6442
35. Siraj MS, Radl S, Glasser BJ, Khinast JG (2011) Effect of blade angle and particle size on powder mixing performance in a rectangular box. Powder Technol 211(1):100–113
36. Bridgwater J (2010) Mixing of particles and powders: where next? Particuology 8 (6):563–567
37. Scheibelhofer O, Balak N, Wahl PR, Koller DM, Glasser BJ, Khinast JG (2013) Monitoring blending of pharmaceutical powders with multipoint NIR spectroscopy. AAPS PharmSciTech 14(1):234–244
38. Liu R, Yin X, Li H, Shao Q, York P, He Y, Xiao T, Zhang J (2013) Visualization and quantitative profiling of mixing and segregation of granules using synchrotron radiation X-ray microtomography and three dimensional reconstruction. Int J Pharm 445 (1–2):125–133
39. Gao JZH, Jain A, Motheram R, Gray DB, Hussain MA (2002) Fluid bed granulation of a poorly water soluble, low density, micronized drug: comparison with high shear granulation. Int J Pharm 237(1–2):1–14
40. Kumar A, Gernaey KV, De Beer T, Nopens I (2013) Model-based analysis of high shear wet granulation from batch to continuous processes in pharmaceutical production–a critical review. Eur J Pharm Biopharm 85(3 Pt B):814–832
41. Osborne JD, Sochon RPJ, Cartwright JJ, Doughty DG, Hounslow MJ, Salman AD (2011) Binder addition methods and binder distribution in high shear and fluidised bed granulation. Chem Eng Res Des 89 (5):553–559
42. Saito Y, Fan X, Ingram A, Seville JPK (2011) A new approach to high-shear mixer granulation using positron emission particle tracking. Chem Eng Sci 66(4):563–569
43. Behzadi SS, Toegel S, Viernstein H (2008) Innovations in coating technology. Recent Pat Drug Deliv Formul 2(3):209–230
44. Burggraeve A, Monteyne T, Vervaet C, Remon JP, De Beer T (2013) Process analytical tools for monitoring, understanding, and control of pharmaceutical fluidized bed granulation: a review. Eur J Pharm Biopharm 83 (1):2–15
45. Dhenge RM, Cartwright JJ, Doughty DG, Hounslow MJ, Salman AD (2011) Twin screw wet granulation: effect of powder feed rate. Adv Powder Technol 22(2):162–166
46. Barrasso D, Ramachandran R (2015) Multiscale modeling of granulation processes: bidirectional coupling of PBM with DEM via collision frequencies. Chem Eng Res Des 93:304–317
47. Dhenge RM, Cartwright JJ, Hounslow MJ, Salman AD (2012) Twin screw wet granulation: effects of properties of granulation liquid. Powder Technol 229:126–136
48. Tu W-D, Ingram A, Seville J (2013) Regime map development for continuous twin screw granulation. Chem Eng Sci 87:315–326
49. Iveson SM, Litster JD (1998) Growth regime map for liquid-bound granules. AIChE J 44 (7):1510–1518
50. Iveson SM, Wauters PAL, Forrest S, Litster JD, Meesters GMH, Scarlett B (2001) Growth regime map for liquid-bound granules: further development and experimental validation. Powder Technol 117(1–2):83–97
51. Hapgood KP, Litster JD, Smith R (2003) Nucleation regime map for liquid bound granules. AIChE J 49(2):350–361
52. Hapgood KP, Amelia R, Zaman MB, Merrett BK, Leslie P (2010) Improving liquid distribution by reducing dimensionless spray flux in wet granulation—a pharmaceutical manufacturing case study. Chem Eng J 164 (2–3):340–349
53. Pohlman DA, Litster JD (2014) Coalescence model for induction growth behavior in high shear granulation. Powder Technol 270:435–444
54. Kano T, Yoshihashi Y, Yonemochi E, Terada K (2014) Clarifying the mechanism of aggregation of particles in high-shear granulation based on their surface properties by using micro-spectroscopy. Int J Pharm 461(1–2): 495–504
55. Kleinebudde P (2004) Roll compaction/dry granulation: pharmaceutical applications. Eur J Pharm Biopharm 58(2):317–326
56. Herting MGM, Kleinebudde P (2007) Roll compaction/dry granulation: effect of raw

material particle size on granule and tablet properties. Int J Pharm 338(1–2):110–118
57. Bozic DZ, Dreu R, Vrecer F (2008) Influence of dry granulation on compactibility and capping tendency of macrolide antibiotic formulation. Int J Pharm 357(1–2):44–54
58. Mansa RF, Bridson RH, Greenwood RW, Barker H, Seville JPK (2008) Using intelligent software to predict the effects of formulation and processing parameters on roller compaction. Powder Technol 181(2):217–225
59. Yu S, Gururajan B, Reynolds G (2012) A comparative study of roll compaction of free-flowing and cohesive pharmaceutical powders. Int J Pharm 428(1–2):39–47
60. Yu S, Adams M, Gururajan B (2013) The effects of lubrication on roll compaction, ribbon milling and tabletting. Chem Eng Sci 86:9–18
61. Liu LX, Marziano I, Bentham AC, Litster JD, White ET, Howes T (2013) Influence of particle size on the direct compression of ibuprofen and its binary mixtures. Powder Technol 240:66–73
62. Wagner CM, Pein M, Breitkreutz J (2015) Roll compaction of granulated mannitol grades and the unprocessed crystalline delta-polymorph. Powder Technol 270:470–475
63. Hadžović E, Betz G, Hadžidedić S, El-Arini SK, Leuenberger H (2011) Investigation of compressibility and compactibility parameters of roller compacted Theophylline and its binary mixtures. Int J Pharm 416(1):97–103
64. Faure A, York P, Rowe RC (2001) Process control and scale-up of pharmaceutical wet granulation processes: a review. Eur J Pharm Biopharm 52(3):269–277
65. Kayrak-Talay D, Dale S, Wassgren C, Litster J (2013) Quality by design for wet granulation in pharmaceutical processing: assessing models for a priori design and scaling. Powder Technol 240:7–18
66. Rambali B, Baert L, Massart DL (2001) Using experimental design to optimize the process parameters in fluidized bed granulation on a semi-full scale. Int J Pharm 220 (1–2):149–160
67. Rambali B (2003) Scaling up of the fluidized bed granulation process. Int J Pharm 252 (1–2):197–206
68. Watano S, Sato Y, Miyanami K (1997) Application of a neural network to granulation scale-up. Powder Technol 90:153–159
69. Nakamura H, Fujii H, Watano S (2013) Scale-up of high shear mixer-granulator based on discrete element analysis. Powder Technol 236:149–156
70. Chaudhury A, Niziolek A, Ramachandran R (2013) Multi-dimensional mechanistic modeling of fluid bed granulation processes: an integrated approach. Adv Powder Technol 24(1):113–131
71. Chitu TM, Oulahna D, Hemati M (2011) Rheology, granule growth and granule strength: application to the wet granulation of lactose – MCC mixtures. Powder Technol 208(2):441–453
72. Dosta M, Heinrich S, Werther J (2010) Fluidized bed spray granulation: analysis of the system behaviour by means of dynamic flowsheet simulation. Powder Technol 204(1): 71–82
73. Chaudhury A, Wu H, Khan M, Ramachandran R (2014) A mechanistic population balance model for granulation processes: effect of process and formulation parameters. Chem Eng Sci 107:76–92
74. Kayrak-Talay D, Litster JD (2011) A priori performance prediction in pharmaceutical wet granulation: testing the applicability of the nucleation regime map to a formulation with a broad size distribution and dry binder addition. Int J Pharm 418(2):254–264
75. Zarate NV, Harrison AJ, Litster JD, Beaudoin SP (2013) Effect of relative humidity on onset of capillary forces for rough surfaces. J Colloid Interface Sci 411:265–272
76. Boerefijn R, Orlovic M, Reimers C, Pogodda M, Jacob M (2013) Application of the flux number approach for the simulation of granulation processes by use of flowsheeting tools. Chem Eng Sci 86:137–145
77. Emady HN, Kayrak-Talay D, Litster JD (2013) Modeling the granule formation mechanism from single drop impact on a powder bed. J Colloid Interface Sci 393: 369–376
78. Poon JM-H, Immanuel CD, Doyle FJ III, Litster JD (2008) A three-dimensional population balance model of granulation with a mechanistic representation of the nucleation and aggregation phenomena. Chem Eng Sci 63(5):1315–1329
79. Smith RM, Litster JD (2012) Examining the failure modes of wet granular materials using dynamic diametrical compression. Powder Technol 224:189–195
80. Peter S, Lammens R, Steffens K (2010) Roller compaction/dry granulation: use of the thin layer model for predicting densities and forces during roller compaction. Powder Technol 199(2):165–175
81. Nesarikar V, Patel C, Early W (2012) Roller compaction process development and scale up

using Johanson model calibrated with instrumented roll data. Int J Pharm 436(1–2): 486–507

82. Miguélez-Morán A, Wu C (2009) Characterisation of density distributions in roller-compacted ribbons using micro-indentation and X-ray micro-computed tomography. Eur J Pharm Biopharm 72(1):173–182
83. Reynolds G, Ingale R, Roberts R, Kothari S, Gururajan B (2010) Practical application of roller compaction process modeling. Comput Chem Eng 34(7):1049–1057
84. Cunningham JJC, Winstead D, Zavaliangos A (2010) Understanding variation in roller compaction through finite element-based process modeling. Comput Chem Eng 34 (7):1058–1071
85. Singh R, Ierapetritou M, Ramachandran R (2012) An engineering study on the enhanced control and operation of continuous manufacturing of pharmaceutical tablets via roller compaction. Int J Pharm 438(1–2): 307–326
86. De Beer T, Burggraeve A, Fonteyne M, Saerens L, Remon JP, Vervaet C (2011) Near infrared and Raman spectroscopy for the in-process monitoring of pharmaceutical production processes. Int J Pharm 417(1–2): 32–47
87. Walker G, Bell SEJ, Vann M, Jones DS, Andrews G (2007) Fluidised bed characterisation using Raman spectroscopy: applications to pharmaceutical processing. Chem Eng Sci 62(14):3832–3838
88. Walker GM, Bell SEJ, Greene K, Jones DS, Andrews GP (2009) Characterisation of fluidised bed granulation processes using in-situ Raman spectroscopy. Chem Eng Sci 64 (1):91–98
89. Burggraeve A, Van den Kerkhof T, Hellings M, Remon JP, Vervaet C, De Beer T (2011) Batch statistical process control of a fluid bed granulation process using in-line spatial filter velocimetry and product temperature measurements. Eur J Pharm Sci 42(5):584–592
90. Burggraeve A, Van Den Kerkhof T, Hellings M, Remon JP, Vervaet C, De Beer T (2010) Evaluation of in-line spatial filter velocimetry as PAT monitoring tool for particle growth during fluid bed granulation. Eur J Pharm Biopharm 76(1):138–146
91. Hansuld EM, Briens L (2014) A review of monitoring methods for pharmaceutical wet granulation. Int J Pharm 472(1–2):192–201
92. Matero S, Poutiainen S, Leskinen J, Järvinen K, Ketolainen J, Reinikainen S, Hakulinen M, Lappalainen R, Poso A (2009) The feasibility of using acoustic emissions for monitoring of fluidized bed granulation. Chemom Intell Lab Syst 97(1):75–81
93. van't Land CM (2012) Drying in the process industry. John Wiley & Sons, Hoboken, NJ
94. Mujumdar AS (ed) (2006) Handbook of industrial drying, 3rd edn. CRC Press, Boca Raton, FL
95. Hamilton P, Littlejohn D, Nordon A, Sefcik J, Slavin P, Andrews J, Dallin P (2013) Investigation of factors affecting isolation of needle-shaped particles in a vacuum-agitated filter drier through non-invasive measurements by Raman spectrometry. Chem Eng Sci 101: 878–885
96. Sahni EK, Chaudhuri B (2013) Numerical simulations of contact drying in agitated filter-dryer. Chem Eng Sci 97:34–49
97. Paudel A, Worku ZA, Meeus J, Guns S, Van den Mooter G (2013) Manufacturing of solid dispersions of poorly water soluble drugs by spray drying: formulation and process considerations. Int J Pharm 453(1):253–284
98. Son Y-J, Worth Longest P, Hindle M (2013) Aerosolization characteristics of dry powder inhaler formulations for the excipient enhanced growth (EEG) application: effect of spray drying process conditions on aerosol performance. Int J Pharm 443 (1–2):137–145
99. Patel BB, Patel JK, Chakraborty S, Shukla D (2013) Revealing facts behind spray dried solid dispersion technology used for solubility enhancement. Saudi Pharm J. Online Dec. 2013
100. Perry RH, Green DW (2012) Fluidized-bed systems. In: Perry's chemical engineer's handbook, 8th edn. McGraw-Hill, New York, p 1–27
101. Geldart D (1972) The effect of particle size and size distribution on the behaviour of gas-fluidised beds. Powder Technol 6 (4):201–215
102. Geldart D (1973) Types of gas fluidization. Powder Technol 7:285–292
103. Mujumdar AS (2015) Handbook of industrial drying, fourth edition, 4th edn. CRC Press, Boca Raton, FL
104. Bhandari B, Bansal N, Zhang M, Schuck P (eds.) (2013) Handbook of food powders—processes and properties. Woodhead Publishing Limited, Cambridge, p 85–104
105. Ganguly A, Alexeenko AA (2012) Modeling and measurements of water–vapor flow and icing at low pressures with application to pharmaceutical freeze-drying. Int J Heat Mass Transf 55(21–22):5503–5513

106. Oetjen GW, Haseley P (2008) Freeze-drying. Wiley-VCH, Weinheim
107. Anwar SH, Kunz B (2011) The influence of drying methods on the stabilization of fish oil microcapsules: comparison of spray granulation, spray drying, and freeze drying. J Food Eng 105(2):367–378
108. Abdelwahed W, Degobert G, Fessi H (2006) Freeze-drying of nanocapsules: impact of annealing on the drying process. Int J Pharm 324(1):74–82
109. Abdelwahed W, Degobert G, Fessi H (2006) Investigation of nanocapsules stabilization by amorphous excipients during freeze-drying and storage. Eur J Pharm Biopharm 63(2): 87–94
110. Abdelwahed W, Degobert G, Stainmesse S, Fessi H (2006) Freeze-drying of nanoparticles: formulation, process and storage considerations. Adv Drug Deliv Rev 58(15): 1688–1713
111. Mcloughlin CM, McMinn WAM, Magee TRA (2000) Microwave drying of pharmaceutical powders. Food Bioprod Process 78 (2):90–96
112. Loh ZH, Liew CV, Lee CC, Heng PWS (2008) Microwave-assisted drying of pharmaceutical granules and its impact on drug stability. Int J Pharm 359(1–2):53–62
113. Chandrasekaran S, Ramanathan S, Basak T (2013) Microwave food processing—a review. Food Res Int 52(1):243–261
114. Hebbar HU, Vishwanathan K, Ramesh M (2004) Development of combined infrared and hot air dryer for vegetables. J Food Eng 65(4):557–563
115. Das I, Das SK, Bal S (2004) Drying performance of a batch type vibration aided infrared dryer. J Food Eng 64(1):129–133
116. Lee G (2002) Pharmaceutical process scale-up. Drugs and the pharmaceutical sciences volume 118. Eur J Pharm Biopharm 54 (3):358
117. Strong J (2009) Scale-up of pharmaceutical manufacturing operations of solid dosage forms. In: Qiu Y, Yisheng C, Zhang GGZ, Liu L, Porter WR (eds) Developing solid oral dosage forms. Elsevier, p 615–636
118. Kemp IC (2011) Drying models, myths, and misconceptions. Chem Eng Technol 34 (7):1057–1066
119. Lekhal A, Girard KP, Brown MA, Kiang S, Glasser BJ, Khinast JG (2003) Impact of agitated drying on crystal morphology: KCl–water system. Powder Technol 132 (2–3):119–130
120. Lekhal A, Girard KP, Brown MA, Kiang S, Khinast JG, Glasser BJ (2004) The effect of agitated drying on the morphology of l-threonine (needle-like) crystals. Int J Pharm 270(1–2):263–277
121. Vauhkonen M, Rimpilainen V, Heikkinen L (2012) Three-dimensional capacitance tomography of a conical fluidized bed reactor. In:2012 I.E. International instrumentation and measurement technology conference proceedings, p 921–923
122. Rimpiläinen V, Heikkinen LM, Vauhkonen M (2012) Moisture distribution and hydrodynamics of wet granules during fluidized-bed drying characterized with volumetric electrical capacitance tomography. Chem Eng Sci 75:220–234
123. Roblegg E, Jäger E, Hodzic A, Koscher G, Mohr S, Zimmer A, Khinast J (2011) Development of sustained-release lipophilic calcium stearate pellets via hot melt extrusion. Eur J Pharm Biopharm 79(3):635–645
124. Khinast JG, Baumgartner R, Roblegg E (2013) Nano-extrusion: a one-step process for manufacturing of solid nanoparticle formulations directly from the liquid phase. AAPS PharmSciTech 14(2):601–604
125. Breitenbach J (2002) Melt extrusion: from process to drug delivery technology. Eur J Pharm Biopharm 54(2):107–117
126. Ghebre-Sellassie I, Martin C (2003) Pharmaceutical extrusion technology. Marcel Dekker Inc., New York
127. Kleinebudde P (2011) Pharmazeutisches Produktdesign: Gezielte Freisetzung von Wirkstoffen durch unterschiedliche Extrusionstechniken. Chemie Ing Tech 83(5): 589–597
128. Douroumis D (ed) (2012) Hot-melt extrusion: pharmaceutical applications. Wiley & Sons, Ltd., Chichester
129. Repka J, Langley MA, DiNunzio N (2013) Melt extrusion - materials, technology and drug product design. Springer, New York
130. Rauwendaal C (2001) Polymer extrusion, 4th edn. Hanser Publishers, Munich, p 781
131. Kohlgrüber K (2008) Co-rotating twin-screw extruders. Carl Hanser Publishers, Munich
132. Kolter K, Karl M, Gryczke A (2012) Hot-melt extrusion with BASF pharma polymers extrusion compendium, 2nd edn. BASF, Ludwigshafen
133. Perdikoulias J, Dobbie T (2003) Die design. Marcel Dekker Inc., New York
134. Han W, Wang X (2012) Optimal geometry design of the coat-hanger die with uniform

outlet velocity and minimal residence time. J Appl Polym Sci 123(4):2511–2516
135. van Laarhoven JAH (2005) Physical-chemical aspects of a coaxial sustained release device based on Poly-EVA. Dissertation, Universiteit van Utrecht
136. Markl D, Wahl PR, Menezes JC, Koller DM, Kavsek B, Francois K, Roblegg E, Khinast JG (2013) Supervisory control system for monitoring a pharmaceutical hot melt extrusion process. AAPS PharmSciTech 14(3): 1034–1044
137. Wahl PR, Treffer D, Mohr S, Roblegg E, Koscher G, Khinast JG (2013) Inline monitoring and a PAT strategy for pharmaceutical hot melt extrusion. Int J Pharm 455(1–2): 159–168
138. Treffer D, Wahl PR, Hörmann TR, Markl D, Schrank S, Jones I, Cruise P, Mürb R-K, Koscher G, Roblegg E, Khinast JG (2014) In-line implementation of an image-based particle size measurement tool to monitor hot-melt extruded pellets. Int J Pharm 466 (1–2):181–189
139. Saerens L, Dierickx L, Quinten T, Adriaensens P, Carleer R, Vervaet C, Remon JP, De Beer T (2012) In-line NIR spectroscopy for the understanding of polymer-drug interaction during pharmaceutical hot-melt extrusion. Eur J Pharm Biopharm 81(1):230–237
140. Treffer D, Wahl P, Markl D, Koscher G, Roblegg E, Khinast J (2013) Hot melt extrusion as a continuous pharmaceutical manufacturing process. In: Repka M (ed) Melt extrusion: equipment and pharmaceutical applications. Springer Publishers. p 363–396
141. Zema L, Loreti G, Melocchi A, Maroni A, Gazzaniga A (2012) Injection molding and its application to drug delivery. J Control Release 159(3):324–331
142. Pawlowski J (1971) Die Ähnlichkeitstheorie in der physikalisch-technischen Forschung, 1st edn. Springer, Berlin/Heidelberg, Germany
143. Rosato D, Rosato D, Rosato M (2000) Injection molding handbook, 3rd edn. Kluwer Academic Publishers Group, Dordrecht, Netherlands
144. Kennedy P, Zheng R (2013) Flow analysis of injection molds, 2nd edn. Carl Hanser Verlag, Munich, p 380
145. Eitzlmayr A, Khinast J, Hörl G, Koscher G, Reynolds G, Huang Z, Booth J, Shering P (2013) Experimental characterization and modeling of twin-screw extruder elements for pharmaceutical hot melt extrusion. AIChE J 59(11):4440–4450
146. Eitzlmayr A, Koscher G, Reynolds G, Huang Z, Booth J, Shering P, Khinast J (2014) Mechanistic modeling of modular, co-rotating twin-screw extruders. Int J Pharm 474(1–2):157–176
147. Eitzlmayr A, Khinast J (2014) Co-rotating twin-screw extruders: detailed analysis based on smoothed particle hydrodynamics. Part 1: hydrodynamics. Chem Eng Sci, May 2015
148. Eitzlmayr A, Khinast J (2014) Co-rotating twin-screw extruders: detailed analysis based on smoothed particle hydrodynamics. Part 2: mixing. Chem Eng Sci, June 2015
149. Sonnergaard J (2001) Investigation of a new mathematical model for compression of pharmaceutical powders. Eur J Pharm Sci 14(2): 149–157
150. Leuenberger H (1982) The compressibility and compactibility of powder systems. Int J Pharm 12(1):41–55
151. Sun CC (2012) Decoding powder tabletability: roles of particle adhesion and plasticity. J Adhes Sci Technol 25(4–5):483–499
152. Palmieri GF, Joiris E, Bonacucina G, Cespi M, Mercuri A (2005) Differences between eccentric and rotary tablet machines in the evaluation of powder densification behaviour. Int J Pharm 298(1):164–175
153. Sinka IC, Motazedian F, Cocks ACF, Pitt KG (2009) The effect of processing parameters on pharmaceutical tablet properties. Powder Technol 189(2):276–284
154. Natoli D, Levin M, Tsygan L, Liu L, Qiu Y, Chen Y, Zhang GGZ, Porter W (2009) Developing solid oral dosage forms. In: Developing solid oral dosage forms. Elsevier, p 725–759
155. Wahl PR, Fruhmann G, Sacher S, Straka G, Sowinski S, Khinast JG (2014) PAT for tableting: inline monitoring of API and excipients via NIR spectroscopy. Eur J Pharm Biopharm 87(2):271–278
156. Kirsch JD, Drennen JK (1999) Nondestructive tablet hardness testing by near-infrared spectroscopy: a new and robust spectral best-fit algorithm. J Pharm Biomed Anal 19(3–4): 351–362
157. Shah RB, Tawakkul MA, Khan MA (2007) Process analytical technology: chemometric analysis of Raman and near infra-red spectroscopic data for predicting physical properties of extended release matrix tablets. J Pharm Sci 96(5):1356–1365

158. Gowen AA, O'Donnell CP, Cullen PJ, Bell SEJ (2008) Recent applications of chemical imaging to pharmaceutical process monitoring and quality control. Eur J Pharm Biopharm 69(1):10–22
159. Wu H, Heilweil EJ, Hussain AS, Khan MA (2008) Process analytical technology (PAT): quantification approaches in terahertz spectroscopy for pharmaceutical application. J Pharm Sci 97(2):970–984
160. Wu H, Heilweil EJ, Hussain AS, Khan MA (2007) Process analytical technology (PAT): effects of instrumental and compositional variables on terahertz spectral data quality to characterize pharmaceutical materials and tablets. Int J Pharm 343(1–2):148–158
161. Serajuddin ATM (1999) Solid dispersion of poorly water-soluble drugs: early promises, subsequent problems, and recent breakthroughs. J Pharm Sci 88:1058–1066
162. Vasconcelos T, Sarmento B, Costa P (2007) Solid dispersions as strategy to improve oral bioavailability of poor water soluble drugs. Drug Discov Today 12 (23–24):1068–1075
163. Ayenew Z, Paudel A, Van den Mooter G (2012) Can compression induce demixing in amorphous solid dispersions? A case study of naproxen-PVP K25. Eur J Pharm Biopharm 81(1):207–213
164. Worku ZA, Aarts J, Van den Mooter G (2014) Influence of compression forces on the structural stability of naproxen/PVP-VA 64 solid dispersions. Mol Pharm 11 (4):1102–1108
165. Au-Yeung KY, Robertson T, Hafezi H, Moon G, DiCarlo L, Zdeblick M, Savage G (2010) A networked system for self-management of drug therapy and wellness. In: Wireless health 2010 on - WH'10, p 1
166. Cole GC(1999) The design and operation of a facility for filling hard shell gelatin capsules. [Online]. http://capsugel.com
167. Augsburger LL (2009) Hard- and soft-shell capsules. In: Florence AT, Siepmann J (eds) Modern pharmaceutics volume 1 basic principles and systems. Informa Healthcare USA, Inc., New York, pp 499–565
168. Ashurst I, Malton A, Prime D, Sumby B (2000) Latest advances in the development of dry powder inhalers. Pharm Sci Technol Today 3(7):246–256
169. Allen LV, Popovich NG, Ansel HC (2011) Caspules. In: Troy DB (ed) Ansel's pharmaceutical dosage forms and drug delivery systems, 9th edn. Lippincott Williams & Wilkins, New York, pp 203–224
170. Podczeck F, Jones BE (2004) Dry filling of hard capsules. In: Jones BE, Podczeck F (eds) Pharmaceutical capsules, 2nd edn. Pharmaceutical Press, London, pp 119–138
171. Watt PR, Armstrong NA (2008) Tablet and capsule machine instrumentation. Pharmaceutical Press, Chicago, pp 1–16
172. Podczeck F, Newton JM (1999) Powder filling into hard gelatine capsules on a tamp filling machine. Int J Pharm 185(2):237–254
173. Edwards D (2010) Applications of capsule dosing techniques for use in dry powder inhalers. Ther Deliv 1(1):195–201
174. Eskandar F, Lejeune M, Edge S (2011) Low powder mass filling of dry powder inhalation formulations. Drug Dev Ind Pharm 37(1): 24–32
175. Faulhammer E, Fink M, Llusa M, Biserni S, Calzolari V, Lawrence SM, Khinast JG (2014) Low dose capsule filling of inhalation products: critical material attributes and process parameters. Int J Pharm 473(1–2):617–626
176. Cole G, Hogan J, Aulton ME, Twitchell AM (2002) Pharmaceutical coating technology. Taylor & Francis, London
177. Toschkoff G, Khinast JG (2013) Mathematical modeling of the coating process. Int J Pharm 457(2):407–422
178. Turton R (2008) Challenges in the modeling and prediction of coating of pharmaceutical dosage forms. Powder Technol 181(2): 186–194
179. Muliadi AR, Sojka PE (2010) A review of pharmaceutical tablet spray coating. Atomization Sprays 20(7):611–638
180. Just S, Toschkoff G, Funke A, Djuric D, Scharrer G, Khinast JG, Knop K, Kleinebudde P (2013) Optimization of inter-tablet coating uniformity for an active coating process at the lab and pilot scale. Int J Pharm 457(1):1–8
181. Brock D, Zeitler JA, Funke A, Knop K, Kleinebudde P (2012) A comparison of quality control methods for active coating processes. Int J Pharm 439(1–2):289–295
182. Dubey A, Boukouvala F, Keyvan G, Hsia R, Saranteas K, Brone D, Misra T, Ierapetritou MG, Muzzio FJ (2012) Improvement of tablet coating uniformity using a quality by design approach. AAPS PharmSciTech 13 (1):231–246
183. Mauritz JMA, Morrisby RS, Hutton RS, Legge CH, Kaminski CF (2010) Imaging

pharmaceutical tablets with optical coherence tomography. J Pharm Sci 99(1):385–391
184. Shen YC (2011) Terahertz pulsed spectroscopy and imaging for pharmaceutical applications: a review. Int J Pharm 417:48–60
185. Ho L, Müller R, Römer M, Gordon KC, Heinämäki J, Kleinebudde P, Pepper M, Rades T, Shen YC, Strachan CJ, Taday PF, Zeitler JA (2007) Analysis of sustained-release tablet film coats using terahertz pulsed imaging. J Control Release 119(3):253–261
186. Koller DM, Hannesschläger G, Leitner M, Khinast JG (2011) Non-destructive analysis of tablet coatings with optical coherence tomography. Eur J Pharm Sci 44(1–2):142–148
187. Ketterhagen W, Am Ende M, Hancock B (2009) Process modeling in the pharmaceutical industry using the discrete element method. J Pharm Sci 98(2):442–470
188. Denis C, Hemati M, Chulia D, Lanne J-Y, Buisson B, Daste G, Elbaz F (2003) A model of surface renewal with application to the coating of pharmaceutical tablets in rotary drums. Powder Technol 130(1–3):174–180
189. Dubey A, Hsia R, Saranteas K, Brone D, Misra T, Muzzio FJ (2011) Effect of speed, loading and spray pattern on coating variability in a pan coater. Chem Eng Sci 66(21): 5107–5115
190. Freireich B, Ketterhagen WR, Wassgren C (2011) Intra-tablet coating variability for several pharmaceutical tablet shapes. Chem Eng Sci 66(12):2535–2544
191. Freireich B, Li J (2013) A renewal theory approach to understanding inter-particle coating variability. Powder Technol 249: 330–338
192. Kalbag A, Wassgren C (2009) Inter-tablet coating variability: tablet residence time variability. Chem Eng Sci 64(11):2705–2717
193. Sandadi S, Pandey P, Turton R (2004) In situ, near real-time acquisition of particle motion in rotating pan coating equipment using imaging techniques. Chem Eng Sci 59(24): 5807–5817
194. Suzzi D, Toschkoff G, Radl S, Machold D, Fraser SD, Glasser BJ, Khinast JG (2012) DEM simulation of continuous tablet coating: effects of tablet shape and fill level on inter-tablet coating variability. Chem Eng Sci 69(1):107–121
195. Mueller R, Kleinebudde P (2007) Prediction of tablet velocity in pan coaters for scale-up. Powder Technol 173(1):51–58
196. Markl D, Zettl M, Hannesschläger G, Sacher S, Leitner M, Buchsbaum A, Khinast JG (2014) Calibration-free in-line monitoring of pellet coating processes via optical coherence tomography. Chem Eng Sci 125:200–208
197. U.S. Food and Drug Administration (2004) Pharmaceutical cGMPs for the 21st century – a risk based approach; Final Report
198. U.S. Food and Drug Administration (2004) Guidance for industry. PAT - a framework for innovative pharmaceutical development, manufacturing, and quality assurance. US Department of Health, Rockville, MD
199. International Conference on Harmonization (2009) ICH Q8(R2), Pharmaceutical Development, Part I: pharmaceutical development, and Part II: annex to pharmaceutical development. FDA, Silver Springs
200. International Conference on Harmonization (2008) ICH Q10, Pharmaceutical Quality System. FDA, Silver Springs
201. Schenck L, Troup G, Lowinger M, Li L, McKelvey C (2011) Achieving a hot melt extrusion design space for the production of solid solutions. Chem Eng Pharm Ind, 819–836
202. Wartewig S, Neubert RHH (2005) Pharmaceutical applications of Mid-IR and Raman spectroscopy. Adv Drug Deliv Rev 57(8): 1144–1170
203. Reich G (2005) Near-infrared spectroscopy and imaging: basic principles and pharmaceutical applications. Adv Drug Deliv Rev 57(8): 1109–1143
204. Krier F, Mantanus J, Sacré P-Y, Chavez P-F, Thiry J, Pestieau A, Rozet E, Ziemons E, Hubert P, Evrard B (2013) PAT tools for the control of co-extrusion implants manufacturing process. Int J Pharm 458(1):15–24
205. Lyndgaard LB, Spångberg R, Gilmour C, Lyndgaard CB, van den Berg F (2014) A process analytical approach for quality control of dapivirine in HIV preventive vaginal rings by Raman spectroscopy. J Raman Spectrosc 45(2):149–156
206. Saerens L, Dierickx L, Lenain B, Vervaet C, Remon JP, De Beer T (2011) Raman spectroscopy for the in-line polymer-drug quantification and solid state characterization during a pharmaceutical hot-melt extrusion process. Eur J Pharm Biopharm 77(1):158–163
207. Tumuluri VS, Kemper MS, Lewis IR, Prodduturi S, Majumdar S, Avery BA, Repka MA (2008) Off-line and on-line measurements of drug-loaded hot-melt extruded films using Raman spectroscopy. Int J Pharm 357(1–2): 77–84
208. Wang Y, Steinhoff B, Brinkmann C, Alig I (2008) In-line monitoring of the thermal

degradation of poly(l-lactic acid) during melt extrusion by UV–vis spectroscopy. Polymer 49(5):1257–1265

209. Barnes SE, Brown EC, Sibley MG, Edwards HGM, Scowen IJ, Coates PD (2005) Vibrational spectroscopic and ultrasound analysis for in-process characterization of high-density polyethylene/polypropylene blends during melt extrusion. Appl Spectrosc 59(5): 611–619
210. Coates PD, Barnes SE, Sibley MG, Brown EC, Edwards HGM, Scowen IJ (2003) In-process vibrational spectroscopy and ultrasound measurements in polymer melt extrusion. Polymer 44(19):5937–5949
211. Domike R, Ngai S, Cooney CL (2010) Light induced fluorescence for predicting API content in tablets: sampling and error. Int J Pharm 391(1–2):13–20
212. Lai CK, Zahari A, Miller B, Katstra WE, Cima MJ, Cooney CL (2004) Nondestructive and on-line monitoring of tablets using light-induced fluorescence technology. AAPS PharmSciTech 5(1):E3
213. Khoptyar D, Subash AA, Johansson S, Saleem M, Sparén A, Johansson J, Andersson-Engels S (2013) Broadband photon time-of-flight spectroscopy of pharmaceuticals and highly scattering plastics in the VIS and close NIR spectral ranges. Opt Express 21(18): 20941–20953
214. Silva AFT, Burggraeve A, Denon Q, Van der Meeren P, Sandler N, Van Den Kerkhof T, Hellings M, Vervaet C, Remon JP, Lopes JA, De Beer T (2013) Particle sizing measurements in pharmaceutical applications: comparison of in-process methods versus offline methods. Eur J Pharm Biopharm 85 (3):1006–1018
215. Keleb EI, Vermeire A, Vervaet C, Remon J-P (2004) Twin screw granulation as a simple and efficient tool for continuous wet granulation. Int J Pharm 273(1):183–194
216. de Boer A, Gjaltema D, Hagedoorn P, Frijlink H (2002) Characterization of inhalation aerosols: a critical evaluation of cascade impactor analysis and laser diffraction technique. Int J Pharm 249(1–2):219–231
217. Li M, Wilkinson D, Patchigolla K (2005) Determination of non-spherical particle size distribution from chord length measurements. Part 2: experimental validation. Chem Eng Sci 60(18):4992–5003
218. Petrak D (2002) Simultaneous measurement of particle size and particle velocity by the spatial filtering technique. Part Part Syst Charact 19(6):391–400
219. Hodzic A, Kriechbaum M, Laggner P (2009) Laboratory SWAXS for application in the pharmaceutical technology. Acta Cryst A 65:120
220. Jednačak T, Hodzic A, Scheibelhofer O, Marijan M, Khinast J, Novak P (2014) Fast real-time monitoring of entacapone crystallization and characterization of polymorphs via Raman spectroscopy, statistics and SWAXS. Acta Pharm 64(1):1–13
221. Hennigan MC, Ryder AG (2013) Quantitative polymorph contaminant analysis in tablets using Raman and near infra-red spectroscopies. J Pharm Biomed Anal 72:163–171
222. Saerens L, Ghanam D, Raemdonck C, Francois K, Manz J, Krüger R, Krüger S, Vervaet C, Remon JP, De Beer T (2014) In-line solid state prediction during pharmaceutical hot-melt extrusion in a 12 mm twin screw extruder using Raman spectroscopy. Eur J Pharm Biopharm 87(3):606–615
223. Bondesson L, Mikkelsen KV, Luo Y, Garberg P, Agren H (2007) Hydrogen bonding effects on infrared and Raman spectra of drug molecules. Spectrochim Acta A Mol Biomol Spectrosc 66(2):213–224
224. Juuti M, Tuononen H, Prykäri T, Kontturi V, Kuosmanen M, Alarousu E, Ketolainen J, Myllylä R, Peiponen K-E (2009) Optical and terahertz measurement techniques for flat-faced pharmaceutical tablets: a case study of gloss, surface roughness and bulk properties of starch acetate tablets. Meas Sci Technol 20 (1):015301
225. May RK, Evans MJ, Zhong S, Warr I, Gladden LF, Shen Y, Zeitler JA (2011) Terahertz in-line sensor for direct coating thickness measurement of individual tablets during film coating in real-time. J Pharm Sci 100 (4):1535–1544
226. Markl D, Hannesschläger G, Sacher S, Leitner M, Khinast JG (2014) Optical coherence tomography as a novel tool for in-line monitoring of a pharmaceutical film-coating process. Eur J Pharm Sci 55:58–67
227. Danzl R, Helmli F, Scherer S (2009) Focus variation—a new technology for high resolution optical 3D surface metrology. J Mech Eng 57(3):245–256
228. Leskinen JTT, Simonaho S-P, Hakulinen M, Ketolainen J (2013) Real-time tablet formation monitoring with ultrasound measurements in eccentric single station tablet press. Int J Pharm 442(1–2):27–34
229. Shlieout G, Lammens RF, Kleinebudde P (2000) Dry granulation with a roller

compactor. Part I: the functional units and operation modes. Pharm Technol Eur 12 (11):24–35

230. Inghelbrecht S (1998) The roller compaction of different types of lactose. Int J Pharm 166 (2):135–144

231. Leuenberger H (2003) Scale-up in the 4th dimension in the field of granulation and drying or how to avoid classical scale-up. Powder Technol 130(1–3):225–230

232. Plumb K (2005) Continuous processing in the pharmaceutical industry: changing the mind set. Chem Eng Res Des 83 (6):730–738

233. Malhotra G (2009) Pharmaceutical processing—batch or a continuous process: a choice. Pharm Process:16–17

234. Mascia S, Heider PL, Zhang H, Lakerveld R, Benyahia B, Barton PI, Braatz RD, Cooney CL, Evans JMB, Jamison TF, Jensen KF, Myerson AS, Trout BL (2013) End-to-end continuous manufacturing of pharmaceuticals: integrated synthesis, purification, and final dosage formation. Angew Chem Int Ed Engl 52(47):12359–12363

235. Vercruysse J, Delaet U, Van Assche I, Cappuyns P, Arata F, Caporicci G, De Beer T, Remon JP, Vervaet C (2013) Stability and repeatability of a continuous twin screw granulation and drying system. Eur J Pharm Biopharm 85(3 Pt B):1031–1038

236. McKenzie P, Kiang S, Tom J, Rubin AE, Futran M (2006) Can pharmaceutical process development become high tech? AIChE J 52 (12):3990–3994

237. Stephanopoulos G, Ng C (2000) Perspectives on the synthesis of plant-wide control structures. J Proc Contr 10(2–3):97–111

238. Garcia T, Cook G, Nosal R (2008) PQLI key topics - criticality, design space, and control strategy. J Pharm Innov 3(2):60–68

239. Boukouvala F, Ramachandran R, Vanarase A, Muzzio FJ, Ierapetritou MG (2011) Computer aided design and analysis of continuous pharmaceutical manufacturing processes, vol 29. Elsevier B.V., p 216–220

240. Schaber SD, Gerogiorgis DI, Ramachandran R, Evans JMB, Barton PI, Trout BL (2011) Economic analysis of integrated continuous and batch pharmaceutical manufacturing: a case study. Ind Eng Chem Res 50(17): 10083–10092

241. Pardeike J, Strohmeier DM, Schrödl N, Voura C, Gruber M, Khinast JG, Zimmer A (2011) Nanosuspensions as advanced printing ink for accurate dosing of poorly soluble drugs in personalized medicines. Int J Pharm 420(1): 93–100

INDEX

Marianthi G. Ierapetritou and Rohit Ramachandran (eds.), *Process Simulation and Data Modeling in Solid Oral Drug Development and Manufacture*, Methods in Pharmacology and Toxicology, DOI 10.1007/978-1-4939-2996-2, © Springer Science+Business Media New York 2016

D

Index

T

U

MIX
Papier aus verantwortungsvollen Quellen
Paper from responsible sources
FSC® C105338

If you have any concerns about our products,
you can contact us on
ProductSafety@springernature.com

In case Publisher is established outside the EU,
the EU authorized representative is:
Springer Nature Customer Service Center GmbH
Europaplatz 3, 69115 Heidelberg, Germany

Printed by Libri Plureos GmbH
in Hamburg, Germany